BUILDER'S
Reference Book

Twelfth Edition

by
Leslie Black

Spon Press
Taylor & Francis Group

LONDON AND NEW YORK

First edition 1951
Twelfth edition 1985
Reprinted 1989, 1991

Transferred to Digital Printing 2003

Spon Press is an imprint of the Taylor & Francis Group

© 1985 E & F N Spon and Leslie Black

Printed and bound in Great Britain by TJI Digital, Padstow, Cornwall

ISBN 0 419 15890 1

Apart from any fair dealing for the purposes of research or private study, or criticism or review, as permitted under the UK Copyright Designs and Patents Act, 1988, this publication may not be reproduced, stored, or transmitted, in any form or by any means, without the prior permission in writing of the publishers, or in the case of reprographic reproduction only in accordance with the terms of the licences issued by the Copyright Licensing Agency in the UK, or in accordance with the terms of licences issued by the appropriate Reproduction Rights Organization outside the UK. Enquiries concerning reproduction outside the terms stated here should be sent to the publishers at the London address printed on this page.

The publisher makes no representation, express or implied, with regard to the accuracy of the information contained in this book and cannot accept any legal responsibility or liability for any errors or omissions that may be made.

A catalogue record for this book is available from the British Library

Preface

This book is not, in itself, a complete text book. It's coverage of the industry is so wide that full texts relating to all the subjects and disciplines mentioned would require the space of many volumes. It should be considered more in the nature of a guide to sources of information relevant to the building industry, legislation affecting it, standards of workmanship and materials, and methods.

It is designed to be a tool, for use in the office or on the site, marshalling facts in a handy compendium style to meet the most common requirements of the busy builder.

Readers are advised always to check the facts quoted here against the latest information on the subject—legislation affecting the industry changes frequently and British Standards and Codes of Practice are continuously under review. At the time of going to press all the information in this totally revised and largely rewritten edition was correct and current. Subsequent editions will be thoroughly revised as the relevant facts alter sufficiently to warrant publication.

Finally the Editor wishes to thank all the various associations and trade sources from which the information contained in this volume was selected, for their assistance and co-operation.

LESLIE BLACK
1980

Contents

		page
	Preface	5
1	General Office Information	9
2	Measuring Data and Preliminaries	50
3	Basic Estimating	75
4	The Standard Method of Measurement of Building Works (SMM6—Sixth Edition)	100
5	Structural Data	130
6	The Excavator	164
7	Builders' Plant and Machinery	175
8	The Concretor	182
9	The Mason	193
10	The Bricklayer	198
11	The Carpenter and Joiner	217
12	Timber Pests and Preservation	274
13	The Roofer	280

		page
14	The Drainlayer	295
15	The Plumber	307
16	The Heating Engineer	322
17	Insulation	338
18	The Electrician	350
19	The Pavior and Tiler	361
20	The Plasterer	371
21	The Painter	382
22	The Glazier	395
Employers' and Unions Organisations		405
Government and Public Organisations		408
Research and Testing Organisations		412
Professional Organisations		419
Trade Organisations		424

1. GENERAL OFFICE INFORMATION

National Working Rule Agreements—Wages and Conditions 10
Building Regulations 20

PAGE 50

NATIONAL JOINT COUNCIL FOR THE BUILDING INDUSTRY

TERMS OF SETTLEMENT—JUNE 1985

NOTICE No. 1

The National Joint Council at its meeting on 5th June received a report from the Building and Civil Engineering Joint Board which had been considering claims for improvements in wages and conditions by the Operatives Parties to the Joint Board.

On the recommendation of the Joint Board, the Council adopted the following terms of settlement of all outstanding claims. The amendments to give effect to these terms are included in Notice No. 2.

Ratification having been declared by the Adherent Bodies as required by Rule 6.1:3.2 of the Rules and Regulations of the Council, the Management Committee was instructed to promulgate these decisions, and so promulgates. The date of promulgation is 6th June, 1985.

1. WAGES

With effect on and from Monday, 24th June, 1985 and subject to the conditions prescribed in the Working Rule Agreement, guaranteed minimum weekly earnings shall be as follows:

    Craft Operatives (Grade A) £107.83½
    Labourers (Grade A)   £ 91.84½

The guaranteed minimum earnings shall be made up as follows:

	Craft Operative £	Labourer £
Standard Basic Grade A Rates of Wages	93.01½	79.36½
Guaranteed Minimum Bonus	14.82	12.48
Guaranteed Minimum Earnings	107.83½	91.84½

Where under the Working Rules, extra payments and other payments are aggregated with the basic rate for the purpose of calculating overtime they shall be adjusted on 24th June, 1985 to maintain the appropriate relationship with the basic rate.

GENERAL OFFICE INFORMATION 11

The weekly amounts payable above the labourers' rate under National Working Rule 22.4—Scaffolders— shall be as follows:

Trainee Scaffolder	£3.12
Basic Scaffolder (on first qualifying as such)	£8.19

2. SICKNESS AND INJURY PAYMENTS

With effect from 24th June, 1985 the weekly rate of payment during absence due to sickness or injury under National Working Rule 16 shall be increased from £36.75 to £38.70.

3. TOOL ALLOWANCES AND EMPLOYERS' LIABILITY

    3.1. With effect from 24th June, 1985 the amount of tool allowances and painters' overall allowance payable under National Working Rule 18 shall be increased as follows:

    Present allowance of 106p per week to become 112p per week
    Present allowance of 82p per week to become 87p per week
    Present allowance of 55p per week to become 58p per week

    3.2. The employers' maximum liability for tools lost through fire or theft under the conditions of National Working Rule 18.2 shall be increased from £185 to £196 with effect from 24th June, 1985.

4. ANNUAL HOLIDAYS SCHEME

By the decision of the four parties to the Annual Holidays Scheme the values of holiday credit stamps shall with effect from 29th July, 1985 be as follows:

Adults	*£12.15
Under 18	£8.50

*This includes a contribution of 90p towards Retirement and Death Benefit under the Benefits Scheme.

5. DEATH BENEFIT

The amount of death benefit payable under the Building and Civil Engineering Industry Benefits Scheme to be increased to £4,500 where death occurs on or after 24th June, 1985. With effect also from this date an additional amount of up to £4,500 will be payable if the cause of death is an accident at the place of work or while travelling to or from work.

6. PERIOD OF SETTLEMENT

The Council shall not be required to consider any application for a major change in Operatives' pay and conditions which would have effect before 30th June 1986.

AMENDMENTS TO NATIONAL WORKING RULES

At its meeting on 5th June, 1985 the Council decided to make the following alterations to National Working Rules. The date of Promulgation is 6th June, 1985.

NATIONAL WORKING RULE 1—GUARANTEED MINIMUM BONUS EARNINGS

    Clause 1.6.6   **Add** 'NWR 2.10' after 'NWR 1.6.3'
    Clause 1.6.8   **Delete** '£14.04' and **Insert** '£14.82'
                          Delete '£11.89½' and **Insert** '£12.48'

NATIONAL WORKING RULE 2—BASIC RATES

    Clause 2.1   **Delete** '£88.33½ (226½p per hour)' and **Insert** '£93.01½ (238½p per hour)'
                         Delete '£75.27 (193p per hour)' and **Insert** '£79.36½ (203½p per hour)'
    Clause 2.2   **Delete** '£7.02' and **Insert** '£7.41'
    Clause 2.9   **Delete** '£37.63½ (96½p)' and **Insert** '£39.58½ (101½p)'
                         Delete '£52.65 (135p)' and **Insert** '£55.57½ (142½p)'
                         Delete '£75.27 (193p)' and **Insert** '£79.36½ (203½p)'

NATIONAL WORKING RULE 3—EXTRA PAYMENTS FOR CONTINUOUS EXTRA SKILL OR RESPONSIBILITY

Key to Code Letters—**Amend** Key to read as follows:

Code Letter	Above Labourers' Rate per week (hourly equivalent in brackets)	
A	£3.90	(10p)
B	£4.29	(11p)
C	£5.07	(13p)
D	£7.41	(19p)
E	£8.19	(21p)
F	£8.58	(22p)
G	£9.36	(24p)
H	£10.14	(26p)
I	£11.70	(30p)
J	£12.09	(31p)
K	£12.87	(33p)
L	£13.65	(35p)
M	£14.43	(37p)
N	£15.99	(41p)

NATIONAL WORKING RULE 13—TERMINATION OF EMPLOYMENT

    Clauses
        13.1.1   **Delete** 'first six normal working days' and
and 13.1.2   **Insert** 'first five normal working days'

Clause 13.3.1 **Amend** existing wording to read:

'In cases of misconduct an operative is liable to dismissal in accordance with the procedures set out in NWR 13.4.'

NATIONAL WORKING RULE 16—SICKNESS AND INJURY PAYMENTS

Clause 16.4 **Delete** '£36.75' and **Insert** '£38.70'

NATIONAL WORKING RULE 18—TOOL AND CLOTHING ALLOWANCES

Clause 18.1 **Delete** opening paragraph and **Insert**

'Tool allowances are paid in respect of the provision, maintenance and upkeep of tools provided by the operative and are NOT to be taken into account for the calculation of overtime, travelling time or of guaranteed minimum weekly earnings nor are they to be offset against guaranteed minimum bonus under NWR 1.6. Where employment starts after Monday, or is terminated in accordance with NWR 13 otherwise than on a Friday, the amount of tool allowance to be paid shall be the appropriate proportion of the weekly allowance for each day on which the operative was available for work in accordance with NWR 1.2.

Note: Tool allowances, although not forming part of an operative's basic rate, are liable to the deduction of tax and National Insurance contributions in the same way as other earnings.'

Clause 18.1 **Delete** existing allowances and **Insert**

Carpenters and Joiners—**Delete** '106p' and **Insert** '112p'
Banker Masons —**Delete** '82p' and **Insert** '87p'
Mason Fixers —**Delete** '55p' and **Insert** '58p'
Plasterers —**Delete** '55p' and **Insert** '58p'
Bricklayers —**Delete** '55p' and **Insert** '58p'
Painters —**Delete** '55p' and **Insert** '58p'
Wall and Floor Tilers —**Delete** '55p' and **Insert** '58p'

Clause 18.2 **Delete** '£185' and **Insert** '£196'

NATIONAL WORKING RULE 19—BENEFITS

Clause 19.1 **Amend** existing wording to read:

'An operative is entitled to be provided by his employer with cover for retirement benefit, of an amount dependent on length of service in the industry, and death benefit of up to £4.500, with an additional amount of us to a further £4,500 payable if the cause of death is an accident at the

Clause 19.3   **Amend** existing wording to read:
place of work or an accident while travelling to or from work, in both cases in accordance with the conditions laid down in the Building and Civil Engineering Benefits Scheme, which is published separately.'

Clause 19.3   **Amend** existing wording to read:
'There is no entitlement to cover in the case of an operative under the age of 18 years or over the age of 65 years, except for death benefit (of up to the sum of the amounts set out in 19.1 above) if the cause of death is an accident at the place of work or an accident while travelling to or from work when the operative is under the age of 18 years, or on or after his 65th birthday.

NATIONAL WORKING RULE 22—SCAFFOLDERS

Clause 22.4   Basic Scaffolder: **Delete** '£7.80' and **Insert** '£8.19'

STANDARD RATES OF WAGES IN ENGLAND, SCOTLAND AND WALES

Under the Terms of Settlement approved by the National Joint Council on 5th June, 1985 and promulgated on 6th June, 1985 the standard basic wages and guaranteed minimum bonus payments (GMB) for building craft operatives and labourers due to be operated with effect on and from MONDAY 24th JUNE, 1985 will be as indicated below.

Note: The rates shown below are weekly rates. The hourly equivalents are shown in brackets.

1. *Craft Operatives and Labourers*
   (i) *London and Liverpool District*

	Rate per week	GMB
Craft Operatives	£93.21 (239p)	£14.82
Labourers	£79.56 (204p)	£12.48

   (ii) *Grade A and Scotland*

	Rate per week	GMB
Craft Operatives	£93.01½ (238½p)	£14.82
Labourers	£79.36½ (203½p)	£12.48

2. *Watchmen*

	Rate per week
London and Liverpool District	£92.04
Grade A and Scotland	£91.84½

Note: Under the provisions of National Working Rule 2.8. the weekly remuneration for watchmen is a sum equivalent to the labourers' guaranteed minimum weekly earnings provided that not less than five shifts (day or night) are worked in the pay week.

3. *Young Labourers*
   (i) London and Liverpool District

Age	% of Labourers' Rate	Rate per week	GMB
16	50	£39.78 (102p)	£6.24
17	70	£55.77 (143p)	£8.77½
18	100	£79.56 (204p)	£12.48

   (ii) *Grade A and Scotland*

Age	% of Labourers' Rate	Rate per week	GMB
16	50	£39.58½ (101½p)	£6.24
17	70	£55.57½ (142½p)	£8.77½
18	100	£79.36½ (203½p)	£12.48

4. *Woodworking Establishment Operatives employed under provisions of NWR 2.10.*

	Rate per week	GMB
London and Liverpool District	£83.85 (215p)	£13.26
Grade A and Scotland	£83.65½ (214½p)	£13.26

NATIONAL WORKING RULE 21 – CONDITIONS OF EMPLOYMENT OF APPRENTICES/TRAINEES

At its meeting on 5th June, 1985 the National Joint Council received a report from its Management Committee which had been considering an Operatives' Side claim for an increase in the amounts payable to apprentices/trainees under National Working Rule 21 who were being trained under the National Joint Training Scheme.

The Council accepted the Report of the Mangement Committee and decided to make the following alterations to the National Working Rule:

NATIONAL WORKING RULE 21—CONDITIONS OF EMPLOYMENT OF APPRENTICES/TRAINEES

Clause 21.2 **Amend** existing wording to read:

'The basic rates and guaranteed minimum bonus of apprentices/trainees (including probationers) shall be as set out below. Payment under the scales is due from the date of entry into employment as an apprentice/trainee, whether the apprentice/trainee is working on site or undergoing full-time training on an approved course, subject to the provisions of NWR 21.3. Payments under the scales are due from the beginning of the payweek during which the specified period starts.

Delete sections 21.2.1 to 21.2.4 in their entirety and **Insert:**

21.2.1 *Entrants under 19 whose employment as apprentices/trainees began on or after 25th June, 1984 and Building Foundation Training Scheme trainees whose employment as apprentices/ trainees began before that date.*

Six-month period	Rate per week	GMB
First	£43.29 (111p)	—
Second	£58.69½ (150½p)	—
Third	£63.96 (164p)	£10.14
*Thereafter, until skills test is passed	£77.22 (198p)	£12.28½
On passing skills test and until completion of training period	£87.16½ (223½p)	£13.84½

*The NJCBI has agreed to the introduction of skills testing under the National Joint Training Scheme. The first practical skills tests will take place in the Spring, 1986.

21.2.2 *Entrants under 19 whose employment as apprentices/trainees began on or after 27th June, 1983 but before 25th June, 1984 (excluding BFTS trainees whose employment as apprentices/ trainees began between those dates, paid under NWR 21.2.1).*

Six month period	Rate per week	GMB
Third	£67.27½ (172½p)	£10.14
Fourth	£77.07½ (197½p)	£12.28½
Fifth	£87.16½ (223½p)	£13.84½
Sixth	£92.04 (236p)	£14.62½

21.2.3 *Entrants under 19 whose employment as apprentices/trainees began before 27th June, 1983.*

Six month period	Rate per week	GMB
Fifth and sixth	£83.07 (213p)	£13.06½

21.2.4 *Entrants from approved CITB Vocational Preparation Course 1982/83.*

Six month period	Rate per week	GMB
Fourth and fifth	£83.07 (213p)	£13.06½

Clause 21.8 – Tool Allowances
Delete present allowance of '106p' and **Insert** '112p'
Delete present allowances of '55p' and **Insert** '58p'

NEW WAGE RATES FROM BATJIC EFFECTIVE MONDAY 24th JUNE 1985

The Building and Allied Trades Joint Industrial Council agreed on 31 May 1985 and promulgated the new Wage Rates to apply from Monday 24 June 1985 for a period of twelve months.

WORKING RULE 1—STANDARD RATE OF WAGES 39 HOUR WEEK

Craftsmen £109.40
Adult General Operatives £93.79½
The hourly rate shall be £2.80½ (Craftsmen) and
£2.40½ (Adult General Operatives)

APPRENTICE/TRAINEES AND YOUNG OPERATIVE RATES

The weekly standard rate for **Apprentices under 19 years of age** shall be these proportions of the Craftsman's Rate:

Age of Entry	1st Year	2nd Year	3rd Year
16	(50%) = £54.70	(70%) = £76.58	(90%) = £98.46
17	(60%) = £65.64	(80%) = £87.52	(90%) = £98.46
18	(80%) = £87.52	(85%) = £92.99	(90%) = £98.46

Entrants over the age of 19, normal period of 3 years

1st Year	2nd Year	3rd Year
(85%) = £92.99	(90%) = £98.46	(95%) = £103.93

Entrants over the age of 19 who undertake full-time off-site training 1st year

1st Year	next 6 months	last 6 months
(85%) = £92.99	(90%) = £98.46	(95%) = £103.93

Trainees who enter industry from Government Training Centre

1st 6 months	2nd 6 months	last 6 months
(85%) = £92.99	(90%) = £98.46	(95%) = £103.93

Standard Rate of Wages for Young Operatives

Age 16 years	(50% of General Operatives Rate)	£46.90
Age 17 years	(70% of General Operatives Rate)	£65.66
Age 18 years	(100% of General Operatives Rate)	£93.79½

WORKING RULE 1C

Intermittent and Consolidated Rates of Pay for Skill

(a) The following are rates of pay per hour which shall be paid in addition to the general operatives' rate of pay to these operatives who apply these skills on an intermittent basis:

7½ pence per hour

Air or electric percussion drill; hammer; rammer etc; cartridge gun operator; compressor driver; concrete-mixer driver; barrow hoist operator; pumpman; handroller operator; mechanical barrow operator; electric operated vibrator and paint sprayer.

14 pence per hour

Drag shovel operator; dumper driver (up to 2000 kg); power roller driver (up to 4000 kg); light tyred tractor driver; pipelayer (up to 300mm); concrete screeder/leveller; forklift/sideloader driver (up to 3000 kg).

21½ pence per hour
Batching plant driver; dumper driver (over 2000 kg); power roller driver (over 4000 kg); banksmen; watchmen; pipelayer (over 300mm); concrete trowel and planthand; forklift/sideloader driver (over 3000 kg).

(b) **The following shall apply to semi-skilled grades with continuous responsibility:**

£100.53 per week
Travelling, overhead, crawler, mobile or tower crane operator (up to 2 tons); wheeled or tracked tractor driver (up to 70 h.p.); trenching machine operator (up to 30 h.p.); excavator driver (up to ⅜ cu.yd.); timberman.

£103.25 per week
Travelling, overhead, crawler, mobile or tower crane operator (over 2 tons); wheeled or tracked tractor driver (over 70 h.p.); trenching machine operator (over 30 h.p.); excavator driver (up to ¾ cu.yd.).

£106.00 per week
Travelling, overhead, crawler, mobile or tower crane operator (over 5 tons); excavator driver (over ¾ cu.yd.).

(c) **Payment for discomfort, inconvenience or risk:**
  (i) **Detached Work**
15m and up to 30m	= 5½ pence per hour
30m and up to 45m	= 7 pence per hour
45m and up to 60m	= 12½ pence per hour
60m and up to 75m	= 19 pence per hour
75m and up to 90m	= 24½ pence per hour

  (ii) **Exposed Work**
Above 40m and up to 50m	= 3½ pence per hour
Above 50m and up to 60m	= 5 pence per hour
Above 60m and up to 75m	= 6½ pence per hour

The extra payment to be increased by 4 pence per hour for each 15 metres above 75 metres. All heights to be calculated from ground level.

  (iii) **Cranes**
Control platform over 15m and up to 30m	= 4 pence per hour
30m and up to 45m	= 5 pence per hour
45m and above	= 6½ pence per hour

  (the above does not apply to Tower Cranes)

  (iv) **Work in Swings** = 14 pence per hour

  (v) **Furnace Firebrick Work and Acid Resisting Brickwork**
Furance or similar work up to 120°F	= 5½ pence per hour
New Firebrick Work	= 14 pence per hour
Repair of Firebrick Work	= 17 pence per hour
Brickwork using acid bonding material	= 7½ pence per hour

GENERAL OFFICE INFORMATION

WORKING RULE 11—TRAVELLING AND LODGINGS
(b) (iii) Daily Traveiling Allowance
This remains at 8p (6p apprentices/trainees and young labourers).
Daily Fare Allowance
Delete scale given and substitute:

Distance (km)	Allowance	Distance (km)	Allowance	Distance (km)	Allowance
1-6	Nil	21	£2.01	36	£2.76
7	£0.20	22	£2.08	37	£2.82
8	£0.39	23	£2.13	38	£2.89
9	£0.57	24	£2.18	39	£2.96
10	£0.74	25	£2.23	40	£3.03
11	£0.90	26	£2.28	41	£3.11
12	£1.05	27	£2.33	42	£3.18
13	£1.20	28	£2.37	43	£3.26
14	£1.34	29	£2.41	44	£3.33
15	£1.47	30	£2.45	45	£3.40
16	£1.58	31	£2.50	46	£3.48
17	£1.69	32	£2.55	47	£3.55
18	£1.78	33	£2.60	48	£3.63
19	£1.87	34	£2.65	49	£3.70
20	£1.94	35	£2.70	50	£3.77

For distances in excess of 50 kilometres a maximum of £3.77 may be paid.

APPENDIX F
GUIDE NOTES TO THE BATJIC AGREEMENT
Lodging Allowance, paragraph 1: **delete** £9.45 **substitute** £10.00.
paragraph 2: **delete** £8.50 **substitute** £10.00.

WORKING RULE 12—SICKNESS AND INJURY BENEFIT
Sick pay to be increased to £7.92 per day with a maximum of £39.60 per week.
Please note that in addition to payment under this Rule you are required to pay Statutory Sick Pay due.

WORKING RULE 13—BENEFITS SCHEMES
Death Benefit Scheme
Line 1 **delete** '£4,000' and **insert** '£4,500'
Line 2 **delete** '£4,000' and **insert** '£4,500'

BATJIC BENEFIT SCHEME CONTRIBUTIONS
The BATJIC National Council have agreed the following new contribution rates for Holiday Pay and Death Benefit. There is no change to the Retirement Benefit contribution.

From Monday 5th August 1985	Under 18	18 or over
Death Benefit	No charge	£1.95 per month
Holiday Pay Scheme	£7.69 per week	£10.17 per week
Retirement Benefit No Change	70p per week	70p per week

The Additional Voluntary Contribution from employers and/or employees were extended in January 1985: 50p, £1.00, £1.50, £2.00, £3.00, £5.00, £8.00 or £12.00.

WORKING RULE 18—TOOL ALLOWANCE

Operatives are required to provide their own tools in order to carry out their duties on behalf of their employers and shall receive a weekly allowance towards the provision and maintenance of their tool-kit as follows:

Carpenters and Joiners .... £1.14 per week
Banker Masons .... 87 pence per week
Mason Fixers .... 59 pence per week
Plasterers .... 59 pence per week
Bricklayers .... 59 pence per week
Wall and Floor Tilers .... 59 pence per week

Note: Apprentices should receive the full tool allowance when they are required to maintain their own tool-kit.

Painters Overall Allowance .... 59 pence per week

Storage of Tools. The employer's maximum liability for loss of tools shall be increased to £198.

BUILDING REGULATIONS

The Building Act of 1984 consolidated various building control statutes enacted over the last 90 years. It is no longer necessary for a builder to search through a confusing array of enactments concerning buildings and related matters; most are now restated in the new Act. Given here is a summary of the new Act, which at the time of going to press had just been officially published.

Powers for local authorities to make by-laws controlling construction, air space and drainage were made available in the Public Health Act 1975. For some 50 years this was the basis of building control supported by model by-laws which were variously adjusted to meet local needs or prejudices. Then the power was taken up by the Public Health Act 1936, but by-laws remained in council's hands as before. Another 24 years and a new Public Health Act in 1961 gave authority for the making of building regulations.

The idea was to achieve uniformity and to enable the requirements to be more readily amended to take account of advancements in knowledge and changes in techniques. The new building regulations were first made in 1965 and have been subjected to numerous amendments and metricated.

GENERAL OFFICE INFORMATION 21

Additional legislation arrived in the form of the Health and Safety at Work etc., Act 1974, which widened building control powers considerably and for the first time gave a definition of 'building'. Further radical alterations came with the passing of the Housing and Building Control Act 1984.

This legislation has been consolidated into the Building Act 1984 made on 31st October, 1984. So instead of searching about in several elderly Acts for various matters relating to building control we now have an Act of 131 sections with seven schedules.

Power to make regulations

The Secretary of State is given power to make regulations to secure the health, safety, welfare and convenience of people using buildings. Regulations may also cover the conservation of fuel and power, and preventing waste, undue consumption, misuse or the contamination of water.

Building regulations may also be made imposing on owners and occupiers of buildings of a prescribed class a continuing duty to maintain fittings and equipment.

Approved documents

In order to provide practical guidance the Secretary of State, or some other body which he may designate, may approve and issue documents saying how the requirements of building regulations may be achieved.

Such approved documents can be revised from time to time and withdrawn or replaced. In effect these documents will replace the deemed-to-satisfy content of existing regulations.

If a builder fails to comply with an approved document he does not necessarily render himself liable to any civil or criminal proceedings. Where, however it is alleged that a person has contravened a regulation, failure to comply with an approved document may tend to establish liability. Where an approved document has been complied with this may tend to negate liability.

Relaxation

The Secretary of State may give a direction dispensing with or relaxing a requirement of the regulations. Building regulations may also allow local authorities to exercise this power, and also for a public body to dispense with an unreasonable regulation affecting work it proposes to carry out.

Applications for relaxation are to be made to the local authority on a form to be obtained from them.

Where the relaxation of a regulation might have wide ranging effect on health or safety generally it must be advertised in local newspapers. No advertisement is required where the effect on public health or safety will be limited, or where the work only affects an internal part of a building.

Type relaxation
The Secretary of State is enabled to dispense with or relax some requirement of a building regulation, not only in a particular case, but generally. Such a relaxation could be subject to conditions and could be given a time limit. Before deciding to give a type relaxation of this nature the Secretary of State is required to consult such bodies as appear to him to represent the interests concerned. He must publish an account of any relaxation issued.

Type approvals
The principle of giving type approvals may be extended to buildings. The Secretary of State is enabled to issue a certificate of approval of a type of building matter as complying either generally or in any class of case with particular requirements of building regulations.

A type aproval may be issued either on application or otherwise, and, where it is granted in consequence of an application a fee may be charged.

This particular power enables some new product or method of construction to be formally assessed for building regulation purposes. It appears that an approval may be made to apply for a limited period only, if necessary.

The availability of a type approval will not operate as a substitute for the normal plan approval process, but will support any plans deposited with a local authority or sent to a person to be known in the future as an approved inspector.

The power to make type approval may be delegated to a person, or body, to such an extent, and subject to such conditions as the Secretary of State may think desirable.

Passing plans
When plans are deposited with a local authority, it has a duty to pass them unless:
- they are defective;
- the proposal would contravene any building regulations.

If the plans are defective, or show that work would contravene any building regulation the council is given two options.
1. The plans may be rejected, or
2. they may be passed subject to conditions.
- specifying modifications to be made to the plans, and
- more plans are to be deposited.

In these circumstances the person depositing plans must have previously requested the council to pass plans subject to conditions or have agreed with the council that they may do so.

When passing plans the council is to state that it operates only for approval under the building regulations and any section of the Building Act 1984 that expressly authorises the council in certain cases to reject plans.

If there is a disagreement between a developer and the council as to whether the proposed work is in conformity with the building regulations the matter may be referred to the Secretary of State for a determination.

There may be delay in this coming into operation, and until it does, any question may be determined by a magistrates' court. Any application to the court must be made before the proposed work is substantially commenced.

Plans deposited with a local authority may be accompanied by a certificate from an approved person. An approved person will act within the limitations of prescribed regulations and they must be specified in his certificate. A person must include evidence of his approval and of proper insurance cover. A local authority is barred from rejecting work covered by one of these certificates.

The council must issue a decision on the passing or rejection of plans within five weeks of deposit.

Consultation

The Secretary of State must continue to appoint and consult with a Building Regulation Advisory Committee before making regulations.

Where it is proposed to dispense with or relax a regulation requirement affecting:

- structural fire precautions;
- means of escape in case of fire;
- the availability of means of escape at all material times;

then the fire authority must be consulted.

Approval of persons to give certificates

We have seen that under the heading 'passing of plans' certain certificates accompanying plans may be accepted as evidence that the matter covered by the certificate is in compliance with the regulations.

Naturally provision is then made for the approval of persons by the Secretary of State or some other body which he appoints. A person who is approved will have limits placed upon the description of work, or the regulations under which he may issue certificates. Fees will be payable to become an approved person, and adequate insurance cover must be provided.

'Approved persons' are not to be confused with 'approved inpectors' who will be described in a later article.

Exemptions

Certain buildings may be exempt from all or any building regulations.

Named in the Act are schools and educational establishments where the plans have been approved by the Secretary of State.

Also included are buildings belonging to:
- statutory undertakers;
- Atomic Energy Authority;
- British Airports Authority;
- Civil Aviation Authority.

Not included in the exemption are:
(i) houses;
(ii) houses or hotels owned by British Airports Authority;
(iii) offices and showrooms not forming part of a railway station;
(iv) offices and showrooms of British Airports Authority or the Civil Aviation Authority not on an aerodrome owned by the authority.

Regulations may also be made to exempt from compliance with building regulations that are not substantive requirements:
- a local authority;
- a county council;
- any other public body not working for its own profit.

A 'substantive requirement' means regulations regarding the design and construction of buildings, including the provision of services, fittings and equipment. The definition separates technical standards from administrative and procedural requirements of the regulations.

Short lived materials
If plans are deposited with a local authority showing that a building is to be constructed of materials which are listed as being short lived in building regulations the plans may be rejected.

The council may pass the plans, if it fixes a period after which the building must be removed, or imposes conditions having regard to the nature of materials to be used.

No condition may conflict with any condition contained within a planning consent for the same building.

Should anyone construct a building of short lived materials without first depositing plans, the local authority may fix a period during which it must be removed or impose conditions when allowing it to remain. Orignal time limits may be extended.

If the owner fails to remove a building subject to a time limit, the council may remove it and recover its costs.

The owner can appeal to the magistrates' court against a council's actions, but if he transgresses, he may be fined and subjected to a further fine of up to £5 per day so long as the offence continues.

Materials unsuitable for permanent building
When this provision is operative, regulations may prescribe:

- a type of material or component;
- a type of service, fitting or equipment likely to be unsuitable in the construction of or provision of services in a permanent building in the absence of conditions with respect to the use of the building, the material or the service.

Buildings to which these regulations may apply are called 'relevant buildings'.

If plans of a relevant building are deposited with a local authority they may be rejected, or if passed contain conditions including a time limit for removal of the building. Other reasonable conditions may be applied but must not conflict with conditions in a planning permission.

Anyone constructing a relevant building without first depositing plans may be required by the local authority to remove the building. A council may allow the building to remain, subject to conditions. Proceedings may be taken against the owner.

Where regulations have not required the deposit of plans with a local authority, but it thinks that the structure is a 'relevant building' it is given 12 months in which to take action. It can require removal of the work within a stated time or agree to retention subject to conditions. Periods may be extended at a local authority's discretion.

A person aggrieved by having his plans rejected, or by the period allowed, or by the conditions may appeal to the Secretary of State.

At the end of the period allowed, and when there has been no extension of time, the owner must remove the work or building.

Anyone contravening the council's requirements may upon conviction be subject to a fine and also a continuing daily penalty of up to £50 for each day the offence continues.

Provision of exits

Plans of buildings listed below must contain such means of ingress and egress and passages or gangways as a local authority, after consultation with the fire authority, says are necessary. If the plans do not comply then they will be rejected.
- Theatre, hall or other building used as a place of public resort.
- Restaurant, shop, store or warehouse to which members of the public are admitted, and in which more than 20 persons are employed.
- Club registered under the Licensing Act 1964.
- School not exempted from building regulations.
- Church, chapel or other place of public worship.

The requirements do not extend to include:
- Private houses to which members of the public are admitted occasionally or exceptionally.
- Buildings used as a church or chapel or other place of worship before section 36. Public Health Acts Amendment Act 1890 came into force.

● A building used prior to October 1 1937 where the 1980 Act, or any similar local provision, never came into operation.

Once again, disputes may be settled by a magistrates' court.

Building over a sewer

When plans deposited with a local authority indicate that a building or an extension will be erected over a sewer or drain on the council's map of sewers, the plans must be rejected. However, the council has discretion to pass the plans if the circumstances are acceptable, or it may pass the plans subject to conditions. Frequently this involves entering into a legal agreement regarding responsibilities for access and damage to the sewer and the building.

If there is a disagreement this may be placed before a magistrates' court for a decision.

Provision of drainage

Buildings and extensions are to be provided with satisfactory drainage unless the local authority decides that it may properly dispense with any provision for drainage.

'Drainage' includes the conveyance by means of a sink and any other necessary appliance of refuse water. It also includes rainwater from roofs.

To be satisfactory, drainage must be made to connect with a sewer, or discharge into a cesspool or some other suitable place.

The council cannot require connection to a sewer unless it is within 100 ft. of the site of the building or extension. Also if the sewer is not 'public', that the developer has a right to connect. Additionally the developer must be entitled to put his drain through any intervening land.

If a sewer is more than 100 ft. away from a building the local authority may require connection thereto, if it is prepared to make a contribution to the cost, not only of the extra pipework, but of maintenance and repairs in the future.

Differences between developers and the council may be settled in a magistrates' court.

Drainage in combination

If a local authority thinks it would be more economical or advantageous that two or more buildings should be drained in combination it may require the owners to do so. Alternatively it may do the work on behalf of the owners.

When making this arrangement the local authority must:

● fix the proportions in which the cost of constructing, maintaining and repairing the private sewer is to be borne by the owners;

● where the distance of the existing sewer from the site exceeds 100 ft., fix the proportion of costs between owners and the council.

A sewer constructed by a local authority under these arrangements is not automatically a public sewer. An owner who is aggrieved by a loal authority's decision may appeal to a magistrates' court.

Refuse facilities
Plans shall be rejected by a local authority if they do not show satisfactory means of access from the building to the street for the purpose of removing refuse. Proper provision must also be made for storing refuse prior to removal.

It is unlawful to block an access provided for the removal of refuse. Any dispute may be determined by a magistrates' court.

Water supply
When plans of a house are deposited they must indicate that a sufficient supply of wholesome water will be available, adequate for domestic purposes.

If the proposals have not been carried into effect and a supply of wholesome water is not available then the local authority may prohibit occupation of the building.

Disagreements may be resolved by recourse to the magistrates' court, and convictions can result in a fine including a continuing fine of £2 per day so long as the offence continues.

Closets
Deposited plans must indicate that sufficient and satisfactory closet accommodation will be provided. Water closets are required unless an adequate supply of water is not available, when earth closets or chemical closets may be used.

In factories or workplaces separate closet accommodation is to be provided for each sex.

Any question arising between the local authority and an applicant may be determined on application to a magistrates' court.

Bathrooms
Plans showing new dwellings or conversions creating dwellings must indicate that each separate dwelling is to be provided with a fixed bath or a shower. There must be suitable provision for hot and cold water to each bath and shower.

Larders
When new dwellings are erected or there is a conversion creating a new dwelling, plans may be rejected unless they show suitable and sufficient accommodation for the storage of food. Alternatively there must be sufficient and suitable space for the provision of such accommodation by the occupier.

Offensive material
When it is proposed to erect a building on ground that has been filled with offensive mater, or ground upon which any faecal, offensive, animal or vegetable matter has been deposited the council must reject the plans.

It is open for the applicant to show that the offensive material has been removed or has become or been rendered, innocuous.

Settling disputes
Disputes between a local authority and a person who has executed, or prooses to execute work may be settled by the Secretary of State.

An application must be made jointly to him and decisions will be made under three heads:
- the application to that work of any building regulation;
- whether the plans are in conformity with the regulations;
- whether the work has been executed in accordance with plans passed by the authority.

Departure from plans
An order may be made to ensure that when plans have been passed it is possible to deposit further plans showing a departure or deviation. The usual procedures will then apply.

Lapse of plans
When work covered by plans which have been approved is not commenced within three years the local authority may serve a notice saying that the plans are of no effect.

Tests
Local authorities are given power to ascertain:
- whether any provision in the building regulations would be contravened by anything done or proposed to be done.

In order to fulfil this power a person may be required to carry out reasonable tests, or the local authority may carry out tests and take samples.

Tests may include:
- soil or subsoil;
- materials or components;
- services, fitting or equipment.

For these purposes a council may require the owner of a building to carry out reasonable tests or itself carry out tests and take samples.

Costs of carrying out the tests are to be borne by the developer although the local authority has a discretion to bear the cost, or part of the cost itself.

Disputes may be referred to a magistrates' court for a ruling.

Classification of buildings

Buildings may be classified by reference to size, description, design, purpose, location or any other characteristic.

Penalty

For contravening building regulations there may, upon conviction, be a fine of up to £1000, and a further fine of £50 per day so long as the default continues.

Independent report

If a local authority has served a notice requiring work to be put right, the recipient may tell the council that he intends to obtain a written report from a suitably qualified person.

When obtained the local authority must consider the independent report. If as a result it withdraws its notice it may pay reasonable expense incurred in obtaining the report.

The action of obtaining a report delays implementation of the council's notice from 28 to 70 days.

Offending work

If any work contravenes building regulations the local authority may take proceedings for a fine, and also may require the owner.

- to pull down or remove the work, or
- if he wishes, to do such alterations as will make it comply with the regulations.

The owner has 28 days in which to comply, after which the council may undertake the work and recover reasonable expenses from him.

A local authority may not require the owner to put work right if:

- the plans were passed by the authority, or
- notice of their rejection was not given within the relevant period from their deposit, and the work has been carried out in accordance with the plans.

This does not prevent the local authority, the Attorney General or any other person from applying for an injunction to have the matter put right. In this event the court may require the council to pay the owner compensation.

Civil liability

A breach of duty imposed by building regulations may be actionable if it causes damage. Compliance with building regulations may provide a defence.

'Damage' in this context includes death of, or injury to any person. This also includes any disease and any impairment of a person's physical or mental health.

Refusal to relax regulations
If a local authority refuses to relax or dispense with the requirements of a building regulation the applicant has one month during which he may appeal to the Secretary of State.

The council must give a decision within two months of the date of the application or such other extended period agreed by the applicant.

An appeal to the Secretary of State must be sent to the local authority who is required then to forward it to the Department of the Environment at once, together with a copy of the application and all the documents furnished by the applicant. The local authority must also state its case, sending a copy to the applicant.

Alteration of offending work
A person aggrieved by a notice requiring him to pull down or remove work may appeal to the magistrates' court.

If the court decides that the council acted correctly it will confirm the notice. In every other case the local authority will be given a direction to withdraw the notice.

An appeal must be made within 28 days, or 70 days, when an independent report is being obtained.

Crown Court
If a person is aggrieved by a decision of a magistrates' court he may appeal to the Crown court.

High Court
An appeal to the High Court on a point of law may be made following a Secretary of State's decision relating to:
- passing of plans;
- materials unsuitable for a permanent building;
- relaxation of regulations;
- plans certificates.

Hearing
When an appeal is made to the Secretary of State about the use of unsuitable materials in a permanent building or a refusal to relax regulations he may give the council and the applicant the opportunity of stating a case before an Inspector.

Crown Buildings
Generally substantive requirements of building regulations apply to work carried out by or on behalf of the Crown, but there are exceptions.

Atomic Energy Authority
Substantive requirements of building regulations may also apply to United Kingdom Atomic Energy Authority Buildings.

Drainage of buildings
A local authority may require an owner to remedy the matter when any of the following situations arise:
- satisfactory provision has not been made for the taking away by means of a sink or other necessary appliance, refuse water or the conveyance of rainwater from roofs;
- a cesspool, soil pipe or any part of a drainage system is insufficient or so defective as to allow subsoil water to penetrate;
- a cesspool, appliance or drainage system is in such a condition as to be prejudicial to health or a nuisance;
- a cesspool or drainage system not now in use, but which is prejudicial to health or a nuisance.

This does not apply to property belonging to statutory undertakers, the British Airports Authority, or the Civil Aviation Authority, unless:
- it is a house;
- it is a house or a hotel owned by the British Airports Authority;
- it is a building used as offices or showrooms not forming part of a railway station;
- it is a building used as offices or showrooms and not on an aerodrome owned by the British Airports Authority or the Civil Aviation Authority.

Soil and vent pipes
A rainwater downpipe is not to be used for discharging the soil or drainage from sanitary conveniences.

WC soil pipes are to be properly ventilated.

A pipe for surface water must not be used as a vent for a drain or sewer carrying foul water.

Drainage repairs
It is not lawful to repair, reconstruct or alter the course of an underground drain that joins a sewer, or a cesspool without giving the council 24 hurs notice of the intention.

This requirement does not apply in the case of genuine emergency. In an emergency the work may be undertaken, but left uncovered for the council to inspect.

Disconnecting drains
When a drain is reconstructed in the same or a new position or the use of a drain is permanently discontinued, the drains which become disused or unnecessary are to be disconnected and sealed.

Questions regarding the reasonableness of a local authority's requirements may be determined by a magistrates' court.

No-one may be required to carry out work on land if he has no right to enter the land for that purpose.

Before undertaking this kind of work at least 24 hours notice must be given to the council.

Non-compliance can attract a fine upon conviction, and also a daily penalty.

Improper construction

If a water closet, drain or soil pipe is so badly constructed or repaired as to give rise to conditions prejudicial to health or a nuisance the person responsible may be charged with an offence and upon conviction, fined.

Provision of closets

Where a building, or a dwelling, is without sufficient closet accommodation, or any closets that have been provided are in such a state as to be prejudicial to health or a nuisance, the local authority may require adequate provision to be made.

Unless an adequate supply of water and a sewer is available, a new water closet may not be required except in substitution of an existing water closet.

This particular requirement of the Building Act does not apply to a factory, a workplace or premises to which the Offices Shops and Railway Premises Act 1963 applies.

Workplaces

A workplace must be provided with sufficient and satisfactory sanitary conveniences for the number of people working in, or in attending at the building.

Separate accommodation is required for the sexes.

This section does not apply to buildings covered by the Offices Shops and Railway Premises Act 1963.

Replacing earth closets

If there is a sufficient water supply and sewer available a local authority may require a water closet to be substituted for an earth closet.

When this happens the council must pay one-half of expenses reasonably incurred.

Loan of toilets

When reconstruction, maintenance or improvement work necessitating disconnection of sanitary conveniences is being carried out, the local authority may supply on loan temporary sanitary conveniences. No charge will be made for the first seven days the toilet is on loan.

Public conveniences

No-one may erect a public sanitary convenience without the prior consent of the local authority, who may require its removal at any time.

Anyone aggrieved by a local authority's refusal, or conditions imposed, may appeal to a magistrates' court.

These limitations do not extend to official bodies including British Rail and Dock undertakings.

Water supply
If an occupied house has not got a supply of wholesome water in pipes in the house, the council may require the house to be connected to a public supply. If a public supply is not available, but another source is, then that water may be taken into the house in pipes.

There is provision for an appeal to a magistrates' court.

Food storage
A local authority may require the provision of sufficient and suitable food storage accommodation in an occupied dwelling. The recipient of a notice requiring the provision of a larder may appeal to a magistrates' court.

Entrances and exits
When a building is not provided with entrances and exits, and passageways or gangways which are necessary, having regard to:
- the activities being carried on in the building, and
- the number of people likely to be present at any one time the local authority may require work to meet their standards.

Before doing so they are obliged to consult the fire authority.

The Fire Precautions Act 1971 section 30 (3) applies.

Buildings affected are:
- theatre, hall or other building used as a place of public resort;
- restaurant, shop, store or warehouse to which members of the public are admitted, and in which more than twenty persons are employed;
- club registered under the Licensing Act 1964;
- school not exempted from building regulations;
- church, chapel or other place of public worship.

Disputes may be settled by a magistrates' court.

Means of escape
If a local authority thinks that a building or proposed building has unsatisfactory means of escape in case of fire it may require the owner to make adequate provision.

Before doing so the fire authority is to be consulted, and the work required is limited to those storeys whose floor is more than 20 feet above the ground.

Requirements apply to a building:
- let in flats or tenement dwellings;
- an inn, hotel, boarding house, hospital, nursing home, boarding school, childrens' home or similar institution;

● restaurant, shop, store or warehouse which has sleeping accommodation on an upper floor.

The Fire Precautions Act 1971 section 30(3) applies.

Raising chimneys

When a taller building is erected alongside, or within six feet of chimneys in an existing building, the developer may be required to raise the height of the existing chimneys. The top of the existing chimneys are to be raised to the same height as the top of the chimneys in the taller building or the top of the taller building, whichever is the higher.

There are penalties upon conviction for failure to comply with a council's notice in this connection. There is also facility for appealing to a magistrates' court against the requirement.

Cellars

Anyone wishing to construct a cellar which is lower than the ordinary level of subsoil water as part of a house, shop, inn, hotel or office must get consent from the local authority.

Exceptions are cellars subject to a justices license or which are part of premises on a railway station.

Contraventions may result in a fine upon conviction.

A local authority consent may contain conditions.

There are various powers available to a local authority in the Public Health Act and the Housing Act 1957 dealing with defective premises.

However, the procedures there are very lengthy and where any premises are in such a defective state as to be prejudicial to health or a nuisance, a much faster procedure is available in the Building Act 1984.

The council may serve a notice on the person it considers responsible, stating that it intends to remedy defects and saying what work it intends to undertake. Nine days after giving this notice the council may execute the necessary work and recover its expenses from the person on whom the notice was served.

It is open for the person receiving one of these notices to reply within seven days saying that he will remedy the defects himself, and the local authority must then take no action.

However, if the repairs are not begun within a reasonable time, or having begun, the work does not proceed at a reasonable rate, the local authority may move in.

When it comes to recovering expenses and the matter gets to court the council must establish that the premises were in a defective state, and that repairs had to be undertaken as a matter of urgency. If the defendent had served a counter notice he must establish that he commenced work promptly and made reasonable progress.

Should the court find that the council was not justified in serving the notice, or that the work had progressed satisfactorily, the local authority may not recover its expenses or any part of them.

The local authority must not require or undertake work which would offend a building preservation order affecting the building.

Dangerous buildings

Legislation has existed in several statutes relating to the procedures to be followed when a building becomes dangerous. They were the Public Health Act 1936, sections 58 and 296 and the Public Health Act 1961 sections 24, 25 and Schedule 5.

Now the requirements have been gathered together in the Building Act 1984. When a council thinks that a building or structure, or part of a building or structure is in such a condition or has to carry such loads as to be dangerous it may apply to a magistrates' court for an order that the danger be removed.

Buildings that are dangerous because of defects in the structure may result in the court requiring repairs to be undertaken, or the owner may, if he wishes, demolish part or all of the building and take away all rubbish resulting from the demolition.

The magistrates may make an order restricting the use of a building which is overloaded.

The owner will be given a period during which work will be completed and if this target is not achieved, the local authority may step in and put matters right.

Expenses reasonably incurred are recoverable from the owner.

Emergencies

There are occasions when emergency action is necessary to make safe a dangerous building. In these cases a local authority may take immediate action to undertake work itself without prior reference to a court. If it is reasonably possible it must give notice of its intentions to the owner and the occupier.

Expenses relating to fencing off the building, or arranging for it to be watched, may be recovered from the owner.

Owners not paying the expenses may be taken to court. The court will not allow expenses if they conclude that the condition of the building did not warrant emergency action.

The local authority is required to take care that no one sustains damage by reason of its action.

Detrimental to amenity

Occasionally a building or structure becomes so ruinous or delapidated that its appearance is detrimental to the neighbourhood.

One would expect that powers to remedy such a situation would be included within the Planning Acts, and not a Building Act, which deals with this problem.

A local authority has powers to require that a building be repaired or restored in the interests of amenity. The owner on receiving a notice to do works of repair or restoration may opt to demolish the property and if he does, all rubbish must be removed from the site and any adjoining land.

Any freedom to express a choice of either repair or demolition must be exercised carefully if the building is listed as being of architectural or historic interest, or is within a conservation area. A separate permission to demolish is required.

Notice of intended demolition

Before a building is demolished the local authority must be notified, together with the gas and electricity undertakings.

This gives the local authority the opportunity of responding with any number or all of the requirements in the Act.

The requirements are:
- shore up any adjacent building;
- weatherproof surfaces on adjacent buildings exposed by the demolition;
- repair damage caused on adjacent buildings;
- clear away rubbish and material arising from demolition;
- disconnect and seal any drain or sewer;
- remove sewers or drains and seal any sewer or drain to which they were connected;
- make good disturbed ground surfaces;
- disconnect gas, water and electricity services;
- arrange any burning of structures or materials with the Health and Safety Executive and the fire authority;
- take steps to make site tidy and protect public amenity.

The person undertaking demolition must give the local authority 48 hours notice when undertaking work on drains and sewers, and 24 hours notice before making good ground surfaces.

Notice of demolition is not required when:
- the work is in consequence of a demolition Order under the Housing Act 1957;
- demolishing an internal part of a building which will continue to be occupied;
- building has less than 1750 cu. ft. or when a greenhouse, conservatory, shed or prefabricated garage forms part of a large building;
- an agricultural building is contiguous to another building that is not an agricultural building.

Appeals

Anyone receiving a notice from a local authority that work is required in consequence of demolishing a building may appeal to a magistrates' court.

Among the reasons for appealing are:

- that the owner of an adjacent building is not entitled to support and ought to pay or contribute towards the cost of shoring;
- that the owner of an adjacent building ought to pay for or contribute towards the cost of waterproofing any exposed surfaces.

Copies of the appeal are to be sent to any person or person named in the appeal.

The most outstanding change in building control legislation centres around the controversial provisions in the Building Act 1984 relating to surveillance of building work otherwise than by local authorities, although they are involved. In the future it will be possible for the supervision of plans and work to be carried out by 'approved inspectors'.

These people will be in private practice and may at the option of the person intending to carry out the work, supervise construction instead of a local authority building control staff. Several new terms are introduced into building control vocabulary such as 'initial notice', 'plans certificate' and 'final notice'.

An 'approved inspector' may achieve his status in one of two ways. He may be approved by the Secretary of State, or by a body or organisation which in accordance with regulations is designated by the Secretary of State for that purpose.

Any such approval may limit the work in relation to which the person concerned is an approved inspector.

The Building Act says that regulations may be made to:

- require the payment of a fee to become an approved inspector;
- provide for the withdrawal of approval and also the removal of designation of an approved body;
- provide that lists of designated bodies and approved inspectors are prepared and kept up to date.

The responsibilities of an approved inspector include the duty of consulting any stipulated person before taking any prescribed step in connection with work or other matters to which building regulations are applicable.

Approved inspectors may make charges for carrying out their responsibilities. Fees charged are to be agreed between an approved inspector and the person who intends to carry out the work.

An approved inspector may arrange for another person to inspect plans or work on his behalf. However he may not delegate the giving of 'plans certificates' or 'final certificates' to any other person. Delegation will also not affect his liability whether civil or criminal if it arises out of functions conferred upon him by the Act or by any building regulation.

An approved inspector is also liable for negligence on the part of anyone carrying on any inspection on his behalf as if it were negligence by a servant acting in the course of his employment.

Regulations say that an approved inspector must:
- ensure that the work will achieve reasonable standards for protection of the health and safety of persons in or about the building and that the work will conform with provisions for the conservation of fuel and power, and facilities for disabled persons;
- ensure that there is satisfactory provision for drainage.

A suitably qualified person wanting to be an approved inspector may apply to the Secretary of State or one of the bodies designated by him to undertake the task of examining or vetting applicants.

In practice it is expected that possibly eight of the existing professional bodies will be designated for this purpose. Great care will be necessary to ensure that the standard of acceptance is co-ordinated to some form of common syllabus. It is expected that in addition to professional qualfication, demonstration of knowledge of building regulations and several years' experience will be necessary before approval can be achieved.

Initial consultation papers from the DoE suggested that there should be several degrees of responsibility depending on the experience of an individual. One suggestion was that some inspectors should be able to certify housing up to three storeys. Others could have responsibility for housing above that level, buildings of framed or crosswall construction, buildings where special fire precautions were necessary and buildings with large span roofs.

It has since been recognised that even those people with the highest qualifications and longest experience might in some cases, not have all the specialist skills relevant to all aspects of building regulations.

Therefore it is vital that an approved inspector should recognise the limitations of his own knowledge, and the need to obtain advice from specialists.

Recognition may be withdrawn from an approved inspector, and provision will also be made for the removal of designation from an approved body.

Responsibilities

Approved inspectors may at the option of a person intending to carry out work, supervise construction instead of a local authority.

The person intending to carry out work and his approved inspector are each required to tell the local authority of the arrangement. This is achieved by completing a form called an 'initial notice'. When these notices are sent to a council they must be accompanied by some plans and evidence of insurance cover relating to the work.

If an initial notice is accepted by the council, the approved inspector may then undertake the prescribed functions with respect to the inspection of plans, supervision of the work and the giving of certificates and notices.

A local authority may not reject an initial notice unless it fails to comply with a list of requirements to be set out in regulations. An initial notice must be rejected if it fails to meet any of the prescribed grounds. The council has no discretion in this matter.

Initial notice rejection
Anticipated grounds for the rejection of an initial notice are as follows:
● the forms, plans information and certificates are incomplete. There is no copy of the document showing the person to be an approved inspector.
● not signed by the approved inspector;
● does not show satisfactory provision for removal of refuse, or building over a sewer;
● drainage is not satisfactory;
● the approved inspector has not signed a declaration that he does not have any direct or indirect financial interest in the work;
● an initial notice has already been given;
● there is not undertaking to consult a fire officer if this is necessary;
● no indication that any local enactment affecting the work is to be complied with.

Certain quite small works are excluded from the requirement that an approved inspector shall not have a direct or indirect professional or financial interests in a project he certifies. These are alterations or extensions of a dwelling house of two storeys or less.

The Public Health Act 1936 enabled a local authority to attach conditions when passing plans, but not actually relating to the implementation of the regulations themselves. They are now listed in the Building Act 1984, sections 18 to 29.

Where, when considering an initial notice, it is apparent to a local authority that they could under the existing plan deposit arrangements have imposed a condition on passing plans, they may impose similar requirements following receipt of an initial notice.

Local authorities are given a minimal time in which to consider an initial notice. This is not stated in the Act but is expected to be prescribed in regulations as ten days. Within this period the local authority must give notice of rejection, specifying the grounds in question to each of the persons by whom the initial notice was given. If they fail to do so within the time allowed the authority shall be conclusively presumed to have accepted the initial notice and have done so without imposing any requirements.

An initial notice comes into effect when it is accepted by a local authority and continues in force until it is cancelled, or until six weeks after completion of the work.

Regulations may empower a local authority to extend the life of an initial notice.

The form of the initial notice has been drafted and both the person undertaking the work and the inspector give the notice and furnish all the information required.

This is:
1. description of the work and its location;
2. use of building;
3. name of approved inspector;
4. name of person intending to carry out the work;
5. to be accompanied by:
   a. block plan;
   b. copy of inspectors approval document;
   c. declaration of insurance;
   d. drainage outfall details;
   e. whether proposal is over a sewer or drain;
   f. relevant local enactments;
   g. whether the work is minor;
   h. declaration that the inspector is aware of his obligations.

The form is to be signed by both the inspector and the person intending to carry out the work.

Regulations are to be made imposing requirements with respect to the provision of insurance cover against the occurrence of defects; in particular:

● prescribe the form and content of policies, and
● make provision for the approval of schemes of insurance.

Professionals in private practice, whether as engineers, architects or surveyors may in the future qualify to become 'approved inspectors' for the purpose of building control. Some may even create practices devoted to building control consultancy. Self certification may only take place where the person is an approved inspector and the works are very minor. According to draft regulations minor works are alterations or extensions to dwelling houses which have two storeys or less.

Earlier we outlined the appointment of approved inspectors and how a person intending to carry out work, and the approved inspector appointed for the job, had to serve an 'initial notice' on the local authority.

An essential ingredient of this scheme is the possession of adequate insurance cover for each job. Proposals are to be prepared by the insurance industry, and must be approved by the Secretary of State before being used by approved inspectors.

Before a scheme is approved it is suggested that the Secretary of State must be satisfied that at least four requirements are met:
- the cover is provided by an insurer authorised by the Insurance Companies Act 1982;
- there is an indemnity for the owner of the works and his successors, in respect of repairing damage attributable to non-compliance with the building regulations. This cover must last for ten years;
- there is also indemnity for the owner and his successors against the consequences of neglect by the person who gave the initial notice if defects appear within fifteen years of giving the final certificate. Liability must not be less than £500,000 plus the value of work to which the cover relates;
- there must be insurance providing for the increase of the amount of cover by reference to an index of building costs or 15 per cent.

Effect of initial notice

So long as an initial notice remains in force the function of enforcing building regulations conferred on a local authority is removed from work which is specified in the notice.

In consequence a local authority may not:
- give notice requiring the removal or alteration of any work which contravenes building regulations;
- institute proceedings for any contravention of the building regulations which arises out of carrying out the work specified in an initial notice.

The giving of an initial notice covers the developer's responsibilities under several enactments, and for this purpose:
- the giving of an initial notice accompanied by plans, is to be treated as the deposit of plans;
- plans accompanying any initial notice are to be treated as deposited plans;
- the acceptance or rejection of an initial notice is regarded as the passing or rejection of plans;
- should an initial notice cease to be in force it is to be treated as a declaration that the plans are of not effect.

The enactments referred to are as follows:
  (a) the power contained in section 36(2) Building Act 1984, which

enables a local authority to require work which has been undertaken without the deposit of plans, or notwithstanding the rejection of plans, to be pulled down or removed;
 (b) the provision whereby a local authority may not require removal of work shown on plans they have approved;
 (c) the power of a local authority, or the Attorney General to obtain an injunction for certain contraventions;
 (d) building over sewers;
 (e) the advance payments code;
section 219 to 225 of the Highways Act 1980.

For the purposes of section 13 of the Fire Protection Act 1971 the acceptance of an initial notice by a local authority is to be treated as a deposit of plans.

Cancellation of initial notice
Where an initial notice is in force and the approved inspector:
● expects to be unable to carry out his functions, or
● is unable adequately to carry out his functions, or
● is of the opinion that there is a contravention of any provision of building regulations with respect to any of the work, he must cancel the initial notice, by informing the local authority and also the person carrying out the work.

In the event of contraventions the approved inspector must, before cancelling the initial notice, give notice of the contravention to the person carrying out the work. If after three months that person has neither pulled down nor removed the work, nor effected such alterations in it as may be necessary to secure compliance with building regulations, then the approved inspector has no option but to cancel the initial notice.

If at any time it appears to the person carrying out or intending to carry out the work specified in an initial notice that the approved inspector is no longer willing or able to carry out his function he must cancel the initial notice. The cancellation must be sent to the local authority, and where practicable, to the approved inspector. Failure to give this notice to a local authority may result in a fine upon summary conviction.

Initial notice ceasing to have effect
There are provisions which apply when an initial notice ceases to be in force. They are quite involved and apply when an initial notice is no longer operative because:
● it is cancelled by the approved inspector saying that he is unable to carry out his functions, or that he is unable to secure a remedy for defective work;

- it is cancelled by the person carrying out the work when it appears to him that the approved inspector is no longer willing or able to carry out his functions;
- the period covered by an initial notice, if one has been prescribed, has expired.

Building regulations may be made to provide that, if:

(a) a plans certificate was given before the day on which the initial notice ceased to be in force, and

(b) that certificate was accepted by the local authority (before, on or after that day), and

(c) before that day, acceptance was not rescinded by a notice that the work had not been commenced within three years of the date when the plans certificate was accepted, then, with respect to the work set out in the certificate the local authority may not give notice for the removal or alteration of work contravening the building regulations or institute proceedings in respect of contraventions.

However, draft building regulations say that a local authority may ask for work to be opened for checking when a final certificate does not exist, when a final certificate has been rejected or the building has been occupied six weeks and no final certificate has been received.

When, on occasions before the initial notice ceases to be in force a final certificate was given in respect of part of the work and accepted by a local authority, before on or after that date, the fact that the initial notice has ceased to be in force does not affect the continuing operation of a requirement that the local authority may not give notice for the removal or alteration of work contravening the regulations or institute proceedings in respect of contraventions.

Notwithstanding anything just outlined, the local authority is given powers referring to any part of the work not specified in a plans certificate, or a final certificate. Building regulations may require the submission of plans which relate not only to that part, but also to the part to which the certificate in question relates.

Any notices which may be served and are normally required to be given within twelve months of a contravention are now to be served within twelve months of the date the initial notice ceased to be in force.

Authority is given for new building regulations to provide that if, before the initial notice ceased to be in force a contravention of building regulations was committed with respect to work specified in the notice summary proceedings may be taken. This must be within six months of the initial notice ceasing to be of effect.

If an initial notice has ceased to be in force and no final certificate has been given and accepted it is possible to give a new initial notice.

However, where:
- a plans certificate relates to any of the work, and
- a plans certificate was given before an initial notice ceased, was accepted and not rescinded, or
- the initial notice was given and accepted the responsibilities of an approved inspector do not apply to work contained in the new initial notice which is specified in the plans certificate.

The whole purpose appears to provide that where no final certificate is issued by the approved inspector supervision of any part of the work which has not been certified becomes the responsibility of the local authority. A maximum fine of £1000 is provided for failure to serve certain notices.

The Building Act 1984 makes provision for building control to be exercised by persons other than building control officers employed by the local authority. Suitably qualified people may become approved inspectors and undertake the checking of plans and inspection of work to secure compliance with the building regulations. They must be adequately insured within schemes which are approved and upon appointment send in 'initial notices' to the local authorities.

Plans certificates

When an approved inspector has examined plans and is satisfied that they are neither defective nor show that the work would contravene building regulations, and when he has complied with any prescribed requirements as to consultation or otherwise, he must, if requested to do so by the person intending to carry out the work, give a 'plans certificate' to the local authority and that person.

It may be possible for an approved inspector to give an initial notice and a plans certificate at the same time on a special combined notice form.

A plans certificate:
- may relate either to the whole or part of the work specified in the initial notice, and
- will not have effect unless it is accepted by the local authority to whom it is given.

However a local authority may not reject a plans certificate except on grounds which are set out in regulations. They must reject the certificate if any of the specified grounds are found to exist.

If the rejection is not sent to the approved inspector and the other person to whom the approved inspector gave the certificate within ten working days then the authority shall be conclusively presumed to have accepted the certificate.

Should a dispute arise between the approved inspector and the person proposing to carry out the work, the matter may be referred to the Secretary of State for a determination. A fee may be charged.

When issuing a plans certificate the approved inspector is required to give the following information:
- a description of the work and its location;
- a declaration that he is the approved inspector appointed for this work and referred to in the initial notice;
- a declaration of insurance;
- a statement that he has examined plans of the work and that he is satisfied that they are not defective. He must also say that work carried out in accordance with the plans would not contravene any building regulations;
- indicate whether the fire authority had been consulted;
- a statement that the work is, or is not, minor work;
- a statement that he has no direct or indirect professional or financial interest;
- a list of plans received giving date and reference number of each plan;
- the certificate must be dated and signed.

Draft regulations say a local authority may reject a plans certificate if:
- it is not on the proper form;
- the certificate does not specify the work to which it relates, or the plans to which it relates;
- there is no initial notice in force;
- it is not signed by the inspector who gave the initial notice, or the person is no longer an approved inspector;
- the certificate does not contain a declaration that the fire officer has been consulted, where it is necessary;
- there is no declaration signed by the insurer that a named scheme of insurance relates to the work;
- there is no declaration that the approved inspector does not have any direct or indirect financial investment.

If, following receipt by a local authority of a plans certificate, the work to which it relates has not commenced within a period of three years the authority may rescind acceptance of the certificate.

Final certificates

Local authority enforcement powers cannot be exercised in respect of work which has been supervised by an approved inspector.

When an approved inspector is satisfied that the work contained in an initial notice has been completed he must give to the local authority and the person for whom the work was carried out, a 'final certificate'. This final certificate will indicate completion of the work and the discharge of his functions by the approved inspector.

The form of a final certificate is prescribed in the regulations, and must contain the following:
- a description of the work and its location;
- a declaration that the person completing the form is an approved inspector and that the work was the subject of an initial notice;
- that the work has been completed, that he is satisfied within the limits of professional skill and care that the work:

   1) complies with the required standards for health and safety;

   2) complies with requirements for fuel and power conservation;

   3) meets requirements for facilities for the disabled;

   4) has satisfactory provision for drainage;

   5) has satisfactory provision for ingress and egress and passages or gangways.
- a declaration of insurance;
- statement that fire authority has been consulted if appropriate;
- whether the work is or is not minor work;
- a certificate that the approved inspector has no direct or indirect professional or financial interest in the work;
- the date, and signature of the approved inspector.

In the same way, and subject to conditions which are to be prescribed, final certificates may be given to and rejected by a local authority. If within ten days which are allowed a local authority does not give notice of rejection the council will be conclusively presumed to have accepted the final certificate.

Draft regulations say a local authority may reject a final certificate if:
- it is not on the proper form;
- the certificate does not specify the work;
- no initial notice applies to the work;
- the certificate was not signed, or the person is no longer an approved inspector;
- the certificate does not contain a declaration that insurance cover exists;
- the certificate does not contain a declaration that the fire officer has been consulted where necessary;
- there is no declaration regarding direct or indirect financial or professional interest.

So long as an initial notice remains in force the local authority may not require the removal of defective work or institute proceedings for contraventions.

Another clause takes this requirement further. It provides that when a final certificate has been given to and accepted by the local authority, the initial notice will cease to apply to that work. Even so, the local authority may not require the removal of defective work or institute proceedings for contraventions.

Changes in building control arrangements will enable a developer to employ an approved inspector to check his plans, inspect the work in progress and finally give a certificate that all is in order.

Running alongside this arrangement is the facility of sending plans to the local authority. But this is to be varied from the way things have been done for many years. A person intending to carry out work, or to make a material change of use, is to have the option of giving a local authority a 'building notice' or deposit full plans.

An outline of the requirements surrounding a building notice is set out below.

However this procedure cannot be used:
● when the regulations require means of escape to be provided in the building in case of fire;
● when the work is on a building intended to be put to a designated use under section 1 of the Fire Precautions Act 1971, (Compulsory fire certificates).

Anyone deciding to use this procedure is required to send a building notice to the council containing the following information.
● a description of the work or material change of use;
● details of location, the number of storeys and proposed use of the building.

When the proposal is for a new building or an extension, there must also be:
● a block plan to a scale of 1:500 showing size and position of building and its relationship to adjoining boundaries;
● the curtilage, and the size, position and use of every other building or proposed building within the curtilage;
● width and position of any adjoining street;
● details of drainage;
● provision being made for entrances and exits, passages and gangways.

The local authority, upon receipt of a building notice, may ask for plans 'as may reasonably be required by them in the discharge of their functions'. They must be provided within a reasonable time.

A building notice and any accompanying plans are not to be treated as having been deposited. In this respect section 16 of the Building Act 1984 says that a local authority on receipt of deposited plans must approve them or reject them or approve them in part, within five weeks. The effect of this arrangement is to put the local authority in a position where it is not given the power to approve or reject plans of a minor character, because they will not have been deposited. Neither will they need the services of an inspector.

Examples embrace many small buildings and extensions, including:
- the erection of a dwelling house of not more than two storeys;
- a building containing flats but not of more than two storeys;
- a material change in the use of a building of not more than two storeys for use as a dwelling house;
- loft conversion of a bungalow.

The building must of course be erected to comply with the regulations. The person carrying out the building work is required to give the usual notices of commencement and at various stages of work and at completion.

The notices to be given are similar to those now in use. However whereas at the present time all notices must be in writing, it will be possible in future for notices to be given by such other means as the builder and local authority may agree.

This idea of an optional deposit of plans, or the giving of a building notice allows developers to choose whether they wish to deposit full plans with the council as they do at the present time, or whether they wish to avoid the routine of getting approval or rejection to their proposal, by providing only outline information.

If a developer chooses to deposit full plans then the authority must deal with them in the same way as in the past. In this respect passed plans give the builder protection against enforcement notices which have been enjoyed under Section 65 Public Health Act 1936; now replaced by Section 36 Building Act 1984.

The builder may think that his project is so straightforward that he does not need the protection of passed plans. He may also be satisfied that informal discussions with the building control officers had solved any possibility of disagreements arising during the progress of his scheme.

In these circumstnces the local authority, on receipt of a building notice, would not be able to insist on the supply of full plans. They will nevertheless be able to ask for any extra information reasonably needed to enable them to discharge their functions in enforcing the regulations, and the listed powers in public and local lesiglation.

The local authority will still be able to take action against contraventions found in the course of the work. It appears that the arrangements are proposed to reduce the need for unnecessary bureaucratic procedures.

This proposal has, in some part, been derived from arrangements in inner London, where for many buildings there is no requirement for prior approval of plans.

Full plans
The alternative to giving a building notice is to deposit full plans with a local authority. Such plans must be approved, rejected or approved in part, as at present.

Full plans may be deposited for any work which could be covered by a building notice, at the builder's option, but there are proposals for which plans must be deposited. They are:
- a dwelling house having more than two storeys;
- addition of a third storey to a two storey dwelling house, and this includes loft conversions;
- buildings having flats in more than two storeys;
- shops;
- offices;
- material change of use in a building or more than two storeys for use as a dwelling;
- work on buildings intended to be put to a use designated by section 1 of the Fire Precautions Act 1971 (compulsory fire certificate).

Full plans are to give a description of the proposed work or material change of use together with: location and number of storeys; proposed use; block plan; size and position of building; boundaries; position of all buildings within the curtilage; details of nearby streets; drainage details; exits; details generally as at present required.

Summary
A person intending to carry out work may engage an approved inspector who is required to:
- send an initial notice to the council;
- check plans and send a plans certificate to the council;
- check the work and send a final notice to the council.

In each instance the council may accept the notice or certificate. If they find any failure to give adequate information it may be rejected.

As an alternative a person intending to carry out work may deal directly with the council.

By sending a building notice he does not get an approval or rejection, but the building control officer will carry out inspections and require compliance with the regulations.

By sending full plans he receives an approval or rejection of his plans and the work is inspected by the building control officer.

2. MEASURING DATA AND PRELIMINARIES

Numeration	51
Imperial System of Weights and Measures	51
Metric System of Weights and Measures	53
Metric Conversion	54
Length	58
Area	60
Capacity	61
Weight	62
Temperature Conversions	62
Geometric Data	66
Trigonometry	67
Map Scales	71
Drawing Paper and Stationery Sizes	71
Abbreviations on Plans and Drawings	73

NUMERATION

With the adoption of the metric system, it is probable that Britain will adopt the Continental billion, trillion, etc., but official circles have not yet reached a decision on this point.

Units	1
Tens	10
Hundreds	100
Thousands	1,000
Tens of thousands	10,000
Hundreds of thousands	100,000
Millions (thousands of thousands)	1,000,000

ROMAN NUMERALS

I	1		LXX	70
II	2		LXXX	80
III	3		LXXXVIII	88
IV	4		XC	90
V	5		XCIX	99
VI	6		C	100
VII	7		CX	110
VIII	8		CXI	111
IX	9		CXC	190
X	10		CC	200
XI	11		CCXX	220
XII	12		CCXXIV	224
XIII	13		CCC	300
XIV	14		CCCXX	320
XV	15		CD	400
XVI	16		D	500
XVII	17		DC	600
XVIII	18		DCCC	800
XIX	19		DCCCLXXVI	876
XX	20		CM	900
XXX	30		CMXCIX	999
XL	40		M	1000
L	50		MD	1500
LV	55		MDCCC	1800
LX	60		MM	2000

IMPERIAL SYSTEM OF WEIGHTS AND MEASURES

This system of weights is derived from Roman, Saxon, Old French and other traditional systems. The traditional basis of TROY WEIGHT (the original English system) is 'one grain from the middle of the ear of wheat', which originally was the Troy grain. APOTHECARIES' WEIGHT is a variant of the TROY system. AVOIRDUPOIS WEIGHT, introduced in mediaeval times, took an ancient Greek coin (the *drachma,* or handful) as its basis.

AVOIRDUPOIS WEIGHT

27½ Grains Troy	=	1 dram or drachm
16 drams	=	1 ounce
1 lb Avoirdupois	=	1.2153 lb Troy

Ounces	Pounds	Stones	Quarters	Cwts	Tons
16	1				
224	14	1			
448	28	2	1		
1,792	112	8	4	1	
35,840	2,240	160	80	20	1

MEASURE OF CAPACITY

Cubic Inch							
8.665 =	1 Gill						
34.659 =	4 =	1 Pint					
69.318 =	8 =	2 =	1 Quart				
138.637 =	16 =	4 =	2 =	1 Pottle			
277.274 =	32 =	8 =	4 =	1 Gallon			
554.548 =	64 =	16 =	8 =	2 =	1 Peck		
2,218.192 =	256 =	64 =	32 =	8 =	4 =	1 Bushel	
17,745.536 =	2,048 =	512 =	256 =	64 =	32 =	8 =	1 Qr.

1 Bushel = 1.284 Feet Cube
(1 Cubic Foot = 6.24 Gallons = 7.4805 US Gallons)

LONG MEASURE

Inches	Feet			
12	1	Yards		
36	3	1	Furlongs	
792	66	22	1	Miles
63,360	5,280	1,760	8	1

SQUARE MEASURE

Square inch	Square feet	Square yard	Rood	Acre	Square mile
144 =	1				
1,296 =	9 =	1			
1,568,160 =	10,890 =	1,210 =	1		
6,272,640 =	43,560 =	4,840 =	4 =	1	
4,024,489,600 =	27,878,400 =	3,097,600 =	2,560 =	640 =	1

ALSO

1 Irish acre = 7,840 sq. yards   1 Scottish acre = 6,104 sq. yards
            yards   years

CUBIC OR SOLID MEASURE

1,728 cubic inches	=	1 cubic foot
27 cubic feet	=	1 cubic yard
1 ton Freight Measure	=	8 Barrels, or 40 cubic feet
1 Load rough or round Timber	=	40 cubic feet
1 Load hewn or sawn Timber	=	50 cubic feet

THE METRIC SYSTEMS OF WEIGHTS AND MEASURES

The basic system was adopted by the French Government in 1795, and is an arbitrary one, in which all units of length area, capacity and weight are derived from the length of a quadrant of the meridian—that is, the distance from the equator to the pole. One ten-millionth of this is *the metre*. From the metre are derived all the other units of measurement: The *are*, which is 100 square metres and is the unit of area; the *litre* which is one cubic decimetre, and is the unity of capacity; and the *gram*, which is one cubic centimetre of water, and is the unit of weight. The metric system is a decimal one—that is, 10 of one unit always equal one of the next highest, as will be seen from the tables below.

The *Systéme International des Unités* (International System of Units) which Great Britain is adopting in the changeover to metric, is a modern extension of the original or basic metric system and contains a large number of closely related units of measurement, some of which differ radically from Imperial and older metric units. It has not been adopted by all countries using the basic metric system.

LINEAR MEASURE

10 millimetres	= 1 centimetre	= 0.3937 in	
10 centimetres	= decimetre	= 3.9370 in	
10 decimetres	= 1 decimetre	= 3 ft 3.3708 in	
10 metres	= 1 decametre	10 yd 2 ft 9.7079 in	
10 decametres	= hectometre	= 109 yd 1 ft 1.0790 in	
10 hectometres	= 1 kilometre	= 1,091 yd 1 ft 10.7900 in	

SQUARE MEASURE

1 centare	= 1.1960 sq. yd	
100 centares	= 1 are	= 119.6033 sq. yd
100 ares	= 1 hectare	= 2 acres, 2280.3326 sq. yd

CUBIC OR SOLID MEASURE

10 centisteres	= 1 decistere	= 3.5317 cu ft
10 decisteres	= 1 stere	= 35.3170 cu ft

CAPACITY

1 litre	= 1.760 pints	
10 litres	= 1 decalitre	= 2 gal 1.607 pints
10 decalitres	= 1 hectolitre	= 22 gal 0.077 pints
10 hectolitres	=´ 1 kilolitre, or stere, or cubic metre	= 220 gal 0.770 pints

WEIGHT

10 centigrams	= 1 gram	= 0.564 drams
10 grams	= 1 decagram	= 5.643 drams
10 decagrams	= 1 hectogram	= 3 oz 6.170 drams
10 hectograms	= 1 kilogram	= 2 lb 3 oz 4.380 drams

CONVERSION FACTORS

Inches	×	2.540 =	cm
Centimetres	×	0.394 =	in
Metres	×	3.280 =	ft
Metres	×	39.370 =	in
Feet	×	0.305 =	m
Square feet	×	929.000 =	sq cm
Cubic feet	×	28.300 =	litres
Feet per second	×	30.500 =	cm per sec
Kilograms	×	2.205 =	lb
Metric tonne	×	0.984 =	ton of 2240 lb
Pounds	×	0.454 =	kg
Pounds per sq in	×	70.300 =	g per sq cm
Pounds per sq ft	×	0.488 =	g per sq cm
Tons per sq in	×	157.500 =	kg per sq cm

METRIC CONVERSION

What follows are no more than elementary calculations: a textbook on decimal arithmetic would be out of place here, but the basic rules are given because decimal calculations have hitherto not figured in building trade calculations. In the metric system, quantities are expressed thus:

```
                        Decimal
                         Point
        Whole numbers      ↓       Fraction
   1    2    3    4    ●    5    6    7
   ↑    ↑    ↑    ↑         ↑    ↑    ↑
   │    │    │    │         │    │    │
   │    │    │    │         │    │    │ Thousandths (1/1000)
   │    │    │    │         │    │ Hundredths (1/100)
   │    │    │    │         │ Tenths (1/10)
   │    │    │    │ Units
   │    │    │ Tens
   │    │ Hundreds
   │ Thousands
```

Written out in full, in words, this is one thousand two hundred and thirty four, plus five tenths, six hundredths and seven thousandths.

WARNINGS

(1) In our new metric system, spaces are used instead of commas to denote groups of three, thus:

     670 592.745 001

        NOT

     670,592.745001, as we previously wrote it.

This is because in Continental usage the comma is used instead of the decimal point and the point is used instead of our former comma and future space, thus:

670.592,745.001

(2) It has been decided that linear measurements shall be in millimetres (thousandths of a metre) and metres, missing out centimetres (hundredths of a metre) and decimetres (tenths of a metre). It is vital to remember that one metre and five millimetres are not written as 1.5m, but as 1.005m. Similarly, the Continental practice (soon to be our own as well) is to omit the decagrams (10 grams) and hectograms (100 grams) and go straight from grams to kilograms (1,000 grams). Therefore, 1.5kg is 1 kilogram and 500 grams while 1 kilogram and 5 grams is 1.005kg.

ADDITION

Place the quantities to be added together below each other, taking care (as in the addition of whole numbers only) that units are under units, tens under tens and so on, and that the decimal points and decimal fractions are under each other, as below:

8 765.432 1
1 234.567 8
---
9 999.999 9

First, put in the decimal point and then add as in normal addition of whole numbers.

SUBTRACTION

Follow the same method of setting out the calculation as in the subtraction of whole numbers. Put in the decimal point and subtract as if the quantities were whole numbers:

8 765.432 1
1 234.567 8
---
7 530.864 3

MULTIPLICATION

The method usually taught is:

1. Ignore the decimal points and multiply as if the number were whole numbers,

2. Add up the number of decimal points in both numbers and insert

the decimal point before that number of figures in the answer, counting from the right.

Multiply 54.321 by 12.345:

```
        54321
        12345

       271605  five times
       217284  four times
       162963  three times
       651852  twelve times

     670592745
```

Inserting the decimal point after the sixth figure, counting from the right, gives 670.592 745 as the answer.

DIVISION

The method usually taught is:

1. Ignore the decimal points and treat as long division of whole numbers.

2. Deduct the number of decimal places in the divisor (see example below) from the number of decimal places in the dividend. Place the decimal point after that number of figures, counting from the right, in the quotient.

Divide 876.54 by 123.4.

```
      Divisor Dividend
      1234  / 87654 /   71 Quotient
              8638

               1274
               1234

                 40 Remainder
```

There are two decimal points in the dividend and one in the divisor. When the latter is subtracted from the former, one remains. Therefore, the decimal point is placed after the first figure of the quotient, counting from the right, and the answer is 7.1.

There are two exceptions to the above rule:

1. If the number of decimal points in the divisor and dividend are equal, the result of the subtraction will be zero, and all the figures in the quotient will be whole numbers.

MEASURING DATA

2. If after subtracting as above (2) there are more decimal points required than figures in the quotient, add zeros to the left of the quotient. For example, if the result of the subtraction is 4 and the quotient is 75, add two zeros to the left of the number, so as to make a four-figure decimal fraction, thus: .007 5.

RECOGNITION POINTS
(Rough, very approximate equivalents in Imperial and metric weights and measures.)

1 oz	30 grams (g)
1 lb	Half a kilogram (kg)
1 cwt	50 kg
1 ton	1 000 kg or 1 metric tonne
1 in	25 millimetres (mm)
1 ft	300 mm
3¼ ft	1 metre (m)
⅝ of a mile	1 kilometre (km)
5 miles	8 km
1 cu.yd	¾ of a m^3
32°F.	0° Celsius (Centigrade) (C.) (water freezes)
60°F.	15.6°C.—background heating temperature
70°F.	21°C.—average central heating temperature
140°F.	60°C.—water temperature at boiler
212°F.	100°C.—water boils

For exact conversion factors and conversion tables, see page 55-59.

In the following tables, look for the number required to be converted in the middle column. If it is in Imperial system, its equivalent in metric will be found under the metric column, and vice versa. The tables can be extended to hundreds by moving the decimal point one place to the right, and to thousands by moving the decimal point two places to the right.

Example: To find the British (Imperial) equivalent of 1 763 millimetres, proceed as follows:

1 000mm	39.3in	(take 10 in table and move decimal point two places to right)
700mm	27.51in	(take 70 in table and move decimal point one place to right)
60mm	2.356in	(as in table)
3mm	0.117in	(as in table)
1 763mm	69.283in	

The *Systéme International des Unités* (International System of Units) called S.I. for short, is a very complex one, and conversion tables only for the units used in building are given below. Special conversion tables, of limited use, are included in the Chapters of this work to which they appertain:

	page
Hours/sq. ft to Hr/m²	95
Hr/cu. ft to Hr/m³	95
Hr/sq. yd to Hr/m²	95
Hr/cu. yd to Hr/m³	95
Lb/ft run to Kg/m. run and Lb/sq. ft to Kg/m²	138
Lb/cu. ft to Kg/m³	138
Cwt/cu. ft to Kg/m³	138
Lb force/sq. ft and kN/m²	139
Ton f/sq. ft to kN/m²	139
Lb f/sq. in to MN/m²	139
In to mm (timber only)	215
Standards (timber) to m³	215
Sq. yd to m²	388
Gallons to litres	388
'K' values—Imperial and metric	342

LENGTH (Long Measure)
FRACTIONS OF ONE INCH IN MILLIMETRES
THIRTY-SECONDS, SIXTEENTHS, EIGHTHS, QUARTERS AND ONE-HALF

In	Mm	In	Mm
1/32	0.794	17/32	13.494
1/16	1.588	9/16	14.288
3/32	2.381	19/32	15.081
1/8	3.175	5/8	15.875
5/32	3.969	21/32	16.668
3/16	4.762	11/16	17.463
7/32	5.556	23/32	18.256
1/4	6.350	3/4	19.050
9/32	7.144	25/32	19.843
5/16	7.938	13/16	20.638
11/32	8.731	27/32	21.431
3/8	9.525	7/8	22.225
13/32	10.318	29/32	23.018
7/16	11.112	15/16	23.812
15/32	11.906	31/32	24.606
1/2	12.700	1 inch	25.400

TWELFTHS, SIXTHS AND THIRDS

In	Mm	In	Mm
1/12	2.11	7/12	14.81
1/16	4.22	2/3	17.00
1/4	6.35	3/4	19.05
1/3	8.50	5/6	21.27
5/12	10.61	11/12	23.31
1/2	12.70	1 inch	25.40

INCHES AND MILLIMETRES

In		Mm	In		Mm
0.0393	1	25.4	0.549	14	355.6
0.0786	2	50.8	0.588	15	381.0
0.1170	3	76.2	0.627	16	406.4
0.1572	4	101.6	0.666	17	431.8
0.1965	5	127.0	0.705	18	457.2
0.2356	6	152.4	0.744	19	482.6
0.2751	7	177.8	0.783	20	508.0
0.3154	8	203.4	0.822	21	533.4
0.3537	9	228.6	0.861	22	558.8
0.393	10	254.0	0.900	23	584.2
0.432	11	279.4	0.939	24	609.6
0.471	12	304.8	0.978	25	635.0
0.510	13	330.2	1.017	26	660.0

FEET AND METRES / YARDS AND METRES

Feet		Metres	Yards		Metres
3.280	1	0.324	1.093	1	0.914
6.561	2	0.648	2.187	2	1.828
9.842	3	0.972	3.280	3	2.743
13.123	4	1.219	4.374	4	3.657
16.404	5	1.524	5.468	5	4.572
19.685	6	1.829	6.561	6	5.486
22.965	7	2.133	7.655	7	6.400
26.246	8	2.484	8.748	8	7.315
29.527	9	2.743	9.842	9	8.229
32.808	10	3.048	10.936	10	9.144
65.616	20	6.096	21.872	20	18.228
98.425	30	9.144	32.808	30	27.432
131.234	40	12.192	43.774	40	36.576
164.042	50	15.240	54.680	50	45.720
196.850	60	18.288	65.610	60	54.864
299.659	70	21.336	76.552	70	64.008
262.467	80	24.384	87.489	80	73.152
295.276	90	27.432	98.425	90	82.296
328.084	100	30.480	109.361	100	91.440

AREA (Square Measure)

Because the numbers would be too large and unwieldy for general use, conversion from inches to square centimetres is given, instead of conversion to square millimetres.

SQUARE INCHES AND SQUARE CENTIMETRES

Sq. in		Cm^2
0.155	1	6.457
0.310	2	12.903
0.465	3	19.355
0.620	4	25.806
0.775	5	32.258
0.930	6	38.709
1.085	7	45.161
1.240	8	51.613
1.395	9	58.064
1.550	10	64.516
3.100	20	129.032
4.650	30	193.548
6.200	40	258.064
7.750	50	332.580
9.300	60	387.096
10.850	70	451.612
12.400	80	516.13
13.95	90	580.64
15.50	100	645.16

SQUARE FEET AND SQUARE METRES

Sq. ft		M^2
10.763	1	0.092
21.527	2	0.186
32.291	3	0.279
43.055	4	0.372
53.820	5	0.464
64.583	6	0.557
75.347	7	0.650
86.111	8	0.743
96.875	9	0.836
107.639	10	0.929
215.278	20	1.858
322.917	30	2.787
430.556	40	3.716
533.20	50	4.645
645.83	60	5.574
753.47	70	6.503
861.11	80	7.432
968.75	90	8.361
1076.39	100	9.290

SQUARE YARDS AND SQUARE METRES

M^2		Sq. yd
0.836	1	1.196
1.672	2	2.392
2.508	3	3.588
3.344	4	4.784
4.181	5	5.980
5.017	6	7.176
5.883	7	8.372
6.689	8	9.568
7.525	9	10.764
8.361	10	11.900
16.723	20	23.92
25.084	30	35.88
33.445	40	47.84
41.806	50	59.80
50.186	60	71.76
58.529	70	83.72
66.890	80	95.68
75.251	90	107.64
83.613	100	119.61

CAPACITY (Cubic Measure)

Because the number would be too large and unwieldy for general use, conversion from cubic inches to cubic centimetres is given, instead of conversion to cubic millimetres.

CUBIC INCHES AND CUBIC CENTIMETRES			CUBIC FEET AND CUBIC METRES		
Cm^3		Cu. in	Cu. ft		M^3
16.38	1	0.06	35.31	1	0.0283
32.77	2	0.12	70.62	2	0.0566
49.16	3	0.18	105.94	3	0.0849
65.54	4	0.24	141.25	4	0.1132
81.93	5	0.30	176.57	5	0.1415
98.32	6	0.36	211.88	6	0.1699
114.70	7	0.42	247.20	7	0.1982
131.09	8	0.48	282.51	8	0.2265
147.48	9	0.54	317.83	9	0.2548
163.87	10	0.61	365.14	10	0.2831
327.74	20	1.22	706.29	20	0.5655
491.61	30	1.83	1059.44	30	0.8549
655.48	40	2.44	1412.59	40	1.1326
819.35	50	3.05	1765.43	50	1.4150
983.22	60	3.66	2118.88	60	1.6990
1147.09	70	4.27	2472.03	70	1.9820
1310.97	80	4.38	2825.17	80	2.2650
1474.84	90	5.49	3178.32	90	2.5480
1638.71	100	6.10	3531.32	100	2.8310

CUBIC YARDS AND CUBIC METRES			GALLONS AND LITRES (LIQUID)		
M^3		Cu. yd	Gallons		Litres
0.76	1	1.30	0.220	1	4.545
1.53	2	2.61	0.440	2	9.091
2.29	3	3.92	0.659	3	13.637
3.05	4	5.23	0.874	4	18.183
3.82	5	6.53	1.099	5	22.729
4.59	6	7.84	1.319	6	27.275
5.35	7	9.15	1.539	7	31.821
6.11	8	10.46	1.759	8	36.367
6.88	9	11.77	1.979	9	40.913
7.64	10	13.07	2.199	10	45.459
15.29	20	26.16	4.399	20	90.918
22.93	30	39.23	6.599	30	136.377
30.58	40	52.31	8.799	40	181.836
38.22	50	65.39	10.998	50	227.395
45.87	60	78.47	13.198	60	272.754
53.51	70	91.55	15.398	70	318.213
61.16	80	104.63	17.598	80	363.662
68.81	90	117.72	19.798	90	409.131
76.46	100	130.80	21.998	100	454.780

WEIGHT

POUNDS AND KILOGRAMS			IMPERIAL TONS AND METRIC TONNES		
Pounds		Kilograms	Imperial tons		Metric tonnes
2.204	1	0.453	0.984	1	1.016
4.409	2	0.907	1.968	2	2.032
6.613	3	1.360	2.952	3	3.048
8.818	4	1.813	3.962	4	4.064
11.321	5	2.267	4.921	5	5.080
13.227	6	2.721	5.095	6	6.096
15.432	7	3.175	6.889	7	7.112
17.637	8	3.028	7.873	8	8.128
19.841	9	4.082	8.857	9	9.144
22.046	10	4.535	9.842	10	10.166
44.092	20	9.071	19.684	20	20.320
66.139	30	13.607	29.526	30	30.481
88.185	40	18.143	39.360	40	40.641
110.231	50	22.679	49.210	50	50.802
132.277	60	27.216	59.050	60	60.902
154.329	70	31.753	68.890	70	71.122
176.370	80	36.290	78.730	80	81.283
198.416	90	40.827	88.570	90	91.442
220.460	100	45.358	98.420	100	101.605

TEMPERATURE CONVERSIONS

It has been decided that Great Britain will use the Centigrade scale of temperature, but will in future call it the Celsius scale (after its inventor, Celsius). The S.I. unit of temperature is the degree Kelvin (after its inventor, Lord Kelvin), which is the same size as the degree Celsius (or Centigrade) but the Kelvin scale begins at minus 273 degrees Celsius, because this is thought to be 'absolute zero'—the lowest possible temperature in the universe. Thus, 100°C equal 373°K. 'Colour temperature' of light sources is measured in degrees K.

Find the temperature that you wish to convert (whether it be Celsius or Fahrenheit) in the central column (in bold type). The Celsius equivalent is then given on the same line, in the left-hand column, and the Fahrenheit equivalent is given on the same line, in the right-hand column.

Examples: (i) What is the Celsius equivalent of 15 F.?
Answer: 9.4 C.

(ii) What is the Fahrenheit equivalent of 15 C.?
Answer: 59.0 F.

TEMPERATURE CONVERSION

C.	F.	C.		F.
−27.2	−17	1.4	−2.2	28	82.4
−26.7	−16	3.2	−1.7	29	84.2
−26.1	−15	5.0	−1.7	30	86.0
−25.6	−14	6.8	−0.6	31	87.8
−25.0	−13	8.6	0	32	89.6
−24.4	−12	10.4	0.6	33	91.4
−23.9	−11	12.2	1.1	34	93.2
−23.3	−10	14.0	1.7	35	95.0
−22.8	−9	15.8	2.2	36	96.8
−22.2	−8	17.6	2.8	37	98.6
−21.7	−7	19.4	3.3	38	100.4
−21.1	−6	21.2	3.9	39	102.2
−20.6	−5	23.0	4.4	40	104.0
−20.0	−4	24.8	5.0	41	105.8
−19.4	−3	26.6	5.6	42	107.6
−18.9	−2	28.4	6.1	43	109.4
−18.3	−1	30.2	6.7	44	111.2
−17.8	0	32.0	7.2	45	113.0
−17.2	1	33.8	7.8	46	114.8
−16.7	2	35.6	8.3	47	116.6
−16.1	3	37.4	8.9	48	118.4
−15.6	4	39.2	9.4	49	120.2
−15.0	5	41.0	10.0	50	122.0
−14.4	6	42.8	10.6	51	123.8
−13.9	7	44.6	11.1	52	125.6
−13.3	8	46.4	11.7	53	127.4
−12.8	9	48.2	12.2	54	129.2
−12.2	10	50.0	12.8	55	131.0
−11.7	11	51.8	13.3	56	132.8
−11.1	12	53.6	13.9	57	134.6
−10.6	13	55.4	14.4	58	136.4
−10.0	14	57.2	15.0	59	138.2
−9.4	15	59.0	15.6	60	140.0
−8.9	16	60.8	16.1	61	141.8
−8.3	17	62.6	16.7	62	143.6
−7.8	18	64.4	17.2	63	145.4
−7.2	19	66.2	17.8	64	147.2
−6.7	20	68.0	18.3	65	149.0
−6.1	21	69.8	18.9	66	150.8
−5.6	22	71.6	19.4	67	152.6
−5.0	23	73.4	20.0	68	154.4
−4.4	24	75.2	20.6	69	156.2
−3.9	25	77.0	21.1	70	158.0
−3.3	26	78.8	21.7	71	159.8
−2.8	27	80.6	22.2	72	161.6

TEMPERATURE CONVERSION—*continued*

C.		F.	C.		F.
22.8	**73**	163.4	47.8	**118**	244.4
23.3	**74**	165.2	48.3	**119**	246.2
23.9	**75**	167.0	48.9	**120**	248.0
24.4	**76**	168.8	49.4	**121**	249.8
25.0	**77**	170.6	50.0	**122**	251.6
25.6	**78**	172.4	50.6	**123**	253.4
26.1	**79**	174.2	51.1	**124**	255.2
26.7	**80**	176.0	51.7	**125**	257.0
27.2	**81**	177.8	52.2	**126**	258.8
27.8	**82**	179.6	52.8	**127**	260.6
28.3	**83**	181.4	53.3	**128**	262.4
28.9	**84**	183.2	53.9	**129**	264.2
29.4	**85**	185.0	54.4	**130**	266.0
30.0	**86**	186.8	55.0	**131**	267.8
30.6	**87**	188.6	55.6	**132**	269.6
31.1	**88**	190.4	56.1	**133**	271.4
31.7	**89**	192.2	56.7	**134**	273.2
32.2	**90**	194.0	57.2	**135**	275.0
32.8	**91**	195.8	57.8	**136**	276.8
33.3	**92**	197.6	58.3	**137**	278.6
33.9	**93**	199.4	58.9	**138**	280.4
34.4	**94**	201.2	59.4	**139**	282.2
35.0	**95**	203.0	60.0	**140**	284.0
35.6	**96**	204.8	60.6	**141**	285.8
36.1	**97**	206.6	61.1	**142**	287.6
36.7	**98**	208.4	61.7	**143**	289.4
37.2	**99**	210.2	62.2	**144**	291.1
37.8	**100**	212.0	62.8	**145**	293.0
38.3	**101**	213.8	63.3	**146**	294.8
38.9	**102**	215.6	63.9	**147**	296.6
39.4	**103**	217.4	64.4	**148**	298.4
40.0	**104**	219.2	65.0	**149**	300.2
40.6	**105**	221.0	65.6	**150**	302.0
41.1	**106**	222.8	66.1	**151**	303.8
41.7	**107**	224.6	66.7	**152**	305.6
42.2	**108**	226.4	67.2	**153**	307.4
42.8	**109**	228.2	67.8	**154**	309.2
43.3	**110**	230.0	68.3	**155**	311.0
43.9	**111**	231.8	68.9	**156**	312.8
44.4	**112**	233.6	69.4	**157**	314.6
45.0	**113**	235.4	70.0	**158**	316.4
45.5	**114**	237.2	70.6	**159**	318.2
46.1	**115**	239.0	71.1	**160**	320.0
46.7	**116**	240.8	71.7	**161**	321.8
47.2	**117**	242.6	72.2	**162**	323.6

TEMPERATURE CONVERSION —continued

C.		F.	C.		F.
72.3	**163**	325.4	97.8	**208**	406.4
73.3	**164**	327.2	98.3	**209**	408.2
73.9	**165**	329.0	98.9	**210**	410.0
74.4	**166**	330.8	99.4	**211**	411.8
75.0	**167**	332.6	100.0	**212**	413.6
75.6	**168**	334.4	100.6	**213**	415.4
76.1	**169**	336.2	101.1	**214**	417.2
76.7	**170**	338.0	101.7	**215**	419.0
77.2	**171**	339.8	102.2	**216**	420.8
77.8	**172**	341.6	102.8	**217**	422.6
78.3	**173**	343.4	103.3	**218**	424.4
78.9	**174**	345.2	103.0	**219**	426.2
79.4	**175**	347.0	104.4	**220**	428.0
80.0	**176**	348.8	105.0	**221**	429.8
80.6	**177**	350.6	105.6	**222**	431.4
81.1	**178**	352.4	106.1	**223**	433.4
81.7	**179**	354.2	106.7	**224**	435.2
82.2	**180**	356.0	107.2	**225**	437.0
82.8	**181**	357.8	107.8	**226**	438.8
83.3	**182**	359.6	108.3	**227**	440.6
83.9	**183**	361.4	108.9	**228**	442.4
84.4	**184**	363.2	109.4	**229**	444.2
85.0	**185**	365.0	110.0	**230**	446.0
85.6	**186**	366.8	110.6	**231**	447.8
86.1	**187**	368.6	111.1	**232**	449.6
86.7	**188**	370.4	111.7	**233**	451.4
87.2	**189**	372.2	112.2	**234**	453.2
87.8	**190**	374.0	112.8	**235**	455.0
88.3	**191**	375.8	113.3	**236**	456.8
88.9	**192**	377.6	113.9	**237**	458.6
89.4	**193**	379.4	114.4	**238**	460.4
90.0	**194**	381.2	115.0	**239**	462.2
90.6	**195**	383.0	115.6	**240**	464.0
91.1	**196**	384.8	116.1	**241**	465.8
91.7	**197**	386.6	116.7	**242**	467.6
92.2	**198**	388.4	117.2	**243**	469.4
92.8	**199**	390.2	117.8	**244**	471.2
93.3	**200**	392.0	118.3	**245**	473.0
93.9	**201**	393.8	118.9	**246**	474.8
94.4	**202**	395.6	119.4	**247**	476.6
95.0	**203**	397.4	120.0	**248**	478.3
95.6	**204**	399.2	120.6	**249**	480.2
96.1	**205**	401.0	121	**250**	482
96.7	**206**	402.8	127	**260**	500
97.2	**207**	404.6	132	**270**	518

GEOMETRIC DATA

CIRCUMFERENCE OR PERIMETERS OF PLANES

Circle............	$3.1416 \times$ Diameter
Ellipse...........	$3.1416 \times \tfrac{1}{2}$ major axis $+ \tfrac{1}{2}$ minor axis
Sector...........	$\dfrac{\text{Radius} \times \text{Degrees in Arc}}{57.3}$

SURFACE AREAS OF PLANES AND SOLIDS

Circle............	π (3.1416) $\times$ Radius^2 or .7854 $\times$ Diameter^2
Cone............	$\tfrac{1}{2}$ Circumference $\times$ Slant Height $+$ Area of Base
Frustum of Cone..	$\pi \times$ Slant Height $\times$ (radius at top $+$ radius at base) $+$ Areas of Top and Base
Cylinder.........	Circumference $\times$ Length $+$ Area of two ends
Ellipse (approx.)..	Product of Axes $\times$ .7854
Parabola.........	Base $\times \tfrac{2}{3}$ Height
Parallelogram....	Base $\times$ Height
Pyramid..........	$\tfrac{1}{2}$ sum of Base Perimeters $\times$ Slant Height $+$ Area of Base
Sector of Circle...	$\dfrac{3.1416 \times \text{Degrees in Arc} \times \text{Radius}^2}{360}$
Segment of Arc...	$\tfrac{2}{3}$ CH $+ \dfrac{H^3}{2C}$ Where C $=$ Chord, H $=$ Rise
Segment of Circle	Area of Sector $-$ Triangle
Sphere...........	Diameter$^2 \times$ 3.1416
Triangle	Half Base $\times$ Perpendicular Height

Note: The above formulae are of general application. They express relationships only, not size. Therefore, they can be used with any system of measurement, whether Imperial, metric or any other. Whatever the system of measurement, numeration remains the same.

TRIGONOMETRY

FUNCTIONS OF AN ANGLE

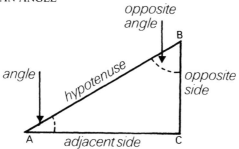

Sine .... = $\dfrac{\text{Perpendicular or Opposite Side}}{\text{Hypotenuse}}$

Cosine .... = $\dfrac{\text{Adjacent Side}}{\text{Hypotenuse}}$

Tangent ...... = $\dfrac{\text{Perpendicular or Opposite Side}}{\text{Adjacent Side}}$

Cotangent .... = $\dfrac{\text{Adjacent Side}}{\text{Perpendicular or Opposite Side}}$

Cosecant ..... = $\dfrac{\text{Hypotenuse}}{\text{Perpendicular or Opposite Side}}$

Secant .... = $\dfrac{\text{Hypotenuse}}{\text{Adjacent Side}}$

Versed Sine = 1 —cosine
Coversed Sine = 1 —sine

DIAMETERS, CIRCUMFERENCES AND AREAS OF CIRCLES
in Imperial or Metric sizes.
(¼ equals 0.25, ½ equals 0.50 and ¾ equals 0.75)

Dia.	Circum.	Area	Dia.	Circum.	Area	Dia.	Circum.	Area
1	3.1416	.7854	¼	7.067	3.90	½	10.995	9.62
¼	3.927	1.23	½	7.854	4.90	¾	11.781	11.04
½	4.712	1.77	¾	8.6394	5.94	4	12.56	12.56
¾	5.499	2.40	3	9.425	7.06	¼	13.351	14.18
2	6.2832	3.14	¼	10.21	8.29	½	14.137	15.9

Dia.	Circum.	Area	Dia.	Circum.	Area	Dia.	Circum.	Area
¾	14.922	17.72	17	53.4	226.98	¼	91.892	671.95
5	15.708	19.64	¼	54.19	233.7	½	92.677	683.49
¼	16.49	21.65	½	54.98	240.53	¾	93.462	695.12
½	17.278	23.76	¾	55.76	247.45	30	94.248	706.86
¾	18.064	25.97	18	56.55	254.47	¼	95.033	718.69
6	18.85	28.27	¼	57.73	261.58	½	95.819	730.61
¼	19.63	30.68	½	58.12	268.8	¾	96.604	742.64
½	20.42	33.18	¾	58.91	276.05	31	97.389	754.76
¾	21.20	35.78	19	59.69	283.53	¼	98.175	766.99
7	21.9	38.48	¼	60.47	291.04	½	98.96	779.31
¼	22.77	41.28	½	61.261	298.64	¾	99.746	791.73
½	23.56	44.18	¾	62.05	306.35	32	100.53	804.24
¾	24.34	47.17	20	62.83	314.16	¼	101.32	816.86
8	25.13	50.26	¼	63.62	322.06	½	102.1	820.57
¼	25.91	53.46	½	64.403	330.06	¾	102.89	842.39
½	26.7	56.74	¾	65.133	338.16	33	103.67	855.3
¾	27.49	60.13	21	65.973	346.36	¼	104.46	868.3
9	28.27	63.61	¼	66.759	354.65	½	105.24	881.41
¼	29.04	67.20	½	67.544	363.05	¾	106.03	894.61
½	29.84	70.88	¾	68.33	371.54	34	106.87	907.92
¾	30.63	74.66	22	69.115	380.13	¼	107.6	921.32
10	31.410	78.54	¼	69.9	388.82	½	108.38	934.82
¼	32.2	82.52	½	70.686	397.6	¾	109.17	948.41
½	32.98	86.59	¾	71.471	406.49	35	109.96	962.11
¾	33.72	90.76	23	72.257	415.47	¼	110.74	975.9
11	34.56	95.03	¼	73.042	424.55	½	111.53	989.8
¼	35.34	99.4	½	73.827	433.73	¾	112.31	1003.8
½	36.128	103.87	¾	74.613	443.01	36	113.1	1017.8
¾	36.914	108.43	24	75.398	452.39	¼	113.88	1032.0
12	37.7	113.09	¼	76.184	461.86	½	114.67	1046.3
¼	38.48	117.86	½	76.696	471.43	¾	115.45	1060.7
½	39.27	122.72	¾	77.754	481.14	37	116.24	1075.2
¾	40.05	127.66	25	78.54	490.87	¼	117.02	1089.8
13	40.84	132.73	¼	79.325	500.74	½	117.81	1104.2
¼	41.62	137.88	½	80.111	510.7	¾	118.6	1119.2
½	42.41	143.14	¾	80.896	520.77	38	119.38	1134.1
¾	43.19	148.49	26	81.681	530.93	¼	120.17	1149.0
14	43.98	153.94	¼	82.467	541.18	½	120.95	1164.1
¼	44.77	159.48	½	83.252	551.54	¾	121.74	1179.3
½	45.55	165.13	¾	84.038	562.0	39	122.52	1194.5
¾	46.34	170.87	27	84.823	572.55	¼	123.31	1209.9
15	47.12	176.72	¼	85.608	583.2	½	124.09	1225.4
¼	47.9	182.65	½	86.394	593.95	¾	124.88	1240.9
½	48.69	188.69	¾	87.179	604.8	40	125.66	1256.6
¾	49.48	194.82	28	87.965	615.75	¼	125.45	1272.4
16	50.26	201.06	¼	88.75	626.79	½	127.23	1288.2
¼	51.05	207.39	½	89.535	637.94	¾	128.02	1304.2
½	51.83	213.82	¾	90.321	649.18	41	123.81	1320.2
¾	52.62	220.35	29	91.106	660.52	¼	129.59	1336.1

MEASURING DATA 69

Dia.	Circum.	Area	Dia.	Circum.	Area	Dia.	Circum.	Area
½	130.38	1352.6	44	138.23	1520.5	½	146.08	1698.2
¾	131.16	1369.0	¼	139.02	1537.8	¾	146.87	1716.5
42	131.95	1385.4	½	139.9	1555.3	47	147.65	1734.9
¼	132.73	1402.0	¾	140.59	1572.8	¼	148.44	1753.4
½	133.52	1418.6	45	141.37	1590.4	½	149.23	1772.0
¾	134.3	1435.3	¼	142.16	1608.1	¾	150.01	1790.7
43	135.09	1452.2	½	142.94	1625.9	48	150.8	1809.5
¼	135.87	1469.1	¾	143.73	1643.9	¼	151.58	1828.4
½	136.66	1486.1	46	144.51	1661.9	½	152.37	1847.4
¾	137.44	1503.3	¼	145.3	1680.0	¾	153.15	1866.5

VOLUMES OF CYLINDERS, PIPES, WELLS, ETC.

Length or height or depth in metres or millimetres or in yards or feet or inches

Diameter		½	¾	1	2	3	4	5	6
0.125	⅛	.04	.07	.09	.17	.26	.36	.44	.52
0.500	½	.1	.15	.2	.39	.59	.79	.98	1.18
0.750	¾	.22	.33	.44	.88	1.33	1.77	2.21	2.65
1.000	1	.39	.59	.79	1.57	2.36	3.14	3.93	4.71
1.250	1¼	.61	.92	1.23	2.45	3.68	4.91	6.14	7.36
1.500	1½	.88	1.32	1.77	3.53	5.3	7.07	8.84	10.6
2.000	2	1.57	2.36	3.14	6.28	9.42	12.57	15.71	18.85
2.500	2½	2.45	3.68	4.91	9.82	14.73	19.64	24.55	29.45
3.000	3	3.53	5.4	7.07	14.14	21.21	28.27	35.34	42.41
3.500	3½	4.81	7.22	9.62	19.24	28.86	38.48	48.11	57.73
4.000	4	6.28	9.42	12.57	25.13	37.7	50.26	62.83	75.4
4.500	4½	7.95	11.93	15.9	31.81	47.71	63.62	79.52	95.42
5.000	5	9.82	14.73	19.64	39.27	58.91	78.54	98.18	117.81
5.500	5½	11.88	17.82	23.76	47.52	71.27	95.03	118.79	142.55
6.000	6	14.14	21.21	28.27	56.55	84.82	113.1	141.37	169.64
6.500	6½	16.59	24.89	33.18	66.37	99.55	132.73	165.92	199.1
7.000	7	19.24	28.86	38.48	76.97	115.45	153.94	192.42	230.9
7.500	7½	22.09	33.14	44.18	88.36	132.54	176.72	220.9	265.07
8.000	8	25.13	37.7	50.27	100.53	150.8	210.06	251.33	301.59
8.500	8½	28.37	42.56	56.75	113.49	170.24	226.98	283.73	340.47
9.000	9	31.81	47.71	63.62	127.23	190.85	254.47	318.09	381.7
9.500	9½	35.44	53.16	70.88	141.76	212.65	283.53	354.41	425.29
	10	39.27	58.01	78.54	157.08	285.62	314.16	392.7	471.24
	15	88.36	132.54	176.72	353.43	530.15	706.86	883.58	1060.29
	20	157.08	235.62	314.16	628.32	942.48	1256.64	1570.8	1884.96
	25	245.44	368.16	490.88	981.75	1472.63	1963.5	2454.38	2945.25
	30	353.43	530.15	706.86	1413.72	2120.58	2827.44	3534.3	4241.16

Diameter		7	8	9	10	15	20	25	30
0.125	⅛	.61	.7	.79	.87	1.31	1.75	2.18	2.62
0.500	½	1.37	1.57	1.77	1.96	2.95	3.93	4.91	5.89
0.750	¾	3.09	3.53	3.98	4.42	6.63	8.84	11.04	13.25
1.000	1	5.5	6.28	7.07	7.85	11.78	15.71	19.64	23.56

Diameter		7	8	9	10	15	20	25	30
1.250	1¼	8.59	9.82	11.04	12.27	18.41	24.54	30.68	36.8
1.500	1½	12.37	14.14	15.9	17.67	26.51	35.34	44.18	53.01
2.000	2	21.99	25.13	28.27	31.42	31.1	62.83	78.54	94.25
2.500	2½	34.36	39.27	44.18	49.09	73.64	98.18	122.73	147.27
3.000	3	49.48	56.55	63.62	70.69	106.03	141.37	176.7	212.06
3.500	3½	67.35	76.97	86.59	96.21	144.32	192.42	240.53	288.63
4.000	4	87.96	100.53	113.1	125.66	188.5	251.32	314.15	377.0
4.500	4½	111.33	127.23	143.14	159.04	238.56	318.08	397.6	477.12
5.000	5	137.45	157.08	176.72	196.35	294.53	392.7	490.88	589.05
5.500	5½	166.31	190.06	213.82	237.58	356.37	475.16	593.95	712.74
6.000	6	197.92	226.19	254.47	282.74	424.11	565.48	706.85	848.22
6.500	6½	232.28	265.46	298.65	331.83	497.75	663.66	829.58	995.49
7.000	7	269.39	307.87	346.36	384.84	577.26	769.68	962.1	1154.52
7.500	7½	309.25	353.43	397.61	441.79	662.69	883.48	1104.48	1325.37
8.000	8	351.86	402.12	452.39	502.65	753.98	1005.3	1256.6	1507.95
8.500	8½	397.22	453.96	510.71	567.45	851.18	1134.9	1418.63	1702.35
9.000	9	445.32	508.94	572.55	636.17	954.26	1272.34	1590.43	1908.51
9.500	9½	496.17	567.06	637.94	708.82	1063.23	1417.64	1772.05	2126.46
	10	549.78	628.32	706.86	785.4	1178.1	1570.8	1963.5	2356.2
	15	1237.01	1413.72	1590.44	1767.15	2650.73	3534.3	4417.88	5301.45
	20	2199.12	2513.28	2827.44	3141.6	4712.4	6283.2	7854.0	9424.8
	25	3436.13	3927.0	4417.88	4908.75	7363.13	9817.5	12271.88	14726.25
	30	4948.02	5654.88	6361.74	7068.6	10602.9	14137.2	17671.5	21205.8

VOLUMES OF SOLIDS OF VARIOUS SHAPES

Cone.......... Area of Base × ⅓ Perpendicular Height
Frustum of Cone ½ Perpendicular Height × π ($R^2 + r^2 + Rr$), where R and r are radii of Base and Top
Cylinder....... 3.1416 × Radius^2 × Height
Pyramid....... Area of Base × ⅓ Perpendicular Height
Sphere .... 4/3 Radius^3 × 3.1416
Wedge .... Area of Base × ½ Perpendicular Height

CAPACITY OF BUILDERS' UTENSILS

Utensil	cu. ft	m³	gallons	litres	bushels
Barrow	3.00	0.085	18.75	81.82	2.33
Bag	2.57	0.065	16.00	72.75	2.00
Hod	0.64	0.017	4.00	18.18	0.50
Pail (approx.)	0.50	0.013	3.00	13.64	0.37

(Based on 6½ gallons equals 1 cu. ft, and 1 bushel equals 8 gallons, converted to S.I. metric units.)

MEASURING DATA 71

MAP SCALES

Scale Ratio	One Inch =	One Mile =
1:63360	1 Mile	1 Inch
1:25000	2083.33 ft	2.53 in
1:12500	1041.66 ft	5.06 in
1:10560	888 ft	6 in
1:5280	440 ft	12 in
1:2500	208.33 ft	25.34 in
1:1250	104.166 ft	50.68 in
1:1056	88 ft	5 ft
1:500	41.66 ft	10 ft-6.72 in

These are old ordnance Survey scales: post-war maps are to differing scales and are marked with National Grid lines, but are not to scales used in building work. The maps most used in building are the 1:2500, commonly call the 'twenty-five inch map', and the 1:500 which is available for all large towns. Copies of these maps are held in every local or municipal engineer's office and are kept up-to-date by his staff. It is usual to allow builders and others to make tracings of these maps, or to provide photo-copies gratis or for a small fee. The 1:2500 map is used for key plans and the 1:500 for drainage plans and site plans that have to be included in deposited plans for all new work. In the old L.C.C. area of the Greater London Council, the 1:1056 map is used instead of the 1:2500. It is probable that these maps will continue to be used for many years to come: the task of transferring all the information to new metric maps, if they were available, would be something no local authority could at present face, on the score of cost alone.

However, in the new editions of the 1:2500 sheets, linear measures and levels will be in metric and sizes of fields in acres and hectares. For conversion factors, see page 50 (acres and hectares).

DRAWING PAPER AND STATIONERY SIZES

All printing paper (except paper in rolls) is in metric measurements and to the sizes laid down by the International Standards Organisation (I.S.O.) as reproduced by B.S. 4000. This makes provision for three series of paper sizes: A, B and C. Drawing and note papers will be made to A series sizes. The B series is intended for posters and the C series for envelopes.

The sizes of the A series papers are based on the A0 size, which is the equivalent of 1 metre square in area. Each size is exactly half the size of the preceding one in area and has

1. The long side identical in length with the short side of the preceding size, and
2. The short size equal to half the length of the long side of the preceding size.

I.S.O. sizes	Millimetres	Inches	Nearest old size from which A size can be cut		Nearest old drawing-board size
			Name	size in inches	
2A	1189 × 1682	46.8 × 66.2	Emperor	48 × 72	—
A0	841 × 1189	33.1 × 46.8	Emperor	48 × 72	—
A1	594 × 841	23.4 × 33.1	Atlas	26 × 34	32 × 54
A2	420 × 594	16.5 × 23.4	Royal	24 × 19	28 × 19
A3	297 × 420	11.7 × 16.5	Foolscap	17 × 13½	23 × 16
A4	210 × 297	8.3 × 11.7	Quarto	8 × 10	—
A5	148 × 210	5.8 × 8.3	Post 4to	7¼ × 9	—
A6	105 × 148	4.1 × 5.8	Small 8vo	4½ × 7¼	—

It is probable that A0 to A4 will be used for drawings, A 3 for specifications and quantities, A4 for commercial note-headings and trade literature (leaflets) and A5 to A6 for private note-paper.

ENVELOPES

The C series is for envelopes and folders. Of this and related series only the two following are within the *Post Office preferred range of sizes*. The use of other C and D series will incur higher postage rates if used for items not more than 2 oz (57 g) in weight.

ENVELOPE SIZES

Type	Millimetres	Inches
C 6	114 × 162	4.49 × 6.38
D L	110 × 220	4.33 × 8.66

PLAN SCALES

Imperial measure	Fraction	Nearest Metric scale, in mm
1/16 in equals 1 ft	1/192	1.200
1/8 in equals 1 ft	1/96	1.100
1/4 in equals 1 ft	1/48	1.50
1/2 in equals 1 ft	1/24	1.20
1 in equals 1 ft	1/12	1.10

Note: Oval scales, giving at least three of the above metric scales, and flat section scales, giving two of the above metric scales, are now available.

ABBREVIATIONS ON PLANS AND DRAWINGS

Recommended in British Standard Specifications

1. PRIMARY UNITS

Centimetre	cm	Millimetre	mm
Chain	ch	Ounce	oz
Cubic Foot	cu. ft	Pound	lb
Cubic Yard	cu. yd	Quarter	qr
Dozen	doz	Square Inch	sq. in
Foot	ft or (')	Square Foot	sq. ft
Hundredweight	cwt	Square Yard	sq. yd
Inch	in or (")	Ton	t
Metres	m	Yard	yd

2. MATERIALS AND GENERAL TERMS

Aggregate	agg.	Galvanized	galv.
Air brick	A.B.	Glazed ware pipe	G.W.P.
Approved	appd.	Grease trap	G.T.
Approximate	approx.	Ground Level	G.L.
Asbestos	asb.	Gully	G.
Asphalt	asph.	Height	ht.
Bench Mark	B.M.	Hose bib	H.B.
Birmingham Gauge	B.G.	Inspection chamber	I.C.
Bitumen	bitn.	Insulated or Insulation	insul.
Brickwork	bwk, or B	Intercepting trap	I.T.
British Standard	B.S.	Internal	int.
Cast-Iron	C.I.	Invert	inv.
Cement	cem.	Lavatory basin	L.B.
Centre Line	C.L. or ^{C}L	Left-hand	L.H.
Centre to centre	c/c.	Macadam	mac.
Checked	ckd.	Manhole	M.H.
Chemical closet	C.C.	Mild Steel	M.S.
Clearing eye	C.E.	Not to scale	N.T.S.
Concrete	conc.	Number	No.
Corrugated	corr.	Petrol interceptor	P.I.
Diameter	dia.	Radius	rad.
Drawing	drg.	Rain water outlet	R.W.O.
Drinking fountain	D.F.	Rain water pipe	R.W.P.
Earth Closet	E.C.	Reinforced concrete	R.C.
Figure	Fig.	Right-hand	R.H.
Fire hydrant	F.H.	Rising main	R.M.
Flushing cistern	F.C.	Round	rd.
Fresh air inlet	F.A.I.	Sink	S.

Sketch	sk.
Sluice or stop valve	S.V.
Soil and vent pipe	S. & V.P.
Soil pipe	S.P.
Specification	spec.
Spigot and socket	S. & S.
Square	sq.
Standard Wire Gauge (Imp)	S.W.G.
Stand Pipe	St. P.
Street Gully	S.G.
Tongued and grooved	t. & g.
Traced	Tcd.
Urinal	U.
Vent pipe	V.P.
Volume	vol.
Waste pipe	W.P.
Waste and vent pipe	W. & V.P.
Water closet	W.C.
Weight	wt.
Yard gully	Y.G.

3. BASIC ESTIMATING

Cost to Employ Labour	77
Cost of Rehabilitation Work on Older Houses	80
Labour Only Constants	94
The Critical Path Method	96
Estimators' Conversion Tables	98
Modular Co-ordination	98

The word 'estimate' is derived from the Latin word meaning a guess: this is exactly what an estimate is. It is a prophecy — an inspired guess, if you will — and the art or science of estimating is to make a guess that approximates very nearly to the future reality. To do this with any certainty, it is necessary to limit the field within which the guesswork must operate. The size of the job can be ascertained from drawings, specifications, bills of quantities or measurements on the job. One of these sources of information is essential. To this must be added the quantities of materials required for a job of any given size. Here this book can help you, first by providing unit sizes of building materials, and then by providing easy means of calculating the number of units required for any particular job. It will also give the approximate standard of workmanship for jobs of moderate to good quality, corresponding to the quality of the materials with which it deals. This is, roughly speaking, the field covered by the Codes of Practice and the British Standards. Every builder knows, however, that there are whole ranges of materials, outside the British Standard range of products, that are much cheaper and often as good. Similarly, a lower, but quite serviceable standard of workmanship is often acceptable. It must be remembered that local authorities and supply undertakings now insist on materials to British Standards quality, and workmanship to the standard prescribed by the Codes of Practice. On any job that is subject to inspection, therefore, such as all new work, structural alterations to existing property, and the installation of all services, including drainage, it pays to estimate for materials to British Standards and workmanship to Code standards, since it is probable that these will be insisted upon. A point worth remembering is that these standards of quality are deemed to satisfy the Building Regulations (except in some rare instances those of the G.L.C.) and that better work cannot be demanded.

Having thus limited the field, so far as materials and quality of workmanship are concerned, there remains labour cost. The simplest way to obtain some firm contact with actuality in this matter is to rely on unit times for building operations, and a list (necessarily incomplete) of hours required to execute various quantities of building work of different kinds is appended to these notes.

All these quantities of labour and materials must be translated into sums of money and then combined to give a price, first for one type of work, and then for all types of work combined.

The *Building Trades Journal* publishes a detailed analysis of the cost of employing one craftsman for one week, including wages and all indirect costs.

COSTS TO EMPLOY BUILDING OPERATIVES AT 'GRADE A' (RATES)

CURRENT

The tables of 'costs to employ' are based on rates of wages in force from 24 June 1985. An additional 30 per cent incentive bonus has been added but it should be adjusted in light of local experience. The National Insurance contribution is based on April 1985 tables. Holidays with pay rates are from 29th July 1985.

The rates are based on annual calculations in accordance with IOB recommendations. They do not include items usually priced in preliminaries nor do they include travelling time or allowances for overheads or profit.

The total annual wages are divided by 'Productive Hours', i.e. basic hours (1911) less an allowance for inclement weather. 'Productive Hours' total 1,873 hrs.

Appendix 'A'

		CRAFTSMAN	LABOURER
Total hours worked per year		1911	1911
Inclement weather 2%		38	38
Total productive hours	A	1873	1873
Non-productive overtime	C	89	89
Days sick per year	D	5	5
Days public holiday	F	8	8
Sick pay per day	E	7.74	7.74
Tool money per week		.87	—
Annual holiday with pay	G	12.15	12.15
Trade supervision per hour	H	.27	.23
Basic hourly rate		238.5	203.5
30% plus rate		71.5	61.0
TOTAL HOURLY RATE	B	310.0	264.5
Basic wages (A×B)		5806.30	4954.09
Non-productive overtime (C×B)		275.90	235.41
Sick pay (D×E)		38.70	38.70
Public holidays (F×B×8)		198.40	169.28
		6319.30	5397.48
National Insurance 10.45%		660.37	564.04
Tool money		38.80	—
Annual holiday with pay (48×G)		583.20	583.20
CITB levy		71.00	18.20

	7672.67	6562.72
Severance pay 2%	153.45	131.25
Employers liability 3rd party insurance 2%	156.52	133.88
Trade supervision (A×H)	505.71	430.79
TOTAL COST	8488.35	7258.64
COST PER HOUR...........	4.53	3.88

COST OF BRICKLAYER WITH ATTENDANT LABOURERS AND SCAFFOLDER

	per hour £ p	rate £ p
Eight bricklayers	4.53	36.24
Three labourers	3.88	11.64
One scaffolder	4.53	4.53
	8	52.41
Per bricklayer hour		6.55

COST OF BRICKLAYER WITH ATTENDANT LABOURERS EXCLUDING SCAFFOLDER

Eight bricklayers	4.53	36.24
Three labourers	3.88	11.64
	8	47.88
Per bricklayer hour		5.99

NJCBI WAGE RATES FOR BUILDING OPERATIVES
Rates of wages operative from June 24 1985

CRAFTSMEN AND LABOURERS

	London and Liverpool District			Scotland and Grade 'A'		
	Per week	Per hour	GMB	Per week	Per hour	GMB
Craftsmen	£93.21	239p	£14.82	£93.01½	238½	£14.82
Labourers	£79.56	204p	£12.48	£79.36½	203½	£12.48

WATCHMEN

	Rate per week
London and Liverpool districts	£92.04
Scotland and Grade 'A'	£91.84½

BASIC ESTIMATING 79

The weekly remuneration for watchmen is a sum equivalent to the labourers' standard weekly rate of wages plus the labourers' GMB provided not less than five shifts (day or night) are worked in the payweek.

APPRENTICES/TRAINEES

Entrants under 19 whose employment as apprentices/trainees began on or after June 25, 1984 and Building Foundation Training Scheme trainees whose employment as apprentices/trainees began before that date

Six-month period	Rate per week	Per hour	GMB
First	£43.29	111p	—
Second	£58.69½	150½p	—
Third	£63.96	164p	£10.14
Thereafter, until skills test is passed	£77.22	198p	£12.28½
On passing skills test	£87.16½	223½p	£13.84½

Entrants under 19 whose employment as apprentices/trainees began on or after June 27, 1983 but before June 25, 1984 (excluding BFTS trainees whose employment as apprentices/trainees began between those dates

Six-month period	Rate per week	Per hour	GMB
Third	£67.27½	172½p	£10.14
Fourth	£77.22	197½p	£12.28½
Fifth	£87.16½	223½p	£13.84½
Sixth	£92.04	236p	£14.62½

Entrants under 19 whose employment as apprentices/trainees began before 27th June, 1983

Age at entry	Six-month period	Rate per week	Per hour	GMB
16	Fifth and sixth	£83.07	213p	£13.06½
17	Fifth and sixth	£83.07	213p	£13.06½
18	Fifth and sixth	£83.07	213p	£13.06½

BATJIC WAGE RATES
From June 24, 1985 Craftsmen £109.40 Adult general operative £93.79½

WAGE RATES FOR PLUMBERS
Rates of wages operative from March 25, 1985.
Basic 37½ hrs/week

	Per week	Per hour
Trained plumber	£116.25	310p
Advanced plumber	£127.50	340p
Technical plumber	£146.25	390p

Allowances:	
Abnormal conditions	139p per day
Lodging allowance	£10.75 per night
Responsibility money	23p per hour
Welding supplement:	
Possession of gas OR arc certificate	15p per hour
Possession of gas AND arc certificate	30p per hour

WAGE RATES FOR ELECTRICIANS
Rates of wages operative from July 27, 1985.

Basic 37½ hrs/week	Per week	per week	London Weighting/hr
Electrical Technician	£162.75	434p	18p
Approved Technician	£140.62½	375p	18p
Qualified Electrician	£129.75	346p	18p
Labourer	£99.00	246p	18p

WAGE RATES FOR HEATING AND VENTILATING OPERATIVES
Rates of wages operative from April 2, 1984.

Basic 38 hrs/week	Per week	Per hour
Foreman	£154.28	406p
Chargehand	£148.20	390p
Advanced Fitter	£130.34	343p
Fitter	£118.56	312p
Improver	£112.48	296p
Assistant	£106.78	281p
Mate over 18	£95.00	250p
Welding supplement:		
Possession of gas OR arc certificate		16p per hour
Possession of gas AND arc certificate		31p per hour

THE COST OF REHABILITATION WORK ON OLDER HOUSES

by RON COOPER F.G.of.S.

The Department of the Environment has authorised even more generous grants for the repairs and rehabilitation of older houses. This move has two purposes; first, to provide more and better housing at an economical cost, and second, to provide a much-needed increase in the volume of work in the construction industry, especially for the small or medium size building firms.

Maximum eligible amounts are laid down for various items or sections of work, i.e., new bath, new sink, conversion of a large house into flats. In housing action areas and general improvement areas grants may be made for repairs only. Generally the amount of the grant is 50 per cent of the eligible amount (not necessarily the maximum) agreed by the local authority. Higher percentages are allowed in general improvement areas 60 per cent and housing action areas 75 per cent.

Local authorities are empowered to agree the amount of the grants and make subsequent payment provided they are satisfied that the estimate of the proposed work is fair and reasonable.

Application for grants are made on local authority questionnaire forms which should be supported by the contractor's estimate. The contractor must be prepared to explain and/or amplify his estimate if necessary. The following schedule shows the analysis of cost of some of the more usual items to be encountered in rehabilitation and improvement grant work.

Man hours against each item shows the total man hours, tradesmen only, tradesmen and labourers or labourers only, irrespective of whether it is a one-man or 'team of six' operation. Man hours per operation take into consideration the physical difficulties to be met in smaller houses, e.g., terrace houses with small backyard and all material wheeled to and from through the hall passage, semi-detached with a metre wide sideway between each pair of houses.

Wages rates will be based on the estimated current costs to employ a tradesman and labourer based on the June 1985 agreement, viz., tradesmen and labourers: £4.53 and £3.88 per hour respectively. Material costs will be based on the latest available prices (published in BTJ). Margin for overheads and profit has been allowed at 20 per cent.

Description of work	Man hours	Nett Labour £	Nett Material £	Unit Price £
EXCAVATION				
Excavate over site average 150 mm deep, including rubbish and hardcore, wheel out, load lorry, m^2	1.20	4.57	—	5.48
Excavate small trenches, etc, for new foundations, part return fill in and ram, remove as last item, m^3	4.50	17.46	—	20.95
Hardcore average 150 mm thick to receive oversite concrete, m^2	0.45	1.75	1.39	3.77
Pulling down existing brickwork, half brick thick, cleaning sound whole bricks for re-use, set aside remainder for hardcore, m^2	0.90	3.49	—	4.19
Ditto one-brick thick, m^2	1.75	6.79	—	8.15

Description of work	Man hours	Nett Labour £	Nett Material £	Unit Price £
CONCRETE WORK				
Forming precast reinforced concrete lintels and hoisting and setting in position—size 113 × 150 mm, lin.m	0.45	1.75	1.70	4.14
Ditto, ditto—size 113 × 225 mm, lin.m	0.65	2.52	2.65	6.20
Ditto, ditto—size 225 × 150 mm, lin.m	0.85	3.30	5.10	10.08
Ditto, ditto—size 225 × 225 mm, lin.m	1.10	4.27	6.54	12.97
Concrete (1-2-6) in founds or pier bases, m³	4.00	15.52	38.00	64.22
Ditto oversite or underfloor 100 mm thick, m²	1.20	4.66	4.00	10.39
Ditto 150 mm thick, m²	1.75	6.79	6.00	15.35
BRICKWORK				
Half-brick walls in Flettons cement-lime mortar, m²	1.90	8.61	8.00	19.93
One brick ditto, ditto, m²	3.50	15.86	16.50	38.83
One-and-a-half-brick, m²	4.90	22.20	26.00	57.84
Cavity brick walls, half-brick inner and outer skins, m²	4.30	19.48	16.50	43.18
Slate damp course in half-brick wall, lin.m	0.60	2.72	1.50	5.06
One-brick wall, lin.m	1.30	5.89	3.20	10.91
Extra-over brickwork for fair-face one side, m²	0.70	3.17	—	—
Extra-over one-brick wall for facing bricks, assuming facing bricks are £50 per m more than commons, m²	1.20	5.44	3.60	10.85
Close cavity wall at openings, including slate damp course, lin.m	0.80	3.62	1.80	6.50
50 mm Clinker concrete—block partition walls, m²	0.80	3.62	2.60	7.46
75 mm ditto, ditto, m²	1.00	4.53	3.70	9.88
100 mm ditto, ditto, m²	1.20	5.44	5.50	13.13

BASIC ESTIMATING

Description of work	Man hours	Nett Labour £	Nett Material £	Unit Price £
Cutting out two courses of brickwork in short lengths, insert slate damp course and build over in walls half-brick thick, lin.m	1.30	5.89	2.50	10.07
Ditto, ditto in wall one-brick thick, lin.m	2.20	9.97	5.20	18.20
Rake out perished mortar, repoint, including scaffolding to two storeys, m^2	1.20	5.44	0.30	6.89
Ditto in walling exceeding two storeys, m^2	1.40	6.34	0.30	7.97
Cut out fractures in wall lacing in new brickwork in half-brick walls, lin.m	2.20	9.97	1.60	13.88
Cutting and forming openings through brick wall, including shoring for new window or door frame half-brick thick, not exceeding 1 m^2, each	4.00	18.12	—	21.74
Ditto one-brick thick, each	7.00	31.71	—	38.05
Ditto, exceeding 1 m^2 half-brick-thick, m^2	3.75	16.99	—	20.39
One-brick-thick, m^2	6.00	27.18	—	32.62
Reforming jambs, reveals in half-brick wall, fair-face, lin.m	2.00	9.06	1.10	12.19
One-brick wall, lin.m	2.50	11.33	2.20	16.24
Hoist and build—in wood or metal door-frame and point externally not exceeding 0.5 m^2, each	1.10	4.98	0.20	6.22
Ditto 0.5-1.0 m^2, each	1.45	6.57	0.30	8.24
Area exceeding 1.5 m^2, each	2.25	10.19	0.35	12.65
Cutting openings in one-brick wall for and including terracotta air brick, size 225 × 225 mm, each	1.50	6.80	3.60	12.48
ROOFWORK				
Strip existing roof slates and clear away, m^2	0.30	1.36	—	1.63
Strip slated roof, clean and set aside sound slates for re-use, clear away, m^2	0.40	1.81	—	2.17

Description of work	Man hours	Nett Labour £	Nett Material £	Unit Price £
Strip slated roof, re-slate using 50 per cent existing sound slates and providing 50 per cent new slates, size 406 × 203 mm, m^2	1.20	5.44	12.60	21.65
Ditto 508 × 254 mm, m^2	0.80	3.62	11.70	18.38
Ditto 610 × 305 mm, m^2	0.60	2.72	17.00	23.66
Take off and renew single slate, including clips size 406 × 203 mm, each slate	0.50	2.27	0.90	3.80
Ditto 508 × 254 mm, each	0.55	2.49	1.90	5.27
Ditto 610 × 305 mm, each	0.60	2.72	3.00	6.86
Ditto up to 30 No. in any one patch, size 406 × 203 mm, each slate	0.22	1.00	0.90	2.28
Ditto 508 × 254 mm, each	0.23	1.04	1.90	3.53
Ditto 610 × 305 mm, each	0.24	1.09	3.00	4.91
Strip and remove tile, slate batten and nails, m^2	0.35	1.59	—	1.91
Strip and remove roof felt and remove all nails, m^2	0.35	1.59	—	1.91
Underfelt laid and fixed on rafters, m^2	0.13	0.59	0.90	1.79
Slate or tile batten to 100 mm gauge, m^2	0.20	0.91	1.90	3.37
Ditto 200 mm gauge, m^2	0.15	0.68	1.05	2.08
Ditto 300 mm gauge, m^2	0.10	0.45	0.88	1.60
Concrete plain tiles to roof slopes, m^2	1.40	6.34	12.00	22.01
Slates size 406 × 203 mm to ditto, m^2	1.10	4.98	20.20	30.22
Ditto 508 × 254 mm, m^2	0.70	3.17	18.90	26.48
Ditto 610 × 305 mm, m^2	0.50	2.27	27.00	35.12
Half-round or angle hip and ridge tiles, lin.m	0.90	4.08	3.80	9.46
Lead stepped flashings 300 mm wide over slates or tiles, lin.m	1.60	7.25	5.60	15.42
Zinc flashings 300 mm wide, lin.m	1.00	4.53	3.90	10.12
Taking-off tiling (or slating) and timbers for removal of water storage cistern, afterwards refixing and reinstating, each	40.00	181.20	10.00	229.44

Description of work	Man hours	Nett Labour £	Nett Material £	Unit Price £
CARPENTRY AND JOINERY				
Take up defective softwood floor and joists, remove, m²	0.66	2.99	—	3.59
Take up existing floor-boards and prepare joists to receive new floorboard, m²	0.32	1.45	—	1.74
50 × 100 mm joists, lin.m	0.75	3.40	—	4.08
50 × 150 mm ditto, lin.m	0.85	3.85	—	4.62
25 × 150 mm plain edge floor supply and fix, m²	0.80	3.62	9.00	15.14
Take up ditto and fix new flooring in patches not exceeding 1 m², each patch	2.20	9.97	9.00	22.76
Ditto in patches 1—3 m², m²	2.00	9.06	9.00	21.67
Take up shrunk or worn softwood plain edge flooring relay and cramp, make good with new flooring, m²	0.90	4.08	1.50	6.70
Take down softwood roof timbers not exceeding 63 × 175 mm, lin.m	0.33	1.49	—	1.79
Ditto exceeding 63 × 175 mm, lin.m	0.40	1.81	—	2.17
Rafters 50 × 100 mm, lin.m	0.20	0.91	1.20	2.53
Ditto 50 × 150 mm, lin.m	0.26	1.18	1.80	3.58
25 mm boards to rafters 50 × 100 mm, lin.m	0.44	1.99	9.80	14.15
Ditto small quantities, m²	0.70	3.17	9.80	15.56
Taking down skeleton stud partition, m²	0.30	1.36	—	1.63
Ditto partition covered with wall board both sides, m²	0.40	1.81	—	2.17
Ditto covered with match board both sides, m²	0.60	2.72	—	3.26
Alter studs to form new door opening size 755 × 1950 mm in partition plaster and lath both sides, m²	1.20	5.44	1.30	8.09
Ditto covered with wall board both sides, m²	1.40	6.34	2.10	10.13
Ditto covered with matchboard both sides, m²	1.60	7.25	1.90	10.98
Stud partition framed with 50 × 100 mm studs left clean for plaster both sides, m²	0.55	2.49	3.45	7.13

Description of work	Man hours	Nett Labour £	Nett Material £	Unit Price £
Ditto, ditto covered with hardboard both sides, m²	1.50	6.79	10.25	20.45
Ditto, ditto covered with matchboard both sides, m²	2.35	10.65	18.45	34.92
Remove skirting, lin.m	0.10	0.45	—	0.54
Stock section softwood skirting up to size 25 × 150 mm and fixing to block wall partition, lin.m	0.15	0.68	1.85	3.04
Ditto to brickwork, lin.m	0.40	1.81	1.85	4.39
Take down door, take out lining or frame, realign and refix. Ease, adjust and re-hang door, re-fix existing architraves, each	5.20	23.56	3.50	32.47
Take down door and set aside for re-use each	0.50	2.27	—	2.72
Take down door and re-hang in new opening, each	2.00	9.06	—	10.87
Take down door, ease, adjust and re-hang, each	1.80	8.15	—	9.78
Take down door, cut and fit fire-resisting board to one side, each	2.30	10.42	9.00	23.30
Ditto, ditto both sides, each	3.30	14.95	18.00	39.54
Take down frame or linning, set aside for re-use, each	0.50	2.27	—	2.72
Take frame or lining and re-fix each	1.25	5.66	—	6.79
Standard door lining fixed in wood partition, each	0.70	3.17	15.90	21.80
In brick partition ditto in brick wall, each	1.10	4.98	15.00	23.98
Standard plywood flush door and hanging including butt hinges, mortice lock and furniture, each	5.00	22.65	22.00	53.58
Standard hardboard flush door and ditto ditto, each	5.00	22.65	16.00	46.38
External part glazed door, hang, including hinges, cylinder lock, door bolt, each	6.50	29.44	47.00	91.73
Take down, ease and adjust casement sash, each	1.20	5.44	—	6.53

BASIC ESTIMATING

Description of work	Man hours	Nett Labour £	Nett Material £	Unit Price £
Take out double lining sashes, ease, re-hang, new cords, each frame	1.60	7.25	4.00	13.50
Standard casement frame 1.23 m × 0.77 m and building in brickwork, each	1.50	6.80	41.00	57.36
Ditto 1.23 × 0.92 m, each	1.80	8.15	43.00	61.38
Ditto 1.23 × 1.38 m, each	2.20	9.97	46.00	67.16
Take down, set aside for re-use staircase item. Spandril framed panelling up to 4 m wide × 3 m, each	2.00	9.06	—	10.87
Newell posts, each	1.00	4.53	—	5.44
Handrail and balusters, each flight	3.00	13.59	—	16.31
Dog-leg stair flight, each	10.00	45.30	—	54.36
Straight stair flight, each	8.00	36.24	—	43.49
Provide and fix new stair flight 1 m wide 3 m rise with balustrade, each	45.00	203.85	347.00	661.02
Ditto, ditto with one-quarter space landing, each	50.00	226.50	387.00	736.20
Ditto, ditto no landing but with three winders, each	53.00	240.09	437.00	812.51
Strengthen handrail and balusters, renew defective balusters, lin.m	1.25	5.66	3.00	10.39
Cutting out defective and worn portion of tread and piecing in new wood, each	1.00	4.53	1.50	7.24
Renew ironmongery to softwood, 75 mm cast iron butt hinges, pair	0.65	2.94	1.60	5.45
100 mm ditto, pair	0.70	3.17	1.90	6.08
75 mm steel butts, pair	0.40	1.81	0.60	2.89
100 mm ditto, pair	0.45	2.04	1.07	3.73
Barrel bolts 200 mm, each	0.60	2.72	2.85	6.68
Casement stay and pegs	0.50	2.27	1.87	4.97
Sash fastener, each	0.75	3.40	1.95	6.42
Sash fasteners (sliding sashes), each	0.75	3.40	2.30	6.84
Rim lock and furniture	0.80	3.62	5.20	10.58
Mortise lock and furniture	1.60	7.25	8.90	19.38
Cylinder night latch	1.40	6.34	9.84	19.42

Description of work	Man hours	Nett Labour £	Nett Material £	Unit Price £
PLUMBING				
Taking down and removing existing gutters or pipes up to two storeys high, lin.m	0.40	1.81	—	2.17
Ditto exceeding two storeys, lin.m	0.50	2.27	—	2.72
100 mm cast-iron half-round gutter including brackets, lin.m	0.75	3.40	6.75	12.18
Extra for angles, each	0.40	1.81	4.50	7.57
Extra for outlets, each	0.40	1.81	4.55	7.63
75 mm ci down pipe, lin.m	0.75	3.40	11.51	17.89
Extra for swan neck, each	0.65	2.94	9.54	14.98
Extra for shoe, each	0.40	1.81	10.16	14.36
100 mm plastic hr gutter and brackets, lin.m	0.50	2.27	2.69	5.95
Extra for angles, each	0.40	1.81	1.83	4.37
Extra for outlets, each	0.40	1.81	1.83	4.37
63 mm plastic RWD, lin.m	0.50	2.27	2.20	5.36
Extra for shoe, each	0.33	1.49	0.88	2.84
Take down, remove ci soil and vent pipes, lin.m	0.55	2.49	—	2.99
100 mm ci soil and vent pipes including caulked lead joints, lin.m	1.70	7.70	16.00	28.44
Extra-over for bends and extra lead joints, each	1.00	4.53	15.00	23.44
Ditto for single junctions	2.00	9.06	24.00	39.67
Ditto for swan neck	2.00	9.06	22.00	37.27
Ditto for sanitary junctions	2.40	10.87	38.00	58.64
Cut into ci stack pipe, insert new junction and extra lead joints, each	4.00	18.12	44.00	74.54
Joint iron pipe to clayware drain socket, each	0.30	1.36	0.10	1.75
Take down and remove lead waste or service pipes up to 50 mm, lin.m	0.40	1.81	—	2.17
Ditto iron ditto, lin.m	0.50	2.27	—	2.72
Cut into iron or copper service pipe and fit new 13 mm stop-valve	1.30	5.89	3.80	11.63

Description of work	Man hours	Nett Labour £	Nett Material £	Unit Price £
13 mm lead service pipe to walls, bends, sockets and one tee (or branch) every 4 lin.m pipe, lin.m	1.20	5.44	4.30	11.69
13 mm galv iron pipe, lin.m	1.00	4.53	1.65	7.42
13 mm copper pipe, lin.m	0.80	3.62	1.43	6.06
13 mm polythene pipe, lin.m	0.90	4.08	0.90	5.98
18 mm lead pipe, lin.m	1.50	6.80	7.20	16.80
18 mm galv iron pipe, lin.m	1.10	4.98	2.80	9.34
18 mm copper pipe, lin.m	0.90	4.08	2.33	7.69
18 mm polythene pipe, lin.m	1.00	4.53	1.43	7.15
Disconnect and take down sanitary fittings, overflows, waste pipes, cap off service pipes, wc suite, each	2.75	12.46	—	14.95
Lavatory basin, each	1.75	7.93	—	9.52
Bath, each	3.00	13.59	—	16.31
Glazed ware sink, each	2.00	9.06	—	10.87
Sanitary fittings supply and fix to copper or iron service and waste pipes, wc suite low level (PC £45.00)	6.00	27.18	48.00	90.22
Lavatory basin (PC £24.00)	5.80	26.27	28.00	65.12
Bath less panels (PC £90.00)	5.80	26.27	100.00	151.52
Ditto with panels (PC £120.00)	8.00	36.24	135.00	205.49
White glazed Belfast sink (PC £35.00), each	4.60	20.84	40.00	73.90
Steel sink and drainer (PC £110.00), each	5.00	22.65	115.00	165.18
Ditto cupboard under (PC £170.00)	7.50	33.96	176.00	251.95
273 litre capacity water storage cistern including ball valve and overflow and connect to services, each	4.50	20.39	101.92	146.77

PLASTERWORK

Description of work	Man hours	Nett Labour £	Nett Material £	Unit Price £
Hack down ceiling plaster, laths and ceiling joists for new plasterboard, m^2	0.85	3.85	—	4.62
Plaster baseboard and set ceiling, m^2	0.70	3.17	2.20	6.44
Hack off wall plaster to brick walls, and rake out joints to form key, m^2	0.70	3.17	—	3.80

Description of work	Man hours	Nett Labour £	Nett Material £	Unit Price £
Render and set on brickwork, m^2	0.85	3.85	1.04	5.87
Expanded metal lathing on wood and plaster finish	1.50	6.80	3.60	12.48
56 mm paramount or similar plasterboard partition walls, m^2	1.05	4.76	14.50	23.11
Hack off wall tiles and screed and rake out brick joints to form key, m^2	0.75	3.40	—	4.08
Cement mortar screed on walls to receive tiling, m^2	0.50	2.27	0.66	3.52
150 × 150 mm wall tiles bedded in mortar, m^2	2.80	12.68	8.48	25.39
Ditto fixed with adhesive, m^2	2.30	10.42	8.48	22.68
19 mm cement mortar screen on concrete to receive other pavings, m^2	0.60	2.72	1.60	5.18
19 mm ditto with dustproofing solution and trowelled to smooth finish, m^2	0.70	3.17	1.75	5.90
Hack off rendering, pebble dash, etc., to outside walls, and rake out joints to form key, m^2	0.60	2.72	—	3.26
Cement mortar render and pebble dash, ditto, m^2	1.40	6.34	1.40	9.29
GLAZING				
Hacking out old glass and cleaning rebates to receive new glass, m^2	2.50	11.33	—	13.60
3.0 mm clear sheet glass ready cut to size in medium squares to wood, m^2	0.90	4.08	10.50	17.50
3.0 mm obscured glass ditto to wood, m^2	0.90	4.08	12.83	20.29
Remove temporary coverings to sashes stopping all nail holes, etc., m^2	1.10	4.98	—	5.98
PAINTWORK				
Strip wall or ceiling paper, stop, prepare surface, m^2	0.30	1.36	—	1.63
Burn off paintwork to wood surfaces, m^2	0.80	3.62	—	4.34

Description of work	Man hours	Nett Labour £	Nett Material £	Unit Price £
Two coats emulsion paint on ceilings, m²	0.30	1.36	0.40	2.11
Ditto on walls, m²	0.25	1.13	0.40	1.84
Two coats oil colour on ceiling, m²	0.40	1.81	0.50	2.77
Ditto on walls, m²	0.35	1.59	0.50	2.51
Ditto wood surfaces, m²	0.40	1.81	0.50	2.77
Ditto on windows in medium squares, m²	0.75	3.40	0.35	4.50
Clean prepare two coats oil colour on ci gutters, down pipes, etc., lin.m	0.50	2.27	0.40	3.20
DRAINAGE				
Excavate trench for new drains average 600 mm deep and refill and cart away surplus, lin.m	2.00	7.26	—	9.31
Extra over last trenches for breaking up concrete paving 100 mm thick and reforming surfaces on completion, lin.m	3.00	11.64	3.90	18.65
Excavating to search for existing clayware drains up to 1 m deep and break out and remove old drains and concrete bed, lin.m	5.00	19.40	—	23.28
150 mm concrete bed under 100 mm drain, lin.m	0.70	2.72	5.85	10.28
150 mm bed and surround to 100 mm drain, lin.m	2.00	7.76	11.70	23.35
100 mm clayware drain laid and jointed, lin.m	0.35	1.36	2.85	5.05
Extra over ditto for junction, each	0.30	1.16	2.75	4.69
Extra over for bed, each	0.15	0.58	1.60	2.61
The following work in providing drain connection from new inside wc to existing manhole.				
Cut hole through 150 mm concrete, excavate about 600 mm deep in hardcore and subsoil from new drain including cutting hole through footing, each	12.00	46.56	—	55.87

Description of work	Man hours	Nett Labour £	Nett Material £	Unit Price £
Build in one vertical length and one horizontal length of 100 mm clayware drain and one bend, make good floor concrete, each	4.50	17.46	9.35	32.17
Cut into 225 mm brick side of manhole to receive new pipe, cut away benching and insert new branch channel, make good, each	4.00	15.52	7.24	27.31

For materials, prices depend very much on the quantity of materials required and any delivery charges, such as cartage and handling, on and off the site. For small quantities of materials, the prices published monthly in the *Building Trades Journal* are a good guide. Handling and delivery charges will vary for every job, and must be ascertained afresh for each job.

Combined, these two figures will give the cost of labour and materials. Three other additions must be made to this price.

1. Cost of site supervision, either by charge-hand, working foreman, walking foreman, site agent and directors. On a medium size to large job, this will amount to *about 18 per cent of labour costs.*

2. Cost of plant for the job, including transport and erection of machinery, store huts and site offices, and a charge for use and waste of timber and depreciation of static plant and machinery, together with fuel costs for the operation of the machines. This can vary so much from job to job that no approximate percentage can be given.

3. A percentage to cover the cost of maintaining the builder's organisation, such as head office and office staff, yards and depots and their staffs, and the materials consumed in the operation of that organisation. In these days, one of the heavy items that must be included is interest on money borrowed to finance works, not only while the jobs are actually going on, but including retention periods.

This is the cost of the work, made up as follows:
1. Materials: including cartage, unloading, stacking, etc. From quotations, catalogue prices or published lists.
2. Labour: Basic agreed rates, plus bonuses, plus overtime. To this add the following indirect costs: VAT, Industrial injuries, NHS, National Insurance contribution, Redundancy fund, Graduated pension, Holidays with Pay, Sick pay scheme, Training levy, Redundancy

disbursement (averaged allowance), Guaranteed week and accident and public liability insurance.

3. Site supervision: As a calculated percentage of actual wages paid, plus indirect costs, as above, where applicable.
4. Plant, etc, (sometimes called 'prelims'): Hire, depreciation, erection, etc.
5. Management, etc (sometimes called 'overheads'): Usually as a percentage.
6. Profit: Usually as a percentage.

Included in materials are *P.C. sums* (prime cost sums), which do not include fixing *provisional sums,* which include labour and materials, the sums should not be exceeded except with written authority

In certain types of work, however, such as sub-contracting for one type of work only, or work in one trade, an experienced man may be able to 'spot price' a job merely by looking at it, and perhaps taking rough overall measurements. For general building work, this is not to be recommended.

Many estimators (but many more quantity surveyors) make extensive use of published books and lists of 'labour and materials' prices. These should be used with caution; there are differences amounting to 25 per cent in the prices for such everyday work as brickwork, block partitions and plastering in the best-known of these publications: the principal trouble being that the reader is never sure what is included in these prices, or—more important—what is omitted. Unloading and stacking, National Insurance, scaffolding and other plant are usually not included and, of course, these prices cannot take account of bonus systems or plus rates. Some of these works include sections giving prices of materials and proprietary goods from suppliers. These are invariably minimum prices, and sometimes extras for cartage, etc., are not included.

Even greater caution will be needed in the next few years, because the metric prices for metric quantities appearing in trade journals and technical works (including this one) are straight conversions from existing rates in Imperial measure and old money. They have not been proved in practice. All works on estimating published before 1969 should be used with great caution. Immense errors can be made when trying to adapt old-money prices for old measurements to new-money prices for metric quantities. Even with modern conversion tables, much extra work must be involved; because there is extra work, it will be done in a hurry—a prolific source of errors. Unfamiliarity with the new coinage and the new measuring units must, nevertheless, cause the work to be done more slowly, and the very element of unfamiliarity must cause some errors. It is vital that one extra check be made of both measurements and prices, and this may be necessary for some years, until everyone has learned to think in the new factors and has begun to forget the old ones.

LABOUR ONLY CONSTANTS

The following are straight conversions from Imperial measure and weight: they should be used with caution because they have not been tried out in practice. Indeed, it will be several years before completely reliable sets of estimating constants or prices are available.

For those who wish to convert estimating data in Imperial sizes and weights to metric, a brief set of conversion tables are given on page 95.

		Hours	Metric unit
Excavator			
Surface, trenches etc.	Labourer	2.275	m^3
Deep trenches	,,	3.245	,,
Load lorries	,,	0.162	,,
Return, fill and ram	,,	1.300	,,
Concretor			
Mixing and depositing by hand	,,	0.200	,,
Mixer and dump trucks		2.600	,,
Bricklayer			
11" cavity wall	{ Bricklayer	2.392	m^2
	Labourer	0.089	,,
4½" wall	{ Bricklayer	1.196	,,
	Labourer	0.347	,,
DPC	{ Bricklayer	0.800	m run
	Labourer	0.200	,,
Breeze partitions:			
2"	{ Bricklayer	0.347	m^2
	Labourer	0.299	,,
3"	{ Bricklayer	0.347	,,
	Labourer	0.591	,,
4"	{ Bricklayer	0.809	,,
	Labourer	0.596	,,
Drainlayer			
Stoneware, laying and jointing			
4" Bricklayer and labourer, each		1.000	4.50m run
6" Bricklayer and labourer, each		1.000	3.15m run
(Testing, 4hr per length not over 100ft)		4.000	30.5m
Agricultural			
Agricultural labourer (cut, lay and cover)		0.600	m run
Asphalter			
(Labour constants are for asphalter and mate, each)			
Fine gritted asphalt, 2-coat work		1.434	m^2
Underfelting		0.038	,,
Vertical work		2.340	,,
Skirtings (6")		0.050	,,
Dampcourse on narrow walls		0.050	m run

BASIC ESTIMATING

				Hours	Metric unit
Mason	(Labour constants are for crafts—men only)				
	Ashlar (1' to 2' cube) 0.28 to 0.56m			3.750	3.75m run
	Ditto (over 2' cube) over 0.56m			3.200	3.20m run
	(The above is for Portland Stone; deduct 50% for Bath Stone; add 50% for medium sandstone, and 100% for hard sandstone, 140% for marbles and 300% for granite.)				
	Hoisting, setting and pointing stonework from ¼ hr to ⅞ hr, according to size, ranging from 6in (150mm) bed to 15in (375mm) bed.				
Roofer	(Labour constants are for roofer and mate, each)............				
	Corrugated sheets	3 hrs			per square of 100ft or 9.3mm^2
	Plain tiling	3 ,,			
	Interlocking tiles	2½ ,,			
	Slating (20" × 10" slates)	2¾ ,,			
	Battening for tiles or slates	2½ ,,			
	Underfelting			0.38	m^2
Carpenter	General work in roofs, floor joists, etc.				
	Carpenter	1 hr	ft cube	1.00	
	Labourer	1/10,,	ft cube	.10	
	Doors—hanging Carpenter	2 hrs	per door	2.00	each
	Gates—hanging ,,	1 hr	per gate	1.00	each
	Doors frames—making ,,	2½ hrs	per ft cube	1.75	each
	Doors frames—fixing ,,	¾ hr	ft cube	0.75	each
	Draining boards ,,	½ ,,	ft super	6.07	m^2
	Sashes—casement, making,,	⅝ ,,	ft super	4.86	
	Sashes, casement, hanging ,,	1/10,,	ft super	1.07	m^2
Ironworker	(Labour constants are for smith and mate, each)				
	Fixing half-round or ogee gutters	1/7 hr	ft run	0.05	m run
	Fixing rainwater pipe..........	¼ ,,	,, ,,	0.25	,, ,,
	Steelwork—up to 10'	2 hrs	cwt	2.00	50kg
	Steelwork—up to 30'	2⅓ ,,	,,	2.50	50kg
				Hours	Metric unit
Plasterer	Fixing plasterboard Craftsman	⅜ hr	yd super	0.347	m^2
	Labourer	⅜ ,,	,, ,,	0.347	m^2
	Metal lathing ,,	1/3,,	,, ,,	0.569	m^2
	Render 1 coat (Portland cement)				
	Craftsman & Labourer, each	1/5,,	,, ,,	0.238	m^2
	Render 2 coats, Portland cement				
	Craftsman & Labourer, each	⅜ ,,	,, ,,	0.347	m^2
	Skim in gypsum plaster				
	Craftsman	⅓ ,,	,, ,,	0.149	m^2
	Labourer	1/6,,	,, ,,	0.098	m^2

				Hours	Metric unit
Plumber	(Labour constants are for plumber and mate, each)				
	Sheet lead in roofwork generally	5 hrs	cwt	5.00	50kg
	ditto in gutters..........	6 ,,	,,	6.00	50kg
	ditto in stepped flashings	7½,,	,,	7.50	50kg
	angles to safes, etc..	1/5hr	each	0.20	each
	soldered dots.......	¼,,	,,	0.25	each
	Lead pipe, fixing, with holdfasts ½in, ¾in and light wastes..........	1/5,,	ft run	0.65	m run
	4in soil pipe	⅓,,	,, ,,	1.60	m run
	ditto bends	1½hrs	each	1.50	each
	Wiped joints ½in, ¾	2/5hr	,,	0.47	each
	1¼in and 1½in	2/3,,	,,	0.76	each
	4in	1 ,,	,,	1.19	each
	Zinc and copper sheet in roofs..	¼,,	ft super	2.63	m²
	ditto ditto in gutters, etc	⅜,,	,, ,,	3.123	m²
Glazier	(Labour constants are for glazier only)				
	Sheet glass in large squares	1/5hr	ft super	1.196	m²
	ditto in small squares	¼,,	,, ,,	0.299	m²
Painter	Burning off Painter	1 ,,	yd super	1.196	m²
	Plain painting, 1st coat ,,	1/5,,	,, ,,	0.238	m²
	2nd coat ,,	1/6,,	,, ,,	0.228	m²
	Finish ,,	1/5,,	,, ,,	0.238	m²
	Distemper ,,	1 ,,	11yds sup	1.000	10m²
Paperhanging	Stripping (labourer)	½ to 1hr	per piece		
	Stopping and preparing (craftsman).....................	½ to ¾ hr	,, ,,		
	Hanging lining paper (craftsman)	1 hr	,, ,,		5 m²
	ditto ordinary papers (craftsman)..............	1.1/5hrs	,, ,,		
	ditto ceiling paper (craftsman)..............	1½hrs	,, ,,		

THE CRITICAL PATH METHOD

The 'critical path method' (called CPM in most of the literature about it) is a highly systematized procedure that aims to ensure the best possible allocations of time and money to each operation (or activity or trade) on a job. If properly carried out, it should enable a job to be completed for the least possible money and in the shortest possible time. Plainly, it involves job planning and job management as well as estimating. This is perhaps more logical than treating estimating as a subject on its own, without much reference to any other factors.

In the first stage of the CPM, the separate activities or operations are arranged in a sequence, in the order that they occur on a job. From this, an arrow diagram is drawn up, showing this sequence. The second stage consists in adding to this arrow diagram estimates of the time each activity will take, and in selecting the 'critical' operations, that must follow in immediate succession if the job is to finish without loss of time. It is these activities that form the 'critical path'. Leeway or 'float' on non-critical operations—that is, the length of time for which they can be delayed without affecting the 'critical path'—is also determined. The general pattern of the job is now revealed. The contractor already knows that, in order to make the best of the job, delays must not occur in the progress of any of the critical operations, nor between the end of one critical activity and the next in sequence. He now knows which are his critical operations, and just where delays must not be allowed to occur: he also knows by how much non-critical operations can be delayed without affecting the finishing time of the job as a whole.

In stage three—the costing stage—several cost estimates are made. The first is based on normal times for the operations, as set out in the CPM schedule. The second is for a 'crash' programme, involving cuts in time allowed for critical operations (even if this means overtime and flooding the job with men) so far as 'float' on non-critical operations will allow. Then, several programmes intermediate between Stage 1 and Stage 2 are estimated. These different combinations of time and cost are plotted as two time-cost curves; one for direct and the other for indirect costs. From these, a total cost curve is constructed, giving the best programme, in time and cost, for the job. With this information in his possession, the contractor can watch the job's progress closely. If he sees signs of a breakdown in the scheduled time for any operation, he will know how much it will cost him to catch up, and will further know whether it would be profitable to initiate a 'crash' programme, or one of the intermediate ones, or to allow the job to fall behind schedule.

It is obvious that there is an enormous number of calculations involved. These are, however, quite straightforward ones, and a computer need only be used if a lot of time and tedious effort cannot be spent. Some authorities estimate that 75 activities is the greatest number that can be dealt with without a computer, while others put the number at 1,500.

This method is now in use on large contracts, and one Government Department specifies its use on a number of its contracts. A fuller introduction to the subject is given in Building Research Station Digest No. 53 (Second Series) available from H.M.S.O. Courses are now available in some evening institutes and technical colleges.

ESTIMATOR'S CONVERSION TABLES

HOURS PER SQUARE FOOT TO HOURS PER SQUARE METRE		HOURS PER CUBIC FOOT TO HOURS PER CUBIC METRE	
Hr per sq. ft	Hr per sq. m	Hr per cu. ft	Hr per cu. m
⅛	1.341	⅛	4.414
1/5	2.142	¼	8.828
¼	2.682	½	17.657
⅜	3.123	¾	26.485
½	4.864	1	35.315
⅝	6.705	2	70.630
¾	7.281	3	105.945
⅞	9.387		
1	10.764		

HOURS PER SQUARE FOOT TO HOUR PER SQUARE METRE		HOURS PER CUBIC FOOT TO HOURS PER CUBIC METRE	
Hr per sq. yd	Hr per sq.m^2	Hr per cu. yd	Hr per m^3
⅛	0.149	⅛	0.162
1/5	0.238	¼	0.325
¼	0.299	½	0.650
⅜	0.347	¾	0.975
½	0.596	1	1.300
⅝	0.745	2	2.600
¾	0.809	3	3.900
1	1.196	4	5.200
2	2.392	5	6.500
3	3.588		
4	4.784		
5	5.980		

MODULAR CO-ORDINATION

'Dimensional co-ordination' or 'modular co-ordination' has been defined as 'rationalization of theoretical dimensions of a structure to ensure that all are as far as possible multiples in whole numbers of a basic unit of measurement known as the module or where necessary rational and limited subdivisions of the module' *(Specification,* vol. 1). Put another way, it means that all components such as windows and doors should be multiples of the basic unit (which in Britain is the brick), in length and height. All lesser units, such as widths of door frame members, should be simple subdivisions of the length of a brick.

A truly 'modular' building should be designed on a modular grid of 300 mm squares, and it follows that the basic measure (or dimension, or module) should be a simple subdivision of 300 mm in all its three dimensions.

At first sight the idea is an attractive one because it seems to bring order out of chaos, substituting simplicity for diversity. The whole system of building would seem to be a planned co-ordinated one, instead of a heterogeneous collection of unrelated traditional sizes. When translated from theory into practice, drawbacks appear and the theory begins to lose its attractiveness. Architects find that planning to a grid limits scope and hinders good planning. On the building site, joint thicknesses, variations in manufactured sizes and tolerances for 'going in' have to be taken into account. These necessitate various clumsy expedients and compromises, such as 'formats' in brickwork and blockwork, which result in numerous departures from the simplicity of the theory. In fact, the result is just another heterogeneous collection of dimensions. In short, the theory does not work in practice.

Nevertheless, it has an attraction for some: it is sufficiently popular for some brickmakers to manufacture special 'modular bricks', but it is never likely to supersede the existing system. Builders and especially their estimators should remember that 'modular' construction is a different sort of work than the normal, and can be very expensive.

THE EUROPEAN ECONOMIC COMMUNITY

Entry into the Common Market imposed some conditions that affect firms tendering for large contracts. *EEC Directive* 71/305, operative since July 1973, lays down that all public bodies (Government, Local Authorities, housing and new town corporations or associations) must advertise the contract to be let in the *Official Journal of the EEC,* if that contract is for £415,000 or over. Firms wishing to tender must submit a 'request to participate' to the public body letting the tender, and this 'request to participate' must be supported by financial and technical evidence of capability to carry out the work. A 'short list' may be selected from the applicants by the public body concerned: a firm already on that authority's approved list need not supply the particulars detailed above. In theory, this throws open a number of large contracts in the UK to foreign competition, but gives an opportunity to British firms to tender for a much greater number of jobs in the European states in the EEC.

4. THE STANDARD METHOD OF MEASUREMENT OF BUILDING WORKS (SMM6-SIXTH EDITION)

Plant Items

Section A	102
Section B	103
Section C	105
Section D	105
Section E	114
Section F	115
Section G	117
Section H	117
Section J and K	117
Section L	118
Section M	118
Section N	118
Section P	120
Section Q	120
Section R	120
Section S	122
Section T	123
Section U	125
Section V	126
Section W	128
Section X	129

This method of measuring building works was agreed between the Royal Institution of Chartered Surveyors and the National Federation of Building Trades Employers, now Building Employers Confederation.

The *Fifth Edition of the Standard Method of Measurement* was published in the Autumn of 1962, and became operative on 1 March 1963. The *Fifth Edition Metric,* published in 1968, was a conversion of the basic Imperial edition and not a revision.

The *Sixth Edition* is an interim edition in which the work sections have been developed in varying degrees, some considerably and some only sufficiently to accommodate them to the general pattern.

SMM6 is accompanied by a non-mandatory *Practice Manual,* which is intended to be read and used in conjunction with the SMM. It has two basic purposes: first, it is intended to give guidance on the communication of information relative to quantities of finished work where their value is modified by position, complexity, simplicity, repetition, eccentric distribution or other cost significant factors: secondly, it is intended to encourage good practice in the measurement of building works. **The sixth edition became operative on 1 March 1979.**

SMM6 deals with methods of measurement and not, with a few exceptions, methods of billing. The general format and design of bills is left to the discretion of the surveyor. However, from the tenderer's point of view it is convenient if all work which will be carried out by one specialist is brought together in one section of the bills of quantities. For example: lead and copper damp-proof courses and flashings could conveniently be grouped with plumbing.

Work such as filler joist floors, thatching, etc, which are types of construction seldom measured nowadays, have been omitted. It is suggested that the surveyor uses his discretion in the measurement of this kind of work. He could refer to previous editions of SMM, or refer to local custom in the area in which the work is to be carried out, or consult with specialists who may be involved.

The same principle applies to new forms of construction or finish or to those which are seldom used. An example would be the formation of simulated beams and timbers in glass fibre.

PLANT ITEMS

The SMM requires items to be given in the bill for the provision and maintenance of plant on site. The reason for giving plant separately from the ordinary items for permanent work is in recognition of the fact that plant costs often depend upon factors other than the quantities and nature of the various items of permanent work in the section.

For example: in erecting precast concrete, the plant cost would include the cost of a crane for erecting precast concrete units. The cost of the crane would depend upon what size of crane was provided and how long it was likely to be required on site. The choice of the crane would depend upon the heaviest duty required, usually the heaviest unit at greatest radius, and this would govern the daily charge for keeping the crane on site and the charges for haulage to and from the site.

The length of time would be governed by other factors such as the rate at which the work becomes available and the number and type of precast units as a whole.

The total plant cost consequently has a fixed element (the cost of providing the plant) and a time-related element (the cost of maintaining the plant on the site for the duration required).

These elements of cost are normally considered separately when estimating and it is helpful if they can also be considered separately when valuing variations.

The plant items enable these cost elements to be priced separately instead of being spread over other groups of items, which should contribute to more realistic pricing generally. It is not the intention that tenderers should be forced to vary their method of pricing. If the items are left unpriced, the plant cost will be assumed to have been included elsewhere.

SECTION A. GENERAL RULES

The major innovation in this section is the definition of drawn information included in clause A5.

The three types of drawn information referred to are location drawings, component details, and bill diagrams.

The location drawings are further defined (A.5. 1a) and are referred to in the majority of work sections which follow. Component details are working drawings (referred to mainly in the 'composite items' section of Section N) and bill diagrams which are used throughout the rules as an adjunct to a brief description.

A.5. DRAWN INFORMATION

1. Drawn information where required to be provided for the full implementation of this document is defined as follows:
    (a) Location drawings.
        (i) Block plan; To identify site and outline of buildings in relation to Town Plan or other wider context.

(ii) Site plan; To locate the position of buildings in relation to setting out point, means of access and general layout of site.

(iii) General locating drawing; To show the position occupied by the various spaces in a building and the general construction and location of principal elements.

(b) Component details. To show all the information necessary for the manufacture and assembly of the component.

(c) Bill diagram. To be drawn information which may be provided with the bills of quantities to aid the description of the item. Information by way of dimensions or detail may be indicated; alternatively such information shall be included in the relevant description that accompanies the bill diagram.

A.7. QUANTITIES

1. Where the unit of billing is the metre, quantities shall be billed to the nearest whole unit. Fractions of a unit less than half shall be disregarded and all other fractions shall be regarded as whole units.
2. Where the unit of billing is the tonne, quantities shall be billed to the nearest two places of decimals.
3. Where the application of clauses A.7. 1, and A.7. 2, would cause an entire item to be eliminated, such an item shall be enumerated, stating the size or weight appropriate.

SECTION B. PRELIMINARIES

A list is now required of the drawings from which the bill of quantities has been prepared (B.3. 1.) and the general location drawings (previously defined in clause A.5. 1(a) (iii) are required to accompany the tender documents.

A check list of example obligations and restrictions which might be imposed by the employer is now included (B.8. 1.). A similar list for the convenience of pricing by the contractor (B.13) indicates that drying the works is not to be the responsibility of the contractor, unless specific temperature and humidity levels are required.

The attendances required on nominated subcontractors have been amplified and now include particulars of the weight and size of significant pieces of equipment (B.9. 3(c)).

B.3. DRAWINGS AND OTHER DOCUMENTS

1. A list shall be given of the drawings from which the bills of quantities have been prepared and which will be available for inspection by the

contractor. The list shall indicate the location drawings as clause A.5. l(a) (i)-(iii) which shall accompany the bills of quantities.

B.8. OBLIGATIONS AND RESTRICTIONS IMPOSED BY THE EMPLOYER

1. Particulars shall be given of any obligations or restrictions to be imposed by the employer in respect of the following, unless they are covered by the schedules given in accordance with clause B.4.

   (a) Access to and possession or use of the site.
   (b) Limitations of working space.
   (c) Limitations of working hours.
   (d) The use or disposal of any materials found on site.
   (e) Hoardings, fences, screens, temporary roofs, temporary name boards and advertising rights.
   (f) The maintenance of existing live drainage, water, gas and other mains or power services on or over the site.
   (g) The execution or completion of the work in any specific order or in sections or phases.
   (h) Maintenance of specific temperatures and humidity levels. Alternatively a provisional or prime cost sum shall be given.
   (j) Temporary accommodation and facilities for the use of the employer including heating, lighting, furnishing and attendance.
   (k) The installation of telephones for the use of the employer and the cost of his telephone calls shall be given as a provisional sum.
   (l) Any other obligation or restriction.

B.13. PRICING

1. For convenience in pricing, items for the following shall be given.

   Maintaining temporary works; adapting, clearing away and making good shall be deemed to be included with the items. Notices and fees to local authorities and public undertakings related to the following items shall be deemed to be included with the items.

   (a) Plant, tools and vehicles.
   (b) Scaffolding.
   (c) Site administration and security.
   (d) Transport for workpeople.
   (e) Protecting the works from inclement weather.
   (f) Water for the works. Particulars shall be given if water will be supplied by the employer.

(g) Lighting and power for the works. Particulars shall be given if current will be supplied by the employer.
(h) Temporary roads, hardstandings, crossings and similar items.
(j) Temporary accommodation for the use of the contractor.
(k) Temporary telephones for the use of the contractor.
(l) Traffic regulations.
(m) Safety, health and welfare of workpeople.
(n) Disbursements arising from the employment of workpeople.
(p) Maintenance of public and private roads.
(q) Removing rubbish, protective c asings and coverings and cleaning the works on completion.
(r) Drying the works.
(s) Temporary fencing, hoardings, screens, fans, planked footways, guardrails, gantries and similar items.
(t) Control of noise, pollution and all other statutory obligations.

SECTION C. DEMOLITION

This section has remained much as SMM5, but alteration work is now measured under the appropriate trade heading.

One innovation, which also occurs throughout the works sections, is the introduction of a requirement to include an item for bringing to, maintaining on, and removing from, site, any plant required for the work (C.2.). This clause is intended to provide for the contractor to price, if he so wishes, the relevant plant in the relevant work section.

C.2. PLANT

1. An item shall be given for bringing to site and removing from site all plant required for this section of the work.
2. An item shall be given for maintaining on site all plant required for this section of the work.

SECTION D. EXCAVATION AND EARTHWORK

Rules for this section have been drafted on the assumption that excavation and earthwork will be carried out by mechanical plant. This explains the disappearance of the term 'get out' in relation to excavation.

Where it would be difficult or impractical to carry out excavation by mechanical means, e.g. alongside existing services or in small pits, the rules have been drawn up to identify these situations.

A soil description (D.3.), is now required, including details of over or underground services. Also, a pre-contract water level must be established from which the bill measurement will be prepared.

The water level will be re-established at the time the excavations are carried out and the measurements adjusted accordingly. Guidance on the application of this rule is given in the *Practice Manual*.

D.3. SOIL DESCRIPTION

1. Particulars of the following shall be given:
   (a) Ground water level and the date when it was established, hereinafter described as the pre-contract water level. The ground water level shall be re-established at the time the various excavation works are carried out and is described hereinafter as the post-contract water level.

   Where ground water levels are subject to periodic changes due to tidal or similar effects they shall be so described and the average of the mean high and low levels given.
   (b) Trial pits or bore holes, stating their location.
   (c) Over or underground services.
(2) If the above information is not available a description of the ground and strata which is to be assumed shall be stated.

D.3.1(a): Such variations in ground water level are not intended to include temporary fluctuations in level which arise as a result of those items at the contractor's risk under Clause D.25. It is not intended that several water levels, with only minor variations, be established for each site but the water level could well differ substantially from one excavation to another (e.g. pits at opposite ends of the site) or from one part to another of an individual excavation (e.g. a very large basement) especially in those instances where springs are encountered. It should be possible, in most cases, to agree a common post-contract ground water level for the whole site.

D.3.1(b): The particulars required to be given will be the information by way of trial hole data that is available. Such information, if given in accordance with the relevant British Standard Code of Practice, would ensure that a reasonable indication of what can be expected has been given.

(Excerpt from Practice Manual)

Excavation is now required to be classified in terms of maximum depth of excavation (D.11) rather than in 1.50 m stages.

STANDARD METHOD OF MEASUREMENT

D.11. DEPTH CLASSIFICATION

Depths of excavation shall unless otherwise required by this document be classified as follows:

Maximum depth not exceeding 0.25 m
,,   ,,   ,,   ,,   1.00 m
,,   ,,   ,,   ,,   2.00 m
,,   ,,   ,,   ,,   4.00 m

and thereafter in 2.00 m stages.

The working space rules (D.12.) are similar in principle to SMM5 but have been clarified and amplified by sketches in the practice manual (reproduced here). Excavation and filling of working space must now be given as a separate item.

D.12. WORKING SPACE

1. Working space (which shall not be subject to adjustment if more or less space is actually required) shall be measured from the face of the finished work as follows;

   (a) Formwork

       (i) 0.25 m from the face of any work which requires formwork where the bottom of the formwork does not exceed 1.00 m below the starting level of the excavation.

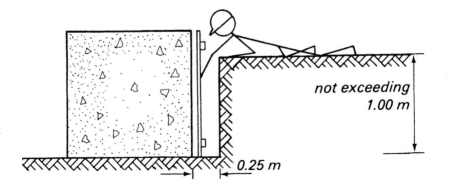

(ii) 0.25 m from the face of any work which requires formwork where the height of the formwork does not exceed 1.00 m and the bottom of the formwork exceeds 1.00 m below the starting level of the excavation.

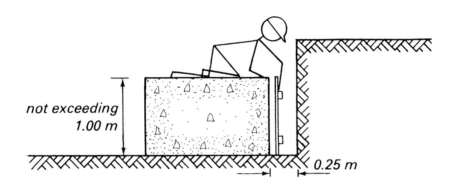

(iii) 0.60 m from the face of any work which requires formwork where the height of the formwork exceeds 1.00 m and where the bottom of the formwork exceeds 1.00 m below the starting level of the excavation.

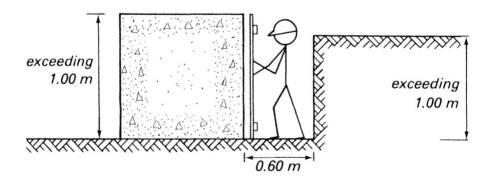

STANDARD METHOD OF MEASUREMENT 109

(b) Work which requires workmen to operate from the outside. 0.60 m from the face of any work which requires workmen to operate from the outside at any depth below the starting level of the excavation.

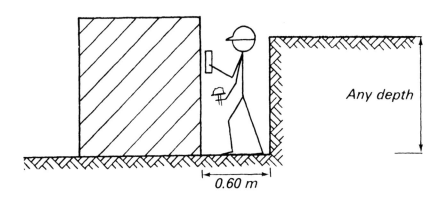

Note: The 'face of any work' would be the outside face of any asphalt tanking or of any brick protective skin if one were provided.

(c) Post tensioned concrete.
1.50 m from the face which requires post tensioning.

2. Excavation and filling of working space shall be classified by depth in accordance with Clause D.11, and given in cubic metres, as a single item, for each of the types of excavation given in clause D.13.

Additional earthwork support and disposal of surface treatment arising from the measurement of working space shall be deemed to be included.

3. Disposal of excavated material from working space shall be described in accordance with clauses D.27 to D.29.

4. Filling of working space shall be described in accordance with clauses D.33 and D.37.

Note: If filling to working space is to be other than excavated material, this should be given in the description stating the material (e.g. type of hardcore or mix of concrete).

5. The number of pits shall be given when measuring working spaces around pits.

D.13. TYPES OF EXCAVATION

This section introduces a new concept of narrow trenches measured lineal (D.13.6(a)) and additional classes of excavation for excavating around pile caps and around services have been included (D.13.7 and D.13.9). Amplification of the meaning of the term 'special plant' in relation to excavation in rock is included in the *Practice Manual*.

6(a) Trenches not exceeding 0.30 m in width, given in metres stating the average depth to the nearest 0.25 m.

7(a) Excavating for pile caps and trenches for ground beams between piles shall be given in cubic metres stating the starting level.

  (b) A location drawing showing the pile sizes and layout shall be provided as clause E.1.1.

9. Excavating around services crossing an excavation shall be enumerated and given as extra over the various descriptions of excavation and working space; details of type, size, depth of service and length exposed shall be given. Any temporary support to services shall be deemed to be included with the item.

D.13.1. Examples of 'special plant' in this context would include:

  (a) Power operated hammers, drills and chisels.

  (b) Special attachments to mechanical plant such as rock buckets, rippers, hammers and chisels.

(Excerpt from Practice Manual)

EARTHWORK SUPPORT

The term 'earthwork support' (D.14 to D.24) is now used instead of 'planking and strutting' in recognition of the wide variety of materials available for this purpose. Earthwork support now includes all shoring, including the special shoring required under SMM5, but does not include interlocking driven sheet steel piling. Earthwork support is now classified by distance between the faces to be supported rather than by types of excavation.

D.14. GENERALLY

1. Interlocking driven sheet steel piling shall be given in accordance with Section E.

2. Curved earthwork support shall be so described irrespective of radius.

D.15. DEFINITIONS

1. Earthwork support shall be deemed to mean providing everything requisite to uphold the sides of excavation by whatever means are necessary (other than interlocking driven sheet steel piling) and shall be measured to all faces of excavation whether or not any is in fact required except to:

    (a) Faces not exceeding 0.25 m in height;

    (b) Sloping face of any excavation where the angle of inclination does not exceed 45 degrees from the horizontal.

    (c) Face of any excavation which abuts an existing wall, pier or other structure:

    (d) Face of any additional excavation which results from the measurement of working space.

D.16. UNIT OF MEASURE AND DEPTH CLASSIFICATION

Earthwork support shall be given in square metres and classified by maximum depth as clause D.11, irrespective of starting level.

D.17. CLASSIFICATION BY DISTANCE BETWEEN FACES

Earthwork support shall be classified by distance between opposing faces, excluding measurement for working space, as follows;
    not exceeding 2.00 m
        2.00—4.00 m
    exceeding 4.00 m.

D.18. TRENCHES ETC., BELOW FACES OF EXCAVATION

Earthwork support to trenches, pits and the like occuring below the face of an excavation should be so described, where the horizontal distance from the face of the excavation above, including any measurement for working space in accordance with clause D.12, is less than the depth of the trench, pit, etc. The height of the excavation face above the top of the trench, pit etc, shall be given.

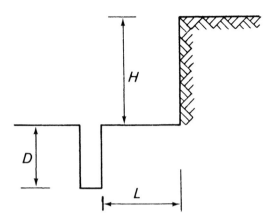

When L (after measurement of working space) is less than D the earthwork support shall be described as below the face of the excavation and the height of the excavated face above (H) shall be given.

D.19. NEXT TO ROADWAY

Earthwork support next to roadways shall be so described if the horizontal distance from the face to be supported to the edge of the roadway or footpath, including any measurement for working space in accordance with clause D.12, is less than the depth of the excavated face below the edge of the roadway or footpath.

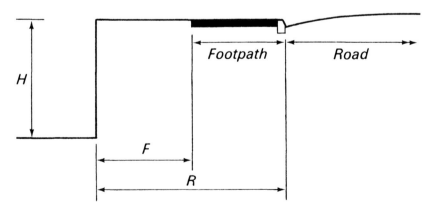

When F or R (after measurement of working space) is less than H the earthwork support shall be described as next roadway.

STANDARD METHOD OF MEASUREMENT   113

D.20. NEXT TO EXISTING BUILDINGS

Earthwork support next to existing buildings shall be so described if the horizontal distance from the face to be supported to the nearest part of the foundations of the building, including any measurement for working space in accordance with clause D.12, is less than the depth of the excavated face below the bottom of the foundations.

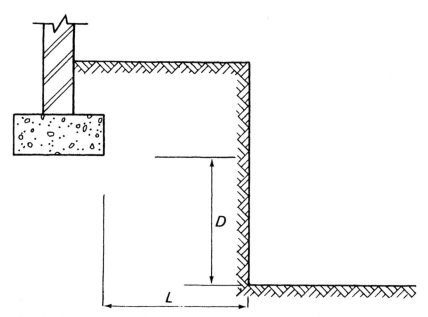

When L (after measurement of working space) is less than D the earthwork support shall be described as next to existing buildings.

D.22. UNSTABLE GROUND

Earthwork support to excavations in running silt, running sand and the like shall be so described and measured from the starting level of the excavation to the full depth.

> The words 'and the like' (D.22) have been added to the *Fifth Edition's* wording of 'running silt or running sand' in an attempt to avoid the many arguments that have arisen by limiting the application of the rule quite literally to running silt or running sand when the same problems pertain with such materials as loose gravel, fly ash, etc. It is suggested, as a guideline, that strata could

be said to fall within the intended category only when the newly excavated face will not remain unsupported sufficiently long to allow the necessary support to be inserted.

(Excerpt from Practice Manual)

The terms 'general water', 'ground water', 'spring or running water', used in SMM5 are replaced by the terms 'surface water' and 'ground water'. Surface water is intended to deal with rainwater, disposal of which is required to be given as an item (D.25). A similar item (D.26) will be given for the disposal of ground water where the pre-contract water level indicates that the excavation work will be affected by ground water. Such an item would be introduced into the variation account, should there be a difference between a change following the recording of the post-contract water level and the pre-contract water level.

SECTION E. PILING AND DIAPHRAGM WALLING

This section has been redrafted to follow the civil engineering standard method of measurement. E.1 requires that location drawings adequately indicate nature of work being measured and what plant items are required (E.3). Rules for measuring diaphragm walling (E.10 to E.13) are included for the first time.

E.13. DIAPHRAGM WALL DESCRIPTION

1. Items for work in diaphragm walls shall be grouped together under an appropriate heading.
2. Excavation for diaphragm walls shall be given in cubic metres stating the thickness. The volume of excavation shall be calculated from the nominal lengths and depths of the walls. The depths shall be taken from the level at which the work of excavation is expected to begin and in accordance with clause E.2.3. Additional excavation to accommodate guide walls or to provide working space and disposal of material so excavated shall be deemed to be included.
3. Excavation for diaphragm walls shall be classified into the following ranges of maximum depth:

   not exceeding 5.00 m
   5.00—10.00 m
   10.00—15.00 m
   15.00—20.00 m
   20.00—25.00 m
   25.00—30.00 m
   Exceeding 30.00 m stating the depth.

4. Excavation in rock and excavation in artificial hard material shall each be given separately in cubic metres stating the nature of the anticipated strata and of the material. Alternatively such work shall be given as extra over the excavation. Standing time and associated labour in connection with such items shall be deemed to be included.
5. Disposal of excavated material shall be given in cubic metres.

Diaphragm walls are concrete walls constructed using Bentonite slurry or other fluids to support a trench which is then filled with concrete to form the wall, the concrete being placed through the support fluid which is thereby displaced.

Excavated material which has been in contact with Bentonite or other support fluid may be contaminated and not fit to be used as filling material. Descriptions for disposal of such excavated material should draw attention to any restrictions imposed in this respect.

(Excerpt from Practice Manual)

SECTION F. CONCRETE WORK

This section has been altered both in the rules for measurement of concrete and formwork. The work is required to be classsified (F.3.) into concrete-framed structures, steel-framed structures and other concrete work.

F.4. IN-SITU CONCRETE

2. Reinforced work shall be so described. Members having a reinforcement content in excess of 5 per cent by volume shall be so described.

The requirement (F.4.2) to distinguish members having a reinforcement content in excess of 5 per cent by volume is intended to indicate which members (slabs, columns, beams, etc) are excessively heavily reinforced resulting in problems in placing concrete. It is the total reinforcement in a complete member which has to be considered and not a small isolated part of a member which is particularly congested, e.g. where the laps occur in the reinforcement.

The selection of 5 per cent is a deliberately high proportion in that it will only apply to exceptional cases and usage will identify the rare instances where this occurs. As a rule of thumb, this will apply where the steel exceeds 0.41 tonnes per cubic metre of the measured member.

(Excerpt from Practice Manual)

F.6. CONCRETE CATEGORIES

1. All members, unless otherwise stated, shall be given in cubic metres and described in the categories set out in clauses F.6. 2—21.

The concrete categories (F.6) follow SMM5 with certain grouping having taken place, i.e. foundations in trenches, pier and column bases will be measurable as one item. Attached beams are no longer separated, but are measured as part of the floor slab.

As retaining walls are not a separate category they will come under walls (F.6.12(a)).

F.8. FINISHES CAST ON TO CONCRETE

Finishes (measured on the exposed face) which are cast on to concrete by lining formwork shall be so described, distinguishing between sides, soffits and upper surfaces and given in square metres as extra over the concrete, stating the mix and the thickness of the finish.

> Examples of finishes which are cast on to concrete would include granolithic, cast stone, terrazzo, mosaic, etc.
> *(Excerpt from Practice Manual)*

Measurement of reinforcement in tonnes follows SMM5, except that maximum lengths and vertical bars are separated.

F.11. BAR REINFORCEMENT

6. Horizontal bars and bars sloping not more than 30 degrees from the horizontal (grouped together) and over 12.00 m long, shall be so described stating the length in further stages of 3.00 m.

7. Vertical bars and bars sloping more than 30 degrees from the horizontal (grouped together) and over 5.00 m long, shall be so described stating the length in further stages of 3.00 m.

The major change in the concrete section occurs in the measurement of formwork (F.13 to F.17).

To highlight the cost significance of this work formwork to soffits (F.15.1-2), and to walls (F.16.1) require the number of separate surfaces to be stated.

Formwork to beams and columns, both isolated and attached, is to be measured lineally (F.15.5 and F.16.6).

In the case of beams and columns of irregular shape the suggestion is made that a bill diagram should accompany the description.

SECTION G. BRICKWORK AND BLOCKWORK

G.4. MEASUREMENT

1. Brickwork and blockwork shall be measured the mean length by the average height. Fair face and facework shall be measured on the exposed face. No deduction shall be made for the following:
   (a) Voids not exceeding 0.10 m^2.
   (b) Flues, lined flues and flue-blocks where the voids and the work displaced do not together exceed 0.25 m^2.
2. Labours on different kinds of work shall be given separately. Labours on existing work shall be so described.
3. Curved work shall be so described stating the mean radius. Curved fair face and curved facework shall be so described stating the radius on face. The provision of curved bricks or curved blocks shall be given in the description of curved work. Rough cutting within the thickness shall be deemed to be included with the curved brickwork or curved blockwork.
4. Work in underpinning shall be given in accordance with Section H.

The major change in Section G is the elimination of reduced brickwork, all walls will now be measured their actual thickness and rules requiring measurement of cutting labours have largely disappeared.

Location drawings are required (G.1) to indicate the nature and extent of the work.

Brickwork and blockwork are to be classified as in foundations, load bearing and non load bearing work (G.3).

Brickwork in deep trenches is now required to be separated (G.5.3) and the categories of brickwork generally have been collected together in one clause (G.5).

The cutting in brickwork is now restricted to fair work (G.15), all other cutting having been eliminated.

SECTION H. UNDERPINNING

Drawings are required (H.1) and plant items (H.2). Other changes follow those in other relevant sections.

SECTION J. RUBBLE WALLING; SECTION K. MASONRY

Sections J and K have remained basically as SMM5. Location drawing items are included (J.1 and K.1), and plant items (J.2 and K.2).

SECTION L. ASPHALT WORK

L.1 to give drawn information includes a need to provide plans at each level of the work and sections if tanking is involved. Other than this, and plant items (L.2), the section remains virtually as SMM5.

SECTION M. ROOFING

Rules for the measurement of roofing follow very closely the requirements of SMM5. Location drawings are required (M.1) to show the extent of the work and the height above ground level. M.6 requires that, when measuring slate or tile roofing, the pitch or pitches (grouped together) shall be stated.

SECTION N. WOODWORK

This section is an amalgam of the old carpentry and joinery sections in SMM5. The rules have been drafted on the assumption that the majority of woodworking is now a shop process using machinery rather than a site craft process.

N.1. INFORMATION

2. All sizes shall be deemed to be basic (nominal) sizes unless stated to finished sizes. Limits on planing margins shall be given. Where deviation from the stated sizes will not be permitted, this shall be stated. The intention is that timber sizes shown on drawings should be the sizes measured. When basic (nominal) sizes are given, the size expected in the finished work will be the basic size subject to planing margins and tolerance.

If, however, finished sizes are given, the contractor will be expected to decide the size of timber required before planing, taking into account either stated planing margins or his own assumptions. The size expected in the finished work will be the finished size subject only to normal tolerance.

Planing margins required to be stated in the bill of quantities should either be specific to the works or refer to a standard specification.

Maximum lengths of timbers (N.1.6) have been changed to align with the maximum marketable lengths currently obtainable.

There is a requirement (N.1.7) that labours shall be related to the items to which they refer. There is also (N.1.11) a requirement to indicate the number of differing cross section shapes, even where the overall sectional dimensions are identical.

The 'carcassing' subsection (N.2 and N.3) replaces the structural element of the former carpentry. There has been considerable grouping of items.

Purlins, rafters, struts, etc., providing they are the same cross section, can be collected together and described 'as in pitched roofs'.

In the 'first fixings' section, boarding and flooring are required to be measured as before with the introduction of a height requirement (N.4.1(c)) and restricted areas (N.4.6).

N.4. BOARDINGS AND FLOORINGS

1 (c) Ceilings and beams, stating whether internal or external, whether horizontal or sloping and where over 3.50 m above floor (measured to ceiling level in both cases), shall each be so described stating the height in further stages of 1.50 m.

6. Areas not exceeding 1.00 m^2 shall be enumerated.

Sheet linings and casings appear in the 'second fixings' section (N.15) and cover woodbased products formerly measured under the rules 'floor, wall and ceiling finishes'.

Examples of such sheeting would include plywood, chipboard, fibreboard, asbestos cement, plastics, and any applied finish (N.15.1).

(Excerpt from Practice Manual)

A major change from SMM5 is the introduction of 'composite' items.

N.17. GENERALLY

1. Composite items, which shall be deemed to mean all items which may be fabricated off-site, shall be described in the form they are likely to be delivered to the work. Unless otherwise stated, their incorporation into the works shall be deemed to be included in the items.

Framed rafters and trusses (N.18)—a bill diagram and an abbreviated description may be the best way to describe the item. A similar suggestion is made for such items as windows, screens and staircases.

Door frames (N.20) although measured lineal, require an indication of the total number, drawing special attention to any repetition.

Fittings are now to be subject of full drawings (N.26) or a provisional sum and the fixing (N.27) to be a separate item.

Unless otherwise stated, spacing of plugs shall be at the contractor's discretion (N.28).

Holes are now given irrespective of size (N.29), the controlling factor being the thickness of the item which contains the hole.

Ironmongery is dealt with as in SMM5, with the requirement that details of any constraints, such as off-site fixing of door fittings, shall be indicated.

SECTION P. STRUCTURAL STEELWORK

Information required to be shown on the location drawings, which will often be the engineer's framing diagrams (P.2), includes:

(a) The position of the work in relation to other parts of the work and of the proposed building.

(b) The types and sizes of structural steel members and their positions in relation to each other.

(c) Details of connections or particulars of the reactions, moments and axial loads at connection points.

In single and built up members, length of units defined (P.7) shall be given in addition to the weight (P.4.3(b)).

Item to be given for the erection of structural steel work in each building or independent structure (P.10).

Off-site painting is now measured in m^2 (P.9.1) but off-site galvanising is measured in tonnes.

SECTION Q. METALWORK

In this section, there is a greater emphasis on enumeration particularly of composite items.

Except for standard items, detail drawings shall be provided and referred to in the description, or they may be subject to a provisional sum (Q.2.2).

Wire mesh coverings over 300 mm wide are to be measured in square metres; but widths under 300 mm are now required to be measured lineal, stating the exact width (Q.8.1) rather than banded as previously.

SECTION R. PLUMBING AND MECHANICAL ENGINEERING INSTALLATIONS

When mechanical engineering is to be measured, a detailed specification and drawings indicating scope of the work should accompany tender documents (R.1.2).

Plant items are required to be given (R.2).

New classifications of work are introduced (R.4).

R.4. CLASSIFICATIONS OF WORK

1. Work, together with its metalwork, supports and the like (unless associated with more than one installation), shall be classified as follows and given under an appropriate heading:

    (a) Rainwater installation.

    (b) Sanitary installation.

STANDARD METHOD OF MEASUREMENT

(c) Cold water installation identifying whether treated water, boosted water, cooling water, chilled water, or domestic cold water.

(d) Fire fighting installation identifying hose reel system, dry riser system, sprinkler system, foam inlets, or hand appliances.

(e) Heated water installation identifying whether steam and condensate, heating (high pressure), heating (medium pressure), heating (low pressure), or domestic hot water.

(f) Fuel oil installation.

(g) Fuel gas installation.

(h) Refrigeration installation.

(j) Compressed air installation.

(k) Hydraulic installation.

(l) Chemical installation.

(m) Special gas installation which shall be deemed to include medical and laboratory installations.

(n) Medical suction installation.

(p) Pneumatic tube installation.

(q) Vacuum installation.

(r) Refuse disposal installation.

(s) Air handling installation, identifying whether ventilation supply, ventilation extract, ventilation with heating (air heated centrally), ventilation with heating (air heated locally), air conditioning (air heated centrally), air conditioning (air heated locally), or foul air extract.

(t) Automatic control installation.

(u) Special equipment identifying whether incinerator and flues, kitchen equipment, laundry equipment, etc, except that where appropriate it may be clarified in accordance with Section S.

(v) Any other installation.

(w) Metalwork, supports and the like associated with more than one installation.

Work is also required to be identified by location, internally, externally and in plant rooms.

R.5. LOCATION OF WORK

1. Work shall be identified in respect of location as follows:

(a) Internally.

(b) Externally.

(c) In plant rooms.

Work internal and external to a building is required to be kept separate as are plant rooms, which would include heating chambers, ventilation machinery rooms, tank rooms, etc. It should be borne in mind that work in existing buildings is required to be separated in accordance with the general rules.

(Excerpt from Practice Manual)

An alternative method of measuring ductwork by weight has been introduced (R.20 to R.21) and for measuring insulation to ductwork in square metres rather than in metres stating the size of the pipe (R.35 to R.36).

R.38. BUILDERS' WORK

1. Unless otherwise stated, all items of builders' work shall be given in accordance with the appropriate sections.
2. Builders' work in connection with plumbing and mechanical engineering installations shall be grouped together under an appropriate heading.

R.39. ELECTRICAL WORK

1. Electrical work in connection with plumbing and mechanical engineering installations shall be given in accordance with:

SECTION S. ELECTRICAL INSTALLATIONS

Similar requirements to Section R, include specification and drawings (S.1.2) and plant (S.2). Classifications of work (S.4) have been extended from those previously required.

A new clause for final sub-circuits (S.21) has been introduced and is a change from measuring in detail to enumeration by points. This is made in response to current practice whereby the layout of the final sub-circuits is often determined by the electrical contractor.

As in the previous section, unless otherwise stated, all items of builders' work (S.26.1-2) are given as in the appropriate sections and grouped together under appropriate headings.

(S.27) BUILDERS' WORK

1. Excavating trenches for cables and pipe ducts not exceeding 55 mm nominal size shall be given in accordance with clause D.13.8. Excavating trenches for pipe ducts exceeding 55 mm nominal size shall

be given in accordance with clause W.3. Sand for bedding cables in trenches shall be given in metres stating the average width and depth of the sand.

D.13.8: NOTES

'Excavating trenches to receive service pipes, cables and the like shall each be given separately in metres, stating the starting level and the average depth to the nearest O.25 m. Details of type and size of service, earthwork support, grading bottoms if required, filling in, compaction and disposal of surplus soil shall be given in the description of such trenches. If the trench is next to a roadway, or next to existing buildings or made in unstable ground, each as defined in clauses D.19, 20 and 22, then these details shall be included in the description.'

W.3: NOTES

'1. Excavating pipe trenches shall be given in metres stating the starting level, the depth-range in increments of 2.00 m and the average depth within the depth range to the nearest 0.25 m.
'2. Excavating around trenches shall be so described irrespective of radius.
'3. Trenches to receive pipes not exceeding 200 mm nominal size may be grouped together and so described. Trenches to receive pipes exceeding 200 mm nominal size shall be given separately stating the size of the pipe.
'4. Earthwork support, treating bottoms, filling in, compaction and disposal of surplus spoil shall be given in the descriptions.
'5. Excavation below ground water shall be given in accordance with clause D.13.13 and any adjustment to earthwork support shall be deemed to be included.
'6. Trenches next to roadway, next to existing buildings or in unstable ground as defined in clauses D.19, 20 and 22 shall each be so described.
'7. Trenches in rock as defined in clause D.13.1 shall be so described. Alternatively rock may be given in metres as extra over the trench in which it occurs.
'8. Breaking up concrete, reinforced concrete, brickwork, tarmacadam and the like shall each be given separately in metres as extra over the trench in which it occurs, stating the pipe diameter as clause W.3.3.'

SECTION T. FLOOR, WALL AND CEILING FINISHINGS

A major change is the disappearance of detailed measurement by bands of work under 300 mm wide. Such work is now required to be kept separate from the general measurement but measured in m^2.

This general rule applies to all the subsections which follow the previous lay-out, except that plain sheet finishings have become flexible sheet finishings, the non-flexible variety are either covered by a new subsection for dry-linings and partitions or are removed to the woodwork section.

Work to walls, ceilings and floors in compartments not exceeding 4.00 m^2 on plan is now required to be given separately, as is work to staircases.

The rules in this section have been drawn up on the assumption that an indication of the type and complexity of the work is more important than the detailed measurement of labours and minor items. It is therefore important to bear in mind that as much information as possible should be given in the bill of quantities.

(Excerpt from Practice Manual)

T.3. GENERALLY

2. Work to walls, ceilings or floors in compartments not exceeding 4.00 m^2 on plan shall be given separately.

   Compartments, in the context of this clause, would include not only rooms but also cupboards, halls, lobbies, walk-in ducts and the like. It would not include areas which are divided up after completion of finishing by patent partitions, etc., e.g. W.C. cubicles.

   (Excerpt from Practice Manual)
3. Work and labour carried out over hand shall be so described.
4. Work shall be grouped according to the kind of material and each group with its associated labours shall be given under an appropriate heading. Mastic asphalt floor finishing shall be given in accordance with Section L.
5. Work to a pattern or in more than one colour shall be so described stating the nature thereof.

   Work to a pattern would include isolated panels within a general wall or other surface notwithstanding the requirement in Clause T.3.10.

   (Excerpt from Practice Manual)
6. The nature of the base to which finishings are to be applied shall be described. Work on differing bases shall be kept separate. Work on existing surfaces shall be so described.
7. Preparatory work to the base shall be measured separately. Preparation of existing work shall be so described.

8. Curved work shall be given separately stating the radius or radii and indicating:
   (a) If conical or spherical.
   (b) Whether concave or convex.
   (c) Whether to more than one radius (elliptical or parabolic).
9. Work to ceilings and beams, except in staircase areas, where over 3.50 m above floors (measured to ceiling level in both cases) shall be so described stating the height in further stages of 1.50 m. Where walls exceed 3.50 m in height and the ceiling is of dissimilar finish then particulars shall be given in the general description of the work as clause T.1 stating the maximum height.

   Work to ceilings and beams in staircase areas has been excluded from the general requirement to keep separate work over 3.50 m above floor level as it is considered that the identification of such work as in staircase areas is sufficient to draw attention to the likelihood of varying heights, particularly over staircase flights.

   (Excerpt from Practice Manual)
10. Work of a repairing nature and to isolated areas not exceeding 1.00 m² shall each be kept separate. Labours as defined in clause T.11 shall be included in the description of such items.
11. Temporary rules, temporary screeds and templets shall be deemed to be included with the items. Temporary support work to the face of risers and the like shall be given in accordance with clause F.15.8.

    This clause is intended to cover such items as granolithic finish to risers and strings of staircases where formwork is required, and is in addition to the formwork required to be measured to the concrete riser or string.

    (Excerpt from Practice Manual)
12. No deduction from the areas of finishings shall be made for grounds.

SECTION U. GLAZING

Little change has been made to this section except that the rules for measuring leaded lights and copper lights have been curtailed as a reflection of modern practice.

Also, there is a greater emphasis on enumeration, particularly of the very thick glasses and the speciality glasses (U.5).

U.5 SPECIAL GLASS

1. The following glass shall be enumerated stating the size of the pane. Where exact thicknesses are required they shall be so described.

(a) Glass less than 10 mm thick where the size of the pane exceeds 4.00 m².

(b) Glass 10 mm thick and over.

(c) Toughened or laminated glass.

(d) Solar control and other speciality glass.

(e) Acrylic, polycarbonate and similar material.

Repetition is recognised by a requirement to keep separate glass where 50 or more panes are identical (U.4.3).

> Repetition of large numbers of identical glass panes provides a significant cost saving and it is considered the indication of any number over 50 provides for this saving to be passed on.
>
> *(Excerpt from Practice Manual)*

The reference to lining up wired glass panes both ways has disappeared, recognising the impracticability of this requirement. Now the rule (V.3.3) is restricted to aligning within the limits of manufacture between adjacent panes.

SECTION V. PAINTING AND DECORATING

The general description of the work required by clause V.1.1, should give an outline of the work—drawing attention to any variety of colour, and such specification items as to whether or not the ironmongery or other fittings are to be removed and refixed, etc.

V.1. GENERALLY

2. Work shall be classified as follows and given under an appropriate heading:

    (a) New work internally.

    (b) New work externally.

    (c) Repainting and redecoration work internally.

    (d) Repainting and redecoration work externally.

    > Work is internal or external according to its position in the finished building, e.g. painting steelwork after fixing but before it is clad is internal work.
    >
    > *(Excerpt from Practice Manual)*

A height requirement has been introduced for ceilings and beams over 3.50 m high and for walls where the finish is dissimilar.

5. Work to ceilings and beams over 3.50 m above the floor (measured to ceiling level in both cases) shall be so described stating the height in further stages of 1.50 m. Where walls exceed 3.50 m in height and the

decoration of the ceiling is dissimilar, particulars shall be given in the general description of the work as clause V.1.1, stating the maximum height.

Unless wallpapering can be fully described, it is to be the subject of a p.c. sum.

V.12. DECORATIVE PAPER, SHEET PLASTIC, OR FABRIC BACKING AND LINING.

1. The supply and delivery to site of paper, fabric or plastic wall coverings, unless they can be fully described shall be included as a prime cost sum and the hanging or fixing shall be given in square metres, stating the number of pieces giving the following particulars:

   (a) Kind of material.
   (b) Nature of base.
   (c) Preparatory work.
   (d) Method of fixing and jointing.

   The width of rolls and type of pattern would both need to be stated before wallpaper could be considered fully described. In describing the method of fixing and jointing (V.12.1d), it would be necessary to draw attention to rolls of paper hung horizontally on walls or other vertical surfaces.

   (Excerpt from Practice Manual)

2. Classification shall be as follows:

   (a) Walls and columns (grouped together).
   (b) Ceilings and beams (grouped together).

3. Border-strips shall be given in metres. Cutting strips to profile shall be given in the description. Mitres and intersections shall be deemed to be included.

4. Corners and motifs shall each be enumerated separately. Cutting units to profile shall be given in the description.

5. Raking and curved cutting (grouped together) shall be given in metres. All cutting and fitting around wants on projections shall be deemed to be included.

   Wants (V.12.5.) would include openings for air bricks, etc., and projections would include pipes, switches, sockets, brackets and the like.

   (Excerpt from Practice Manual)

SECTION W. DRAINAGE

Rules for measuring drainage have been redrafted to align with similar work in other sections. Plant items have been introduced (W.2) and rules for the disposal of water are repeated from the excavation section.

W.6. PIPEWORK

1. Pipes (measured overall pipe fittings) shall each be given in metres stating the kind and quality of pipe, the nominal size and the method of jointing. Any pipes not laid in the bottom of the trench shall be given separately, and the depth stated in accordance with the depth-range stages as in W.3.1.

   Note: W.3.1 states; 'Excavating pipe trenches shall be given in metres stating the starting level, the depth-range in increments of 2.00 m and the average depth within the depth-range to the nearest 0.25 m.'

2. Suspended pipes shall be so described and the supports enumerated separately. Classification shall be as follows:

   (a) In ducts in the ground or below a floor.

   (b) Bracketed off walls, stating the average height above the floor.

   (c) Suspended from soffits, stating the average height above the floor.

3. Pipes in runs not exceeding 3.00 m long shall be so described stating the number.

4. Vertical pipes shall be so described.

5. Pipe fittings shall each be described and enumerated as extra over the pipes in which they occur. Cutting and jointing pipes to fittings and providing everything necessary for jointing shall be deemed to be included.

6. Accessories shall each be described and enumerated stating the kind, quality and size of the accessory and the nominal size of each inlet and outlet. Jointing pipes to such accessories, bedding in concrete and providing gratings, sealing plates and the like shall be given in the description of the item.

   Pipes not laid in the bottom of the trench (W.6.1) are usually those found in a multi-purpose trench. The requirements to give depths does not apply to those laid on a bed in the bottom of the trench.

   Types of fittings (W.6.5) should be clearly described and in cast iron work special consideration should be given to cost significant items; e.g. long radius and short radius bends being kept separate.

   It could be considered good practice to identify items by reference to a manufacturer's catalogue number. Examples of accessories (W.6.6) would include gullies, traps, inspection shoes, fresh air inlets, non-return flaps, etc.

(Excerpt from Practice Manual)

SECTION X. FENCING

A general description (X.1.1) is required unless location drawings indicate scope of the work, otherwise the section follows the previous rules.

Note: Copies of the *Standard Method of Measurement of Building Works* 6th Edition and the *Practice Manual* are available from the Royal Institution of Chartered Surveyors, 12 Great George Street, London W1.

5. STRUCTURAL DATA

Codes of Practice	131
British Standard	132
Approximate Weights of Material	133
Structural Steelwork	136
Conversion Tables	141
Bending Moments and Deflection Formulae	151
Sizes: ISO Metric Black Hexagon Bolts, Screws and Nuts B.S. 4190	152
Bearing Pressures of Materials	154
Safe Loads on Brickwork	154
Nailing Timbers	158
Timber Connectors	158
Joist Hangers	160
Framing Anchors	161
Concrete Lintels	163

This section is intended to facilitate very simple calculations, such as simple beams in timber and steel. Builders should not attempt anything more ambitious without much more information than is contained in this section. In any event, it is quicker, safer, and usually less expensive to consult a civil or structural engineer, if more than a simple beam or lintel is required.

CODES OF PRACTICE

CP 3: Code of basic data for the design of buildings:

Chapter I Lighting

Chapter II Thermal insulation in relation to the control of the environment

Chapter III Sound insulation and noise reduction

Chapter IV Precautions against fire (in three parts)

Chapter V Loading (in two parts)

Chapter VII Engineering and utility services

Chapter VIII Heating and thermal insulation

Chapter IX Durability

Chapter X Precautions against vermin and dirts

CP 110: The structural use of concrete:

Part 1: design, materials and workmanship.

Part 2: design charts for singly reinforced beams, doubly reinforced beams and rectangular columns.

Part 3: design charts for circular columns and prestressed beams.

CP 111: Structural recommendations for loadbearing walls: for loadbearing walls of brickwork and blockwork.

CP 112: The structural use of timber:

Part 1: (Imperial units, now withdrawn).

Part 2: deals with design of structures in timber and plywood.

Part 3: Trussed rafters for roofs of dwellings: recommends materials, functional requirements, manufacture, design, testing, permissible spans, etc.

CP 114: Structural use of reinforced concrete in buildings: deals with reinforced concrete design applied to beams, slabs, columns, flat slab construction, walls and bases in buildings, also floors, roofs and stairs.

CP 115: Structural use of prestressed concrete in buildings: covers design and construction of prestressed concrete as applied to buildings; also manufacture of precast, prestressed concrete units.

CP 116: The structural use of precast concrete: reinforced and prestressed concrete units.

Addendum No. 1: large panel structures and structural connections in precast concrete.

CP 117: Composite construction in structural steel and concrete:
   Part 1: simply supported beams in building.
   Part 2: beams for bridges.

BRITISH STANDARDS

B.S. 4: Structural steel sections: deals with channels, channel beams, equal and unequal angles, and "T" bars. Sizes of ranges, thicknesses of members and weights are given.

B.S. 449: The use of structural steel in building:
   Part 1: 1970: Imperial units (obsolescent).
   Supplement No. 1, Recommendations for design.
   Part 2: 1969: Metric units: relates to the use in building of hot rolled steel sections and plates, and normalised tubular shapes.
   Addendum No. 1, The use of cold formed steel sections in building: relates to the use in building of cold formed steel sections, formed from plate, sheet and strip steel 6 mm thick and under.

B.S. 6323: Steel tubes for mechanical, structural and general engineering purposes: covers eight types; hot finished welded, hot finished seamless, cold drawn seamless, electric resistance welded, cold drawn electric resistance welded, oxy-acetylene welded, hydraulic lap welded, electric fusion welded. Straightness, lengths, galvanising, methods of inspection and test are described.

B.S. 2994: Specification for cold rolled steel sections: gives dimensions and properties (moments of inertia, section moduli, radii of gyration) of the basic ranges of sections. Covers angles, channels, tees, zeds and compound sections made from channels.

B.S. 4483: Steel fabric for the reinforcement of concrete: preferred dimensions and properties. Quality of steel, tolerances on mass, size of mesh, sheet and roll (supersedes B.S. 1221).

B.S. 4848: Hot rolled structural steel sections:
   Part 2: Hollow sections: in B.S. range of metric sizes.
   Part 4: Equal and unequal angles: dimensions, rolling tolerances and properties of a co-ordinated metric range of equal and unequal angles.
   For British Standards concerning concrete, see 'Concretor'; and for timber see 'Carpenter and Joiner'.

APPROXIMATE WEIGHTS OF MATERIALS

This table is divided into three parts, marked A, B and C, according to the unit of measurement used. The section indexes give references to tables in other sections of this work, where weights of other materials are given.

PART A: WEIGHTS IN POUNDS PER CUBIC FOOT AND IN KILOGRAMS PER CUBIC METRE

Material	lb per cu. ft	kg per m^3
Ashes	50	800
Aluminium	162	2 559
Asphalt:		
Block, without grit	111	1 922
,, fine gritted	196	3 139
coarse gritted	147	2 254
Brass	525	8 129
Brickwork:		
Common fletton	125	1 822
Glazed brick	130	2 080
Staffordshire blue in cement mortar	135	2 162
Red engineering	140	2 240
Bronze	524	8 113
Cement, Portland	90	1 441
Chalk	140	2 240
Chippings (stone)	110	1 762
Clay (dry)	100	1 601
Clay (wet)	125	1 822
Clinker (furnace)	50	800
Coal	50	800
Coke breeze	40	640
Concrete:		
Reinforced steel 2%	150	2 400
Ballast or stone	140	2 241
Brick	115	1 841
Clinker	90	1 441
Pumice	70	1 121
Copper	550	8 810
Cork bark	5	80
Gravel	110	1 762
Gunmetal	528	8 475
Iron:		
cast	450	7 207
wrought	480	7 687
Lead	708	11 260
Lime:		
chalk (lump)	44	704
ground	60	961
quick	55	880
Peat (hard)	55	880

Material	lb per cu. ft	kg per m³
Pitch	72	1 152
Sand:		
dry	100	1 601
wet	88	1 281
Snow	3 to 10	48 to 160
Stone:		
artificial	140	2 242
Bath	140	2 242
Blue Pennant	168	2 682
Cragleith	145	2 322
Darley Dale	148	2 370
Forest of Dean	149	2 386
Granite	166	2 642
Marble	170	2 742
Portland	135	2 170
Slate	180	2 882
York	150	2 402
Terra-cotta	132	2 116
Timber:		
Ash	50	800
Baltic spruce	30	480
Beech	51	816
Birch	45	720
Box	60	961
Cedar	30	480
Chestnut	40	640
Ebony	76	1 217
Elm	39	624
Greenheart	60	961
Jarrah	51	816
Maple	47	752
Mahogany:		
Honduras	36	576
Spanish	66	1 057
Oak:		
English	53	848
American	45	720
Austrian and Turkish	44	704
Pine:		
Pitchpine	50	800
Red Deal	36	576
Yellow Deal	33	528
Spruce	31	496
Sycamore	38	530
Teak:		
African	60	961
Indian	41	656
Moulmein	46	736
Walnut:		
English	41	496
Black	45	720

STRUCTURAL DATA 135

Material	lb per cu. ft	kg per m³
Tin	465	7 448
Water	62	933
White lead:		
dry	400	6 407
in oil	250	4 003
Zinc	466	7 464

PART B: POUNDS PER SQUARE FOOT AND KILOGRAMS PER SQUARE METRE

Material	lb per sq. ft	kg per m²
Asbestos flat sheet	2·30	11·22
Blockboard per inch (25·4mm) of thickness	2·50	12·20
Cement screed per ½in (13mm) of thickness	6·00	29·29
Flooring:		
hardwood	3·30	16·10
softwood	2·30	11·22
Plasterboard ⅜in (9·5mm)	2·00	9·76
Plasterboard plus setting coat	3·00	14·64
Plywood per mm of thickness	0·14	0·67
Plaster (fibrous)	6·00	29·29
Roofing materials:		
Asbestos corrugated sheet:		
'Trafford Tile'	2·48	12·04
'Big Six'	2·76	13·46
Standard	2·50	12·20
Asphalt per in (25·4mm) of thickness	11·00	53·70
Glazed roofing	6·00	29·29
Copper sheet (24g)	1·33	6·34
Felt	1·00	4·88
Slating:		
Cornish	7·50	36·61
Welsh	11·50	56·14
Westmorland	15·50	75·67
Asbestos (diamond)	3·30	16·10
Asbestos (rectangular)	2·90	14·05
Tiling (plain)	14·50	70·78
Weatherboarding, ¾in (19mm)	1·50	7·29
Zinc sheet (0·025 in)	0·94	4·48
Zinc sheet (0·031in)	1·19	9·17
Zinc sheet (0·041in)	1·56	7·58

PART C: POUNDS PER FOOT RUN AND KILOGRAMS PER LINEAR METRE

Pipes		lb per ft run	kg per linear run
Asbestos:			
Soil and vent	3½in (95mm)	3·75	0·58
Rainwater	2½in (69mm)	2·00	0·28
,,	2½in (69mm)	2·00	0·28

Pipes			lb per ft run	kg per linear m
Rainwater	4in	(100mm)	4·25	0 59
,,	2½in	(69mm)	3·25	0·49
,,	3in	(75mm)	4·00	0·56
,,	4in	(100mm)	6·00	0·84
Soil, heavy	3in	(75mm)	7·00	0·98
,,	4in	(100mm)	9·00	1·26
Soil, medium	2in	(50mm)	4·00	0·56
,,	3in	(75mm)	6·00	0·84
,,	4in	(100mm)	8·00	1·12
,,	6in	(150mm)	12·00	1·68

WEIGHTS OF MATERIALS IN HUNDREDWEIGHTS PER CUBIC FOOT: AS ASSUMED BY THE LONDON DISTRICT SURVEYORS ASSOCIATION AND METRIC EQUIVALENTS

Materials	Cwt/cu. ft	tonnes/m^3
Stone, blue and glazed brickwork	1½	2·7
Ordinary brickwork	1	1·8
Concrete		
Stone ballast aggregate	1¼	2·2
Brick aggregate	1	1·8
Clinker aggregate	½	0·89
Reinforced concrete	1⅜	1·9
Timber	½	0·89

STRUCTURAL STEELWORK

NOTES

1. Generally, tabular loads are based on a flexural stress of 165 N/mm^2 assuming adequate lateral support. Tabular loads printed in bold face type exceed the buckling capacity of the unstiffened web without allowance for actual length of bearing.

2. Tables on pages 134-137 are reproduced from the *Handbook on Structural Steelwork* by permission of the British Constructional Steelwork Association Ltd. and the Constructional Steelwork Research and Development Association, Copies of the *Handbook* can be had from the BCSA, 1 Vincent Square, London SW1P 2PJ.

STRUCTURAL DATA   137

UNIVERSAL BEAMS

SAFE LOADS FOR GRADE 55 STEEL

Serial size	Mass per metre	Safe distributed loads in kilonewtons for spans in metres and deflection coefficients													Critical span L_c
		4.00	5.00	6.00	7.00	8.00	9.00	10.00	11.00	12.00	13.00	14.00	15.00	16.00	
mm	kg	28.00	17.92	12.44	9.143	7.000	5.531	4.480	3.702	3.111	2.651	2.286	1.991	1.750	m
914 × 419	388		**6729**	**5819**	**4988**	**4364**	**3879**	3491	3174	2909	2686	2494	2328	2182	6.96
	343		**6012**	**5111**	**4381**	**3833**	**3408**	:3067	2788	2556	2359	2191	2045	1917	6.71
914 × 305	289	**6089**	**4872**	**4060**	**3480**	**3045**	*2706*	2436	2214	2030	1874	1740	1624	1522	4.66
	253	**5314**	**4252**	**3543**	**3037**	**2657**	**2362**	2126	1933	1771	1635	1518	1417	1329	4.51
	224	**4615**	**3692**	**3077**	**2637**	**2307**	**2051**	:1846	1678	1538	1420	1319	1231	1154	4.32
	201	**4028**	**3222**	**2685**	**2301**	**2014**	**1790**	:1611	1465	1343	1239	1151	1074	1007	4.10
838 × 292	226	**4464**	**3571**	**2976**	**2551**	**2232**	:1984	1786	1623	1488	1373	1275	1190	1116	4.41
	194	**3714**	**2972**	**2476**	**2123**	**1857**	:1651	1486	1351	1238	1143	1061	991	929	4.15
	176	**3292**	**2634**	**2195**	**1881**	**1646**	:1463	:1317	1197	1097	1013	941	878	823	3.98
762 × 267	197	**3485**	**2788**	**2323**	**1991**	**1742**	1549	1394	1267	1162	1072	996	929		4.05
	173	**3009**	**2408**	**2006**	**1720**	:1505	1338	1204	1094	1003	926	860	803		3.86
	147	**2504**	**2003**	**1669**	**1431**	:1252	1113	1002	910	835	770	715	668		3.65
686 × 254	170	**2745**	**2196**	**1830**	1569	1373	1220	1098	998	915	845				3.93
	152	**2444**	**1955**	**1629**	:1396	1222	1086	977	889	815	752				3.82
	140	**2228**	**1783**	**1485**	:1273	:1114	990	891	810	743	686				3.71
	125	**1944**	**1555**	**1296**	:1111	: 972	864	778	707	648	598				3.54
914 × 419	388		**6729**	**5830**	**4997**	**4372**	**3887**	3498	3180	2915	2691	2499	2332	2186	7.20
	343		**6012**	**5123**	**4391**	**3842**	**3415**	:3074	2794	2561	2364	2196	2049	1921	6.97
914 × 305	289	**6099**	**4879**	**4066**	**3485**	**3049**	*2711*	2440	2218	2033	1877	1743	1626	1525	4.78
	253	**5324**	**4259**	**3549**	**3042**	**2662**	**2366**	2130	1936	1775	1638	1521	1420	1331	4.64
	224	**4625**	**3700**	**3083**	**2643**	**2313**	**2056**	:1850	1682	1542	1423	1321	1233	1156	4.47
	201	**4038**	**3230**	**2692**	**2307**	**2019**	**1794**	:1615	1468	1346	1242	1154	1077	1009	4.28
838 × 292	226	**4472**	**3578**	**2981**	**2556**	**2236**	:1988	1789	1626	1491	1376	1278	1193	1118	4.55
	194	**3723**	**2978**	**2482**	**2127**	**1861**	:1655	:1489	1354	1241	1146	1064	993	931	4.32
	176	**3301**	**2641**	**2200**	**1886**	**1650**	:1467	:1320	1200	1100	1016	943	880	825	4.16
762 × 267	197	**3491**	**2793**	**2327**	**1995**	1746	1552	1396	1269	1164	1074	997	931		4.17
	173	**3016**	**2412**	**2010**	**1723**	:1508	1340	1206	1097	1005	928	862	804		4.00
	147	**2510**	**2008**	**1674**	**1435**	:1255	:1116	1004	913	837	772	717	669		3.81
686 × 254	170	**2750**	**2200**	**1833**	1572	1375	1222	1100	1000	917	846				4.06
	152	**2448**	**1959**	**1632**	:1399	1224	1088	979	890	816	753				3.95
	140	**2233**	**1787**	**1489**	:1276	:1117	993	893	812	744	687				3.86
	125	**1949**	**1559**	**1300**	:1114	:975	866	780	709	650	600				3.71
610 × 305	238	**3673**	**2938**	*2449*	2099	1837	1632	1469	1336	1224					5.70
	179	**2750**	**2200**	**1833**	1572	1375	1222	1100	1000	917					5.29
	149	**2290**	**1832**	**1527**	:1309	:1145	1018	916	833	763					5.09
610 × 229	140	**2031**	**1624**	**1354**	1160	1015	900	812	738	677					3.72
	125	**1804**	**1443**	**1203**	1031	902	802	722	656	601					3.61
	113	**1612**	**1290**	**1075**	:921	806	717	645	586	537					3.51
	101	**1408**	**1127**	:939	:805	704	626	563	512	469					3.37
533 × 210	122	**1567**	*1254*	1045	896	784	697	627							3.50
	109	**1385**	**1108**	924	792	693	616	554							3.38
	101	**1287**	**1030**	858	735	643	572	515							3.32
	92	**1163**	**930**	:775	664	581	517	465							3.25
	82	**1007**	**806**	:671	575	503	448	403							3.11

For explanation of table see page 16.
Values in the shaded area relate to Universal Beams with tapered flanges. See page 9.
The loads listed are based on a bending stress of 280 N/mm², and assume adequate lateral support. Without such support the span must not exceed

L_c unless the compressive stress is reduced in accordance with clause 19a of BS 449.
Loads printed in bold type may cause overloading of the unstiffened web, the capacity of which should be checked. See page 14.
See also footnotes to page 233.

UNIVERSAL BEAMS

SAFE LOADS FOR GRADE 55 STEEL

Serial size	Mass per metre	Safe distributed loads in kilonewtons for spans in metres and deflection coefficients													Critical span L_c
		2.00	2.50	3.00	3.50	4.00	4.50	5.00	5.50	6.00	7.00	8.00	9.00	10.00	
mm	kg	112.0	71.68	49.78	36.57	28.00	22.12	17.92	14.81	12.44	9.143	7.000	5.531	4.480	m
457 × 191	98	†1812	1753	1460	1252	1095	*974*	876	797	730	626	548	487	438	3.29
	89	†1671	1586	1322	1133	991	*881*	793	721	661	566	496	441	396	3.19
	82	†1549	1444	2104	1032	903	802	722	657	602	516	451	401	361	3.11
	74	†1415	1309	1091	935	818	727	‡655	595	545	468	409	364	327	3.04
	67	†1311	1161	968	829	726	645	‡581	‡528	484	415	363	323	290	2.96
457 × 152	82	†1692	1395	1163	996	*872*	*775*	698	634	581	498	436	388		2.50
	74	†1553	1260	1050	900	787	*700*	630	573	525	450	394	350		2.42
	67	1400	1120	933	800	700	*622*	560	509	467	400	350	311		2.34
	60	†1237	1004	836	717	627	558	‡502	‡456	418	358	314	279		2.33
	52	1063	850	709	607	531	‡472	‡425	‡387	354	304	266			2.21
406 × 178	74	†1361	1186	989	847	*741*	659	593	539	494	424	371			3.01
	67	†1225	1064	887	760	665	591	532	484	444	380	333			2.94
	60	†1078	948	790	677	592	‡527	‡474	431	395	339	296			2.88
	54	1036	829	691	592	518	‡461	415	377	345	296	259			2.75
406 × 140	46	871	697	581	498	436	‡387	‡348	317	290	249	218			2.17
	39	702	562	468	401	‡351	‡312	‡281	‡255	234	201				2.04
356 × 171	67	†1126	961	801	687	601	534	481	437	401	343				3.05
	57	†975	803	669	574	502	446	402	365	335	287				2.90
	51	†883	713	594	510	‡446	396	357	324	297	255				2.82
	45	769	615	513	440	‡385	342	308	280	256	220				2.71
356 × 127	39	640	512	427	366	‡320	285	256	233	213	183				1.94
	33	527	422	351	‡301	‡264	234	211	192	176					1.84
305 × 165	54	†814	675	562	‡482	422	375	337	307	281					3.02
	46	†700	581	484	‡415	‡363	323	290	264	242					2.91
	40	628	503	419	‡359	‡314	279	251	229	209					2.82
305 × 127	48	686	*549*	*457*	392	343	305	274	249	229					2.12
	42	595	*476*	*397*	340	297	264	238	216	198					2.03
	37	528	422	*352*	302	264	235	211	192	176					1.97
305 × 102	33	465	372	*310*	266	232	207	186	169	155					1.58
	28	393	315	*262*	225	197	175	157	143	131					1.50
	25	322	258	*215*	184	161	143	129	117	107					1.38
254 × 146	43	566	453	377	323	283	252	226							2.76
	37	486	389	324	278	243	216	194							2.64
	31	395	316	264	226	198	176	158							2.46
254 × 102	28	345	*276*	230	197	172	153	138							1.66
	25	297	*238*	198	170	149	132	119							1.56
	22	253	*202*	169	144	126	112	101							1.47
203 × 133	30	313	250	209	179	156									2.47
	25	260	208	173	148	130									2.32

Loads printed in italic type do not cause overloading of the unstiffened web, and do not cause deflection exceeding span/360.
Loads printed in ordinary type should be checked for deflection. See page 15.

† Load is based on allowable shear of web and is less than allowable load in bending.
‡ Load exceeds buckling capacity of unstiffened web.
See also footnotes to page 232.

STRUCTURAL DATA

JOISTS

SAFE LOADS FOR GRADE 55 STEEL

Nominal size mm	Mass per metre kg	Safe distributed loads in kilonewtons for spans in metres and deflection coefficients													Critical span L_c m
		1.00	1.25	1.50	1.75	2.00	2.25	2.50	2.75	3.00	3.25	3.50	4.00	4.25	
		448	287	199	146	112	88.5	71.7	59.2	49.8	42.4	36.6	28.0	24.8	
254 × 203	81.85						†881	848	771	706	652	606	530	499	4.46
254 × 114	37.20		†656	599	513	449	399	359	327	299	276	257	225	211	1.90
203 × 152	52.09			†615	603	528	469	422	384	352	325	302	264	248	3.43
203 × 102	25.33		†401	337	289	253	225	202	184	169	156	144	126	119	1.79
178 × 102	21.54	†320	306	255	219	191	170	153	139	128	118	109	96	90	1.79
152 × 127	37.20	535	428	356	305	267	238	214	194	178	164	153	134	126	2.88
152 × 89	17.09	†254	207	173	148	130	115	104	94	86	80	74	65	61	1.61
152 × 76	17.86	257	205	171	147	128	114	103	93	86	79	73	64	60	1.40
127 × 114	29.76	345	276	230	197	173	153	138	126	115	106	99	86	81	2.69
127 × 114	26.79	†320	267	222	190	167	148	133	121	111	103	95	83	78	2.74
127 × 76	16.37	201	161	134	115	100	89	80	73	67	62	57	50	47	1.59
127 × 76	13.36	168	134	112	96	84	75	67	61	56	52	48	42	39	1.45
114 × 114	26.79	288	231	192	165	144	128	115	105	96	89	82	72	68	2.74
102 × 102	23.07	214	172	143	123	107	95	86	78	71	66	61	54	50	2.61
102 × 64	9.65	96	77	64	55	48	43	38	35	32	30	27	24	23	1.24
102 × 44	7.44	67	54	45	38	34	30	27	24	22	21	19	17	16	0.76
89 × 89	19.35	155	124	103	88	77	69	62	56	52	48	44	39	36	2.48
76 × 76	14.67	101	81	67	58	51	45	40	37	34	31	29	25	24	2.17
76 × 76	12.65	93	75	62	53	47	41	37	34	31	29	27	23	22	2.20

For explanation of table see page 16.

Sections with mass shown in italics are, although frequently rolled, not in BS 4. Availability should be checked with BSC Sections Product Unit. Flanges of BS 4 joists have a 5° taper; all others taper at 8°.

The loads listed are based on a bending stress of 280 N/mm², and assume adequate lateral support. Without such support the span must not exceed L_c unless the compressive stress is reduced in accordance with clause 19 a of BS 449.

Loads printed in bold face type may cause overloading of the unstiffened web, the capacity of which should be checked (see p. 14).

Loads printed in italic type do not cause overloading of the unstiffened web, and do not cause deflection exceeding span/360.

Loads printed in ordinary type should be checked for deflection (see p. 15).

† Load is based on allowable shear of web and is less than allowable load in bending.

CHANNELS

SAFE LOADS FOR GRADE 55 STEEL

Nominal size	Mass per metre	Safe distributed loads in kilonewtons for spans in metres and deflection coefficients												Critical span L_C
		1.50	2.00	2.50	3.00	3.50	4.00	4.50	5.00	5.50	6.00	7.00	8.00	9.00	
mm	kg	199.1	112.0	71.68	49.78	36.57	28.00	22.12	17.92	14.81	12.44	9.143	7.000	5.531	m
432 × 102	65.54	*1480*	*1110*	888	740	634	555	493	444	404	370	317	278		2.156
381 × 102	55.10	*1168*	876	701	584	500	438	389	350	318	292	250			2.293
305 × 102	46.18	805	604	483	402	345	302	268	241	220	201				2.405
305 × 89	41.69	692	519	415	346	297	259	231	208	189	173				1.999
254 × 89	35.74	523	392	314	261	224	196	174	157						2.196
254 × 76	28.29	396	297	238	198	170	148	132	119						1.696
229 × 89	32.76	443	332	266	221	190	166	148							2.299
229 × 76	26.06	341	256	205	170	146	128	114							1.810
203 × 89	29.78	366	275	220	183	157	137								2.412
203 × 76	23.82	287	215	172	143	123	107								1.924
178 × 89	26.81	294	221	177	147	126									2.532
178 × 76	20.84	225	168	135	112	96									1.973
152 × 89	23.84	228	171	137	114										2.678
152 × 76	17.88	*167*	125	100	83										1.978
127 × 64	14.90	113	85	68											1.840
102 × 51	10.42	61	46												1.474
76 × 38	6.70	29													1.248

For explanation of table see page 16.

The loads listed are based on a bending stress of 280 N/mm², and assume adequate lateral support. Without such support the span must not exceed L_C unless the compressive stress is reduced in accordance with clause 19a of BS 449.

Loads printed in bold face type may cause overloading of the unstiffened web, the capacity of which should be checked (see p. 16).

Loads printed in italic type do not cause overloading of the unstiffened web, and do not cause deflection exceeding span/360.

Loads printed in ordinary type should be checked for deflection (see p. 17).

CONVERSION TABLES

POUNDS PER FOOT RUN TO KILOGRAMS PER METRE RUN

lb/ft run	kg/lin. m
1	1·488
2	2·976
3	4·464
4	5·952
5	7·443
6	8·929
7	10·417
8	11·905
9	13·393
10	14·886

POUNDS PER SQUARE FOOT TO KILOGRAMS PER SQUARE METRE

lb/sq. ft	kg/m^3
1	4·882
2	9·764
3	14·647
4	19·528
5	24·412
6	29·294
7	34·177
8	38·116
9	42·998
10	48·824

POUNDS PER CUBIC FOOT TO KILOGRAMS PER CUBIC METRE

lb/cu. ft	kg/m^3	lb/cu. ft	kg/m^3
1	16·015	90	1 441·660
2	32·037	100	1 601·850
3	48·052	110	1 762·030
4	64·074	120	1 922·220
5	80·092	125	2 002·312
6	96·110	130	2 084·404
7	112·127	135	2 164·496
8	128·148	140	2 242·580
9	134·166	145	2 322·672
10	160·184	150	2 402·770
20	140·368	200	3 203·700
30	480·530	300	4 805·350
40	640·739	400	6 407·140
50	800·923	500	8 009·234
60	961·108	550	8 810·150
70	1 121·290	600	9 071·084
80	1 281·480	700	11 132·318

HUNDREDWEIGHTS PER CUBIC FOOT TO TONNES PER CUBIC METRE

cwts/cu. ft	tonnes/m^3
$\frac{1}{4}$	0·45
$\frac{3}{8}$	0·67
$\frac{1}{2}$	0·89
$\frac{3}{4}$	1·35
1	1·8
$1\frac{1}{4}$	2·24
$1\frac{1}{2}$	2·7
2	3·6

KILONEWTONS PER SQUARE METRE AND POUNDS FORCE PER SQUARE FOOT

kN/m²	1	lb f/sq. ft	kN/m²	1	lb f/sq. ft
0·05	1	20	1·00	20	400
0·10	2	40	1·50	30	600
0·15	3	60	2·00	40	800
0·20	4	80	2·50	50	1 000
0·25	5	100	3·00	60	1 200
0·30	6	120	3·50	70	1 400
0·35	7	140	4·00	80	1 600
0·40	8	160	4·50	90	1 800
0·45	9	180	5·00	100	2 000
0·50	10	200			

APPLICATION—Floor loading

KILONEWTONS PER SQUARE METRE AND TONS FORCE PER SQUARE FOOT

tons f/sq. ft	kN/m²	tons f/sq. ft	kN/m²
1	107·25	7	749·75
2	214·50	8	848·00
3	321·75	9	965·25
4	429·00	10	1 072·50
5	535·25	20	2 145·00
6	642·50		

APPLICATION—Safe bearing capacity of soils, etc.

POUNDS FORCE PER SQUARE INCH MEGANEWTONS PER SQUARE METRE

lb f/sq. in	MN/m²	lb f/sq. in	MN/m²
50	0·33	1 200	8·00
100	0·66	1 350	9·00
150	1·00	1 500	10·00
300	2·00	3 000	20·00
450	3·00	4 500	30·00
600	4·00	6 000	40·00
750	5·00	7 500	50·00
900	6·00	9 000	60·00
1 050	7·00		

APPLICATION—reinforced concrete, etc.

STRUCTURAL DATA

STEEL CHANNELS TO B.S. 4: Part 1; 1972

DESIGNATION		Mass/unit length		Depth of section D	Width of section B	Thickness	
Nominal size						Web t	Flange T
mm	(in)	kg/m	(lb/ft)	mm	mm	mm	mm
432 × 102	(17 × 4)	66.54	(44.0)	431.8	101.6	12.2	16.8
381 × 102	(15 × 4)	55.10	(37.0)	381.0	101.6	10.4	16.3
305 × 102	(12 × 4)	46.18	(31.0)	304.8	101.6	10.2	14.8
305 × 89	(12 × 3½)	41.69	(28.0)	304.8	88.9	10.2	13.7
254 × 89	(10 × 3½)	35.74	(24.0)	254.0	88.9	9.1	13.6
254 × 76	(10 × 3)	28.29	(19.0)	254.0	76.2	8.1	10.9
229 × 89	(9 × 3½)	32.76	(22.0)	228.6	88.9	8.6	13.3
229 × 76	(9 × 3)	26.06	(17.5)	228.6	76.2	7.6	11.2
203 × 89	(8 × 3½)	29.78	(20.0)	203.2	88.9	8.1	12.9
203 × 76	(8 × 3)	23.82	(16.0)	203.2	76.2	7.1	11.2
178 × 89	(7 × 3½)	26.81	(18.0)	177.8	88.9	7.6	12.3
178 × 76	(7 × 3)	20.84	(14.0)	177.8	76.2	6.6	10.3
152 × 89	(6 × 3½)	23.84	(16.0)	152.4	88.9	7.1	11.6
152 × 76	(6 × 3)	17.88	(12.0)	152.4	76.2	6.4	9.0
127 × 64	(5 × 2½)	14.90	(10.0)	127.0	63.5	6.4	9.2
102 × 51	(4 × 2)	10.42	(7.0)	101.6	50.8	6.1	7.6
76 × 38	(3 × 1½)	6.70	(4.5)	76.2	38.1	5.1	6.8

STEEL ANGLES (EQUAL SIDED) TO B.S.4: Part 1; 1972

DESIGNATION			Leg lengths A	Thickness t	Mass/unit length	
Nominal size		Nominal thickness				
mm	(in)	mm	(in)	mm	mm	kg/m	(lb/ft)
203 × 203 (8 × 8)		25	(1)	203.2	25.3	76.00	(51.02)
		24	(15/16)		23.7	71.51	(48.01)
		22	(7/8)		22.1	67.05	(45.01)
		21	(13/16)		20.5	62.56	(42.00)
		19	(3/4)		18.9	57.95	(38.90)
		17	(11/16)		17.3	53.30	(35.78)
		16	(5/8)		15.8	48.68	(32.68)
152 × 152 (6 × 6)		22	(7/8)	152.4	22.1	49.32	(33.11)
		21	(13/16)		20.5	46.03	(30.90)
		19	(3/4)		19.0	42.75	(28.70)
		17	(11/16)		17.3	39.32	(26.40)
		16	(5/8)		15.8	36.07	(24.22)
		14	(9/16)		14.2	32.62	(21.90)
		13	(1/2)		12.6	29.07	(19.52)
		11	(7/16)		11.0	25.60	(17.18)
		9	(3/8)		9.4−	22.02	(14.79)
127 × 127 (5 × 5)		19	(3/4)	127.0	19.0	35.16	(23.61)
		17	(11/16)		17.4	32.47	(21.80)
		16	(5/8)		15.8	29.66	(19.91)
		14	(9/16)		14.2	26.80	(18.00)
		13	(1/2)		12.6	23.99	(16.10)
		11	(7/16)		11.0	21.14	(14.19)
		10	(3/8)		9.5−	18.30	(12.29)

DESIGNATION			Leg lengths A	Thickness t		Mass/unit length	
Nominal size		Nominal thickness					
mm	(in)	mm	(in)	mm	mm	kg/m	(lb/ft)
102 × 102 (4 × 4)		19 17 16 14 13 11 9 8	(¾) (11/16) (⅝) (9/16) (½) (7/16) (⅜) (5/16)	101.6	19.0 17.4 15.8 14.2 12.6 11.0 9.4 7.8	27.57 25.48 23.37 21.17 18.91 16.69 14.44 12.06	(18.51) (17.10) (15.69) (14.21) (12.70) (11.21) (9.69) (8.10)
89 × 89 (3½ × 3½)		16 14 13 11 9 8 6	(⅝) (9/16) (½) (7/16) (⅜) (5/16) (¼)	89.9	15.8 14.2 12.6 11.0 9.4 7.9 6.3	20.10 18.31 16.38 14.44 12.50 10.58 8.49	(13.50) (12.29) (11.00) (9.70) (8.39) (7.10) (5.70)
76 × 76 (3 × 3)		14 13 11 9 8 6	(9/16) (½) (7/16) (⅜) (5/16) (¼)	76.2	14.3 12.6 11.0 9.4 7.8 6.2	15.50 13.85 12.20 10.57 8.93 7.16	(10.41) (9.30) (8.19) (7.10) (5.99) (4.81)
64 × 64 (2½ × 2½)		12 11 9 8 6	(½) (7/16) (⅜) (5/16) (¼)	63.5	12.5 11.0 9.4 7.9 6.2	11.31 10.12 8.78 7.45 5.96	(7.60) (6.79) (5.89) (5.00) (4.00)

DESIGNATION			Leg lengths A	Thickness	Mass/unit length	
Nominal size	Nominal thickness					
mm (in)	mm	(in)	mm	mm	kg/m	(lb/ft)
57 × 57 (2¼ × 2¼)	9 8 6 5	(⅜) (⁵⁄₁₆) (¼) (³⁄₁₆)	57.2	9.3 7.8 6.2 4.6	7.74 6.55 5.35 4.01	(5.19) (4.40) (3.59) (2.69)
51 × 51 (2 × 2)	9 8 6 5	(⅜) (⁵⁄₁₆) (¼) (³⁄₁₆)	50.8	9.4 7.8 6.3 4.6	6.85 5.80 4.77 3.58	(4.60) (3.90) (3.20) (2.40)
44 × 44 (1¾ × 1¾)	8 6 5	(⁵⁄₁₆) (¼) (³⁄₁₆)	44.5	7.9 6.1 4.7	5.06 4.02 3.13	(3.40) (2.70) (2.10)
38 × 38 (1½ × 1½)	8 6 5	(⁵⁄₁₆) (¼) (³⁄₁₆)	38.1	7.8 6.3 4.7	4.24 3.50 2.68	(2.85) (2.35) (1.80)
32 × 32 (1¼ × 1¼)	6 5 3	(¼) (³⁄₁₆) (⅛)	31.8	6.2 4.6 3.1	2.83 2.16 1.49	(1.90) (1.45) (1.00)
26 × 26* (1 × 1)	6 5 3	(¼) (³⁄₁₆) (⅛)	25.4	6.4 4.7 3.1	2.23 1.72 1.19	(1.50) (1.15) (0.80)

* This metric designation of 26 mm × 26 mm has been given for the 1 in × 1 in size to avoid confusion with the metric size of 25 mm × 25 mm chosen for the co-ordinated metric range of sizes.

NOTE 1. Some of the thicknesses given in this table are obtained by raising the rolls. (Practice in this respect is not uniform throughout the industry. In such cases the legs will be slightly longer and the backs of the toes will be slightly rounded.
NOTE 2. Finished sections in which the angle between the legs is not less than 89° and not more than 91° shall be deemed to comply with the requirements of this British Standard.

STEEL ANGLES (UNEQUAL SIDED) TO B.S. 4: Part 1; 1972

DESIGNATION				Leg lengths		Thickness	Mass/unit length	
Nominal size		Nominal thickness		A	B			
mm	(in)	mm	(in)	mm	mm	mm	kg/m	(lb/ft)
229 × 102 (9 × 4)		22	(⅞)	228.6	101.6	22.1	53.77	(36.10)
		21	(13⁄16)			20.6	50.21	(33.71)
		19	(¾)			18.9	46.45	(31.18)
		17	(11⁄16)			17.4	42.87	(28.78)
		16	(⅝)			15.8	39.20	(26.32)
		14	(9⁄16)			14.2	35.43	(23.78)
		13	(½)			12.6	31.56	(21.18)
203 × 152 (8 × 6)		22	(⅞)	203.2	152.4	22.1	58.09	(39.00)
		21	(13⁄16)			20.5	54.22	(36.40)
		19	(¾)			18.9	50.32	(33.78)
		17	(11⁄16)			17.3	46.30	(31.09)
		16	(⅝)			15.8	42.32	(28.41)
		14	(9⁄16)			14.2	38.29	(25.71)
		13	(½)			12.6	34.10	(22.89)
203 × 102 (8 × 4)		19	(¾)	203.2	101.6	19.0	42.75	(28.70)
		17	(11⁄16)			17.3	39.32	(26.40)
		16	(⅝)			15.8	36.07	(24.22)
		14	(9⁄16)			14.2	32.62	(21.90)
		13	(½)			12.6	29.07	(19.52)
178 × 89 (7 × 3½)		16	(⅝)	177.8	88.9	15.8	31.30	(21.01)
		14	(9⁄16)			14.2	28.28	(18.98)
		13	(½)			12.6	25.31	(16.99)
		11	(7⁄16)			11.1	22.36	(15.01)
		9	(⅜)			9.4	19.22	(12.90)

DESIGNATION				Leg lengths		Thickness t	Mass/unit length	
Nominal size		Nominal thickness		A	B			
mm	(in)	mm	(in)	mm	mm	mm	kg/m	(lb/ft)
152 × 102	(6 × 4)	19	(3/4)	152.4	101.6	19.0	35.16	(23.61)
		17	(11/16)			17.4	32.47	(21.80)
		16	(5/8)			15.8	29.66	(19.91)
		14	(9/16)			14.2	26.80	(17.99)
		13	(1/2)			12.6	23.99	(16.10)
		11	(7/16)			11.0	21.14	(14.19)
		9	(3/8)			9.5	18.30	(12.29)
152 × 89	(6 × 3½)	16	(5/8)	152.4	88.9	15.7	27.99	(18.79)
		14	(9/16)			14.2	25.46	(17.09)
		13	(1/2)			12.6	22.77	(15.28)
		11	(7/16)			11.1	20.12	(13.51)
		9	(3/8)			9.4	17.26	(11.59)
		8	(5/16)			7.8	14.44	(9.70)
152 × 76	(6 × 3)	16	(5/8)	152.4	76.2	15.8	26.52	(17.80)
		14	(9/16)			14.2	23.99	(16.10)
		13	(1/2)			12.6	21.45	(14.40)
		11	(7/16)			11.0	18.92	(12.70)
		9	(3/8)			9.5	16.39	(11.00)
		8	(5/16)			7.8	13.69	(9.19)
127 × 89	(5 × 3½)	16	(5/8)	127.0	88.9	15.8	24.86	(16.69)
		14	(9/16)			14.2	22.64	(15.20)
		13	(1/2)			12.6	20.26	(13.60)
		11	(7/16)			11.1	17.89	(12.01)
		9	(3/8)			9.4	15.35	(10.31)
		8	(5/16)			7.9	12.94	(8.69)
127 × 76	(5 × 3)	14	(9/16)	127.0	76.2	14.2	21.17	(14.21)
		13	(1/2)			12.6	18.91	(12.70)
		11	(7/16)			11.0	16.69	(11.21)
		9	(3/8)			9.4	14.44	(9.69)
		8	(5/16)			7.8	12.06	(8.10)

STRUCTURAL DATA

DESIGNATION				Leg lengths		Thickness	Mass/unit length	
Nominal size		Nominal thickness		A	B	t		
mm	(in)	mm	(in)	mm	mm	mm	kg/m	(lb/ft)
102 × 89 (4 × 3½)		16 14 13 11 9 8	(⅝) (9/16) (½) (7/16) (⅜) (5/16)	101.6	88.9	15.8 14.2 12.6 11.0 9.5 7.8	21.75 19.67 17.72 15.62 13.55 11.31	(14.60) (13.21) (11.89) (10.49) (9.10) (7.59)
102 × 76 (4 × 3)		14 13 11 9 8	(9/16) (½) (7/16) (⅜) (5/16)	101.6	76.2	14.2 12.6 11.0 9.4 7.9	18.31 16.38 14.44 12.50 10.58	(12.29) (11.00) (9.70) (8.39) (7.10)
102 × 64 (4 × 2½)		11 9 8 6	(7/16) (⅜) (5/16) (¼)	101.6	63.5	11.0 9.5 7.8 6.3	13.40 11.61 9.69 7.89	(8.99) (7.79) (6.51) (5.30)
89 × 76 (3½ × 3)		14 13 11 9 8 6	(9/16) (½) (7/16) (⅜) (5/16) (¼)	88.9	76.2	14.2 12.7 11.0 9.5 7.8 6.3	16.83 15.20 13.40 11.61 9.69 7.89	(11.30) (10.20) (8.99) (7.79) (6.51) (5.30)
89 × 64 (3½ × 2½)		11 9 8 6	(7/16) (⅜) (5/16) (¼)	88.9	63.5	11.0 9.4 7.8 6.2	12.20 10.57 8.93 7.16	(8.19) (7.10) (5.99) (4.81)

DESIGNATION

Nominal size		Nominal thickness		Leg lengths		Thickness t	Mass/unit length	
mm	(in)	mm	(in)	A mm	B mm	mm	kg/m	(lb/ft)
76 × 64 (3 × 2½)		11 9 8 6	(7/16) (3/8) (5/16) (1/4)	76.2	63.5	11.0 9.4 7.9 6.2	11.17 9.68 8.19 6.56	(7.50) (6.50) (5.50) (4.40)
76 × 51 (3 × 2)		11 9 8 6 5	(7/16) (3/8) (5/16) (1/4) (3/16)	76.2	50.8	11.0 9.4 7.9 6.2 4.7	10.12 8.78 7.45 5.96 4.62	(6.79) (5.89) (5.00) (4.00) (3.10)
64 × 51 (2½ × 2)		9 8 6 5	(3/8) (5/16) (1/4) (3/16)	63.5	50.8	9.3 7.8 6.2 4.7	7.74 6.55 5.35 4.01	(5.19) (4.40) (3.59) (2.69)
64 × 38 (2½ × 1½)		8 6 5	(5/16) (1/4) (3/16)	63.5	38.1	7.8 6.3 4.6	5.80 4.77 3.58	(3.90) (3.20) (2.40)
51 × 38 (2 × 1½)		8 6 5	(5/16) (1/4) (3/16)	50.8	38.1	7.9 6.1 4.7	5.06 4.02 3.13	(3.40) (2.70) (2.10)

NOTE 1. Some of the thicknesses given in this table are obtained by raising the rolls. (Practice in this respect is not uniform throughout the industry.) In such cases, the legs will be slightly longer and the backs of the toes will be slightly rounded.
NOTE 2. Finished sections in which the angle between the legs is not less than 89° and not more than 91° shall be deemed to comply with the requirements of this British Standard.

STRUCTURAL DATA

BENDING MOMENTS AND DEFLECTION FORMULAE

Loading Diagram	Maximum Bending Moment	Maximum Shear	Deflection
Cantilever with point load at free end	$W.l$	W	$\dfrac{1}{3} \dfrac{W.l^3}{E.I}$
Cantilever with uniformly distributed load	$\dfrac{W.l}{2}$	W	$\dfrac{1}{8} \dfrac{W.l^3}{E.I}$
Simply supported beam with central point load	$\dfrac{W.l}{4}$	$\dfrac{W}{2}$	$\dfrac{1}{48} \dfrac{W.l^3}{E.I}$
Simply supported beam with triangular load	$\dfrac{W.l}{6}$	$\dfrac{W}{2}$	$\dfrac{1}{60} \dfrac{W.l^3}{E.I}$
Simply supported beam with uniformly distributed load	$\dfrac{W.l}{8}$	$\dfrac{W}{2}$	$\dfrac{5}{384} \dfrac{W.l^3}{E.I}$
Propped cantilever with uniformly distributed load	$\dfrac{W.l}{8}$	$\dfrac{5}{8}W$	$0 \cdot 0054 \dfrac{W.l^3}{E.I}$
Fixed-ended beam with central point load	$\dfrac{W.l}{8}$	$\dfrac{W}{2}$	$\dfrac{1}{192} \dfrac{W.l^3}{E.I}$
Fixed-ended beam with uniformly distributed load	$\dfrac{W.l}{12}$	$\dfrac{W}{2}$	$\dfrac{1}{384} \dfrac{W.l^3}{E.I}$

w	=	Total Load	l =	Length
▽	=	Point load	E =	Modulus of elasticity
▭	=	Distributed load	I =	Moment of inertia
△	=	Free support	⫢ =	Fixed support

SIZES: ISO METRIC BLACK HEXAGON BOLTS, SCREWS AND NUTS, B.S. 4190

1	2	3	4	5	6	7	8	9	10
Nominal size and thread diameter	Pitch of thread	Diameter of unthreaded shank		Width across flats		Width across corners		Height of head	
d	(Coarse pitch series)	d max.	min.	s max.	min.	e max.	min.	k max.	min.
M5	0.8	5.48	4.52	8.00	7.64	9.2	8.63	3.875	3.125
M6	1	6.48	5.52	10.00	9.64	11.5	10.89	4.375	3.625
M8	1.25	8.58	7.42	13.00	12.57	15.0	14.20	5.875	5.125
M10	1.5	10.58	9.42	17.00	16.57	19.6	18.72	7.45	6.55
M12	1.75	12.70	11.30	19.00	18.48	21.9	20.88	8.45	7.55
M16	2	16.70	15.30	24.00	23.16	27.7	26.17	10.45	9.55
M20	2.5	20.84	19.16	30.00	29.16	34.6	32.95	13.90	12.10
(M22)	2.5	22.84	21.16	32.00	31.00	36.9	35.03	14.90	13.10
M24	3	24.84	23.16	36.00	35.00	41.6	39.55	15.90	14.10
(M27)	3	27.84	26.16	41.00	40.00	47.3	45.20	17.90	16.10
M30	3.5	30.84	29.16	46.00	45.00	53.1	50.85	20.05	17.95
(M33)	3.5	34.00	32.00	50.00	49.00	57.7	55.37	22.05	19.95
M36	4	37.00	35.00	55.00	53.80	63.5	60.79	24.05	21.95
(M39)	4	40.00	38.00	60.00	58.80	69.3	66.44	26.05	23.95
M42	4.5	43.00	41.00	65.00	63.80	75.1	72.09	27.05	24.95
(M45)	4.5	46.00	44.00	70.00	68.80	80.8	77.74	29.05	26.95
M48	5	49.00	47.00	75.00	73.80	86.6	83.39	31.05	28.95
(M52)	5	53.20	50.80	80.00	78.80	92.4	89.04	34.25	31.75
M56	5.5	57.20	54.80	85.00	83.60	98.1	94.47	36.25	33.75
(M60)	5.5	61.20	58.80	90.00	88.60	103.9	100.12	39.25	36.75
M64	6	65.20	62.80	95.00	93.60	109.7	105.77	41.25	38.75
(M68)	6	69.20	66.80	100.00	98.60	115.5	111.42	44.25	41.75

STRUCTURAL DATA

1	2	3	4	5	6	7	8	9	10	11	12
Nominal size and thread diameter	Pitch of thread (Coarse pitch series)	Width across flats s		Width across corners e		Thickness of nut m				Thickness of thin nut t (faced both sides)	
		max.	min.	max.	min.	Black		Faced one side			
						max.	min.	max.	min.	max.	min.
M5	0.8	8.00	7.64	9.2	8.63	4.375	3.625	4.0	3.52	—	—
M6	1	10.00	9.64	11.5	10.89	5.375	4.625	5	4.52	5.00	4.52
M8	1.25	13.00	12.57	15.0	14.20	6.875	6.125	6.5	5.92	6.00	5.52
M10	1.5	17.00	16.57	19.6	18.72	8.45	7.55	8	7.42	7.00	6.42
M12	1.75	19.00	18.48	21.9	20.88	10.45	9.55	10	9.42		
M16	2	24.00	23.16	27.7	26.17	13.55	12.45	13	12.30	9.00	8.42
M20	2.5	30.00	29.16	34.6	32.95	16.55	15.45	16	15.30	9.00	8.42
(M22)	2.5	32.00	31.00	36.9	35.03	18.55	17.45	18	17.30	10.00	9.42
M24	3	36.00	35.00	41.6	39.55	19.65	18.35	19	18.16	10.00	9.42
(M27)	3	41.00	40.00	47.3	45.20	22.65	21.35	22	21.16	12.00	11.30
M30	3.5	46.00	45.00	53.1	50.85	24.65	23.35	24	23.16	12.00	11.30
(M33)	3.5	50.00	49.00	57.7	55.37	26.65	25.35	26	25.16	14.00	13.30
M36	4	55.00	53.80	63.5	60.79	29.65	28.35	29	28.16	14.00	13.30
(M39)	4	60.00	58.80	69.3	66.44	31.80	30.20	31	30.0	16.00	15.30
M42	4.5	65.00	63.80	75.1	72.09	34.80	33.20	34	33.0	16.00	15.30
(M45)	4.5	70.00	68.80	80.8	77.74	36.80	35.20	36	35.0	18.00	17.30
M48	5	75.00	73.80	86.6	83.39	38.80	37.20	38	37.0	18.00	17.30
(M52)	5	80.00	78.80	92.4	89.04	42.80	41.20	42	41.0	20.00	19.16
M56	5.5	85.00	83.60	98.1	94.47	45.80	44.20	45	44.0	—	—
(M60)	5.5	90.00	88.60	103.9	100.12	48.80	47.20	48	47.0	—	—
M64	6	95.00	93.60	109.7	105.77	51.95	50.05	51	49.80	—	—
(M68)	6	100.00	98.60	115.5	111.42	54.95	53.05	54	52.80	—	—

NOTE. Sizes shown in brackets are non-preferred.

BEARING PRESSURES OF MATERIALS

The following may be taken as the average safe pressures that various materials will withstand:

Description in Bearing	Average Safe Pressure Tons per sq. ft	Tonnes/ m³
Ordinary brick in lime mortar	2 – 4 (very variable)	3·6–7·2
Ordinary brick in cement mortar	5	10·8
Hard brick in cement mortar	8	14·4
Blue brick in cement mortar	12	21·5
Lime concrete (good)	8	14·4
Portland cement concrete (1:2 4)	12	21·5
Granolithic bearing blocks	12	21·5
Freestone bearing blocks	12	21·5
Freestone ashlar masonry in cemt	15	26·9
Granite ashlar masonry in cemt	25	34·1

SAFE LOADS ON BRICKWORK

No disengaged pier in brick should have a height of more than 6 times its least width without lateral supports. With such proper supports, a pier may have a height between such supports of not more than 12 times its least width, the latter to be not less than one and a half bricks.

Brickwork in English bond is generally accepted as being stronger than Flemish or other bonds.

Three courses of engineering bricks in cement mortar are accepted in lieu of hard stone for padstones.

Reinforcing brickwork with 'Brickforce' or other mesh greatly increases its strength and allows thinner walls. To avoid cracking below window openings in domestic work, the three courses immediately above the dampcourse should be reinforced and cement mortar should always be used. Reinforcing meshes can also be used instead of lintels over small openings and the manufacturers will be glad to supply details on application. Quarter-inch steel rods (generally two are sufficient) can also be used instead of steel mesh for lintels, but the bottom course should always be brick-on-end and stirrups of $\frac{1}{8}$ in wire should be inserted in every perpend.

STRUCTURAL DATA

SAFE DISTRIBUTED LOADS P.S.I. ON ORDINARY CARCASSING TIMBERS, 2" WIDE (SOFTWOOD, CLASS "B", STRESSED TO 800 LB PER SQUARE INCH)

Depth of beam		CLEAR SPAN						
mm	inches	1·5m / 5ft	1·8m / 6ft	2·1m / 7ft	2·4m / 8ft	2·7m / 9ft	3·0m / 10ft	3·6m / 12ft
75	3	320	—	—	—	—	—	—
100	4	569	474	—	—	—	—	—
125	5	888	740	635	555	—	—	—
150	6	1280	1066	914	800	711	640	—
175	7	1742	1452	1244	1089	965	871	—
200	8	2275	1895	1625	1420	1265	1137	950
225	9	2880	2408	2057	1800	1600	1440	1200

For 1 in timber, divide by 2: for 3 in timber, take $1\frac{1}{2}$ times the above values. For a point load, at midspan, take half the above values.

The formula used in calculating the above table is

$$DB^2 = \frac{3WL}{4f}$$

When B equals breadth, D equals depth, W equals weight in pounds, L equals span and f is a constant, which for class B timber is 800 lb per sq. in. For Class A timber, such as Douglas fir or western hemlock (CLS) f equals 1,000 lb per sq. in.

(Basis for conversion of length of span and depth of beam: 25 mm = 1 in).

METRIC VERSION IN KILONEWTONS PER M²

Beam depth in mm	CLEAR SPAN IN METERS						
	1·5	1·8	2·1	2·4	2·7	3·0	3·6
75	16·00	—	—	—	—	—	—
100	28·45	23·70	—	—	—	—	—
125	44·40	37·00	36·75	27·70	—	—	—
150	61·60	53·30	40·70	40·00	40·00	32·00	—
175	89·10	92·60	62·20	54·50	48·20	43·80	—
200	113·75	100·00	81·21	91·00	63·20	56·80	47·50
225	144·00	120·40	101·30	90·00	80·00	92·00	60·00

NAIL SIZES FOR DIFFERING THICKNESSES OF TIMBER

	Thickness of Timbers		Length of Nail	Guage of Nail
	A	B		
	*	$\frac{3}{4}"$	$\frac{3}{4}"$	16
		1"	1"	14

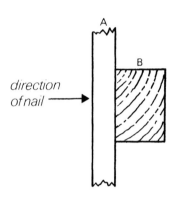

direction of nail →

1. Thickness A is the maximum, and thickness B the minimum which can be used if the full strength of the nails is to be developed in a joint.

2. *These two sizes of nails are only suitable for the structural purpose of fixing thin metal gusset plates (say 20g) to timber.

3. Where a metal component or plate is nailed to a timber member the value of the recommended load may be increased by 25%.

4. Where the minimum penetration (shown in previous table) cannot be obtained, the load per nail should be reduced by the factor:

$$\frac{\text{Actual Penetration}}{\text{Minimum Penetration}}$$

Thickness of Timbers		Length of Nail	Guage of Nail
A	B		
$\frac{1}{8}"$	$\frac{5}{8}"$	$\frac{3}{4}"$	17
	$\frac{7}{8}"$	$1"$	15
$\frac{1}{4}"$	$\frac{1}{2}"$	$\frac{3}{4}"$	18
	$\frac{3}{4}"$	$1"$	16
	$1"$	$1\frac{1}{4}"$	14
	$1\frac{1}{4}"$	$1\frac{1}{2}"$	12
$\frac{3}{8}"$	$\frac{7}{8}"$	$1\frac{1}{4}"$	15
	$1\frac{1}{8}"$	$1\frac{1}{2}"$	13
$\frac{1}{2}"$	$1"$	$1\frac{1}{2}"$	14
	$1\frac{1}{4}"$	$1\frac{3}{4}"$	12
	$1\frac{1}{2}"$	$2"$	10
$\frac{5}{8}"$	$1\frac{1}{8}"$	$1\frac{3}{4}"$	13
	$1\frac{3}{8}"$	$2"$	11
$\frac{3}{4}"$	$1"$	$1\frac{3}{4}"$	14
	$1\frac{1}{4}"$	$2"$	12
	$1\frac{1}{2}"$	$2\frac{1}{4}"$	10
	$1\frac{3}{4}"$	$2\frac{1}{2}"$	9
$\frac{7}{8}"$	$1\frac{3}{8}"$	$2\frac{1}{4}"$	11
$1"$	$1\frac{1}{2}"$	$2\frac{1}{2}"$	10
	$2"$	$3"$	8
$1\frac{1}{8}"$	$1\frac{3}{8}"$	$2\frac{1}{2}"$	11
$1\frac{1}{4}"$	$1\frac{3}{4}"$	$3"$	9
	$2\frac{1}{4}"$	$3\frac{1}{2}"$	7
$1\frac{3}{8}"$	$2\frac{5}{8}"$	$4"$	6
$1\frac{1}{2}"$	$1\frac{3}{4}"$	$3"$	10
	$2"$	$3\frac{1}{2}"$	8
$1\frac{3}{4}"$	$2\frac{1}{4}"$	$4"$	7
$1\frac{7}{8}"$	$2\frac{5}{8}"$	$4\frac{1}{2}"$	6
$2"$	$3"$	$5"$	5
$2\frac{3}{8}"$	$2\frac{5}{8}"$	$5"$	6
$2\frac{3}{4}"$	$3\frac{1}{4}"$	$6"$	4

STRUCTURAL DATA

(By courtesy of the Timber Research and Development Association)

Thickness of Timbers		Length of Nail	Guage of Nail
A	B		
3"	3"	6"	5
$3\frac{3}{8}"$	$3\frac{5}{8}"$	7"	3
$3\frac{5}{8}"$	$4\frac{3}{8}"$	8"	$\frac{5}{16}$
$4\frac{5}{8}"$	$4\frac{3}{8}"$	9"	$\frac{5}{16}$
$4\frac{3}{4}"$	$5\frac{1}{4}"$	10"	$\frac{3}{8}$

THE SPACING OF NAILS IN NAILED JOINTS In Softwood

Guage of Nails	Minimum Penetration	End Distance		Edge Distance		Spacing across the grain		Spacing between Nails along the grain		Permissible Lateral Load per Nail (lb)		Permissible Withdrawal Load (lb) per inch of penetration	
		(1)	(2)	(3)	(4)	(5)	(6)	(7)	(8)	(9)	(10)	(11)	(12)
		Un-bored	Pre-bored	Un-bored	Pre-bored	Un-bored	Pre-bored	Un-bored	Pre-bored	Group* 1	2 & 3	Group* 1	2 & 3
18	$\frac{1}{2}"$	1"	$\frac{1}{2}"$	$\frac{1}{4}"$	$\frac{1}{4}"$	$\frac{1}{2}"$	$\frac{3}{16}"$	1"	$\frac{1}{2}"$	17	14	12	7
17	$\frac{5}{8}"$	$1\frac{1}{8}"$	$\frac{9}{16}"$	$\frac{5}{16}"$	$\frac{5}{16}"$	$\frac{9}{16}"$	$\frac{3}{16}"$	$1\frac{1}{8}"$	$\frac{9}{16}"$	22	18	14	8
16	$\frac{3}{4}"$	$1\frac{5}{16}"$	$\frac{5}{8}"$	$\frac{5}{16}"$	$\frac{5}{16}"$	$\frac{3}{8}"$	$\frac{3}{16}"$	$1\frac{5}{16}"$	$\frac{5}{8}"$	27	22	16	9
15	$\frac{7}{8}"$	$1\frac{7}{16}"$	$\frac{3}{4}"$	$\frac{3}{8}"$	$\frac{3}{8}"$	$\frac{3}{4}"$	$\frac{1}{4}"$	$1\frac{7}{16}"$	$\frac{3}{4}"$	32	26	18	10
15	1"	$1\frac{5}{8}"$	$\frac{13}{16}"$	$\frac{7}{16}"$	$\frac{7}{16}"$	$\frac{13}{16}"$	$\frac{1}{4}"$	$1\frac{5}{8}"$	$\frac{13}{16}"$	37	31	19	11
13	$1\frac{1}{8}"$	$1\frac{7}{8}"$	$\frac{15}{16}"$	$\frac{1}{2}"$	$\frac{1}{2}"$	$\frac{15}{16}"$	$\frac{15}{16}"$	$1\frac{7}{8}"$	$\frac{15}{16}"$	46	38	22	13
12	$1\frac{1}{4}"$	$2\frac{1}{16}"$	$1\frac{1}{16}"$	$\frac{1}{2}"$	$\frac{1}{2}"$	$1\frac{1}{16}"$	$\frac{3}{8}"$	$2\frac{1}{16}"$	$1\frac{1}{16}"$	55	45	25	15
11	$1\frac{3}{8}"$	$2\frac{3}{8}"$	$1\frac{3}{16}"$	$\frac{9}{16}"$	$\frac{9}{16}"$	$1\frac{3}{16}"$	$\frac{3}{8}"$	$2\frac{3}{8}"$	$1\frac{3}{16}"$	65	53	28	16
10	$1\frac{1}{2}"$	$2\frac{9}{16}"$	$1\frac{1}{4}"$	$\frac{5}{8}"$	$\frac{5}{8}"$	$1\frac{1}{4}"$	$\frac{7}{16}"$	$2\frac{9}{16}"$	$1\frac{1}{4}"$	76	62	31	18
9	$1\frac{3}{4}"$	$2\frac{7}{8}"$	$1\frac{7}{16}"$	$\frac{3}{4}"$	$\frac{3}{4}"$	$1\frac{7}{16}"$	$\frac{1}{2}"$	$2\frac{7}{8}"$	$1\frac{7}{16}"$	90	74	35	20
8	2"	$3\frac{3}{16}"$	$1\frac{5}{8}"$	$\frac{13}{16}"$	$\frac{13}{16}"$	$1\frac{5}{8}"$	$\frac{9}{16}"$	$3\frac{3}{16}"$	$1\frac{5}{8}"$	106	86	39	22
7	$2\frac{1}{4}"$	$3\frac{1}{2}"$	$1\frac{3}{4}"$	$\frac{7}{8}"$	$\frac{7}{8}"$	$1\frac{3}{4}"$	$\frac{9}{16}"$	$3\frac{1}{2}"$	$1\frac{3}{4}"$	122	100	43	25
6	$2\frac{5}{8}"$	$3\frac{7}{8}"$	$1\frac{15}{16}"$	$\frac{15}{16}"$	$\frac{15}{16}"$	$1\frac{15}{16}"$	$\frac{9}{16}"$	$3\frac{7}{8}"$	$1\frac{15}{16}"$	139	114	47	27
5	3"	$4\frac{1}{4}"$	$2\frac{1}{8}"$	$1\frac{1}{16}"$	$1\frac{1}{16}"$	$2\frac{1}{8}"$	$\frac{5}{8}"$	$4\frac{1}{4}"$	$2\frac{1}{8}"$	161	132	52	30
4	$3\frac{1}{4}"$	$4\frac{5}{8}"$	$2\frac{5}{16}"$	$1\frac{3}{16}"$	$1\frac{3}{16}"$	$2\frac{5}{16}"$	$1\frac{1}{16}"$	$4\frac{5}{8}"$	$2\frac{5}{16}"$	184	151	57	32
3	$3\frac{5}{8}"$	$5\frac{1}{16}"$	$2\frac{9}{16}"$	$1\frac{1}{4}"$	$1\frac{1}{4}"$	$2\frac{9}{16}"$	$\frac{3}{4}"$	$5\frac{1}{16}"$	$2\frac{9}{16}"$	209	171	61	35
$\frac{5}{16}$	$4\frac{3}{8}"$	$6\frac{1}{4}"$	$3\frac{1}{8}"$	$1\frac{9}{16}"$	$1\frac{9}{16}"$	$3\frac{1}{8}"$	$\frac{15}{16}"$	$6\frac{1}{4}"$	$3\frac{1}{8}"$	288	236	76	44
$\frac{3}{8}$	$5\frac{1}{4}"$	$7\frac{1}{2}"$	$3\frac{3}{4}"$	$1\frac{7}{8}"$	$1\frac{7}{8}"$	$3\frac{3}{4}"$	$1\frac{1}{8}"$	$7\frac{1}{2}"$	$3\frac{3}{4}"$	379	319	91	52

*Note—Group 1—Douglas fir. Group 2—All other softwoods.
(By courtesy of the Timber Research and Development Association)

NAILING TIMBERS

The following points should be taken into consideration:
1. The holding power of nails, as thus calculated, depends on the density and dryness of the wood. A timber that has, for its species, large rings of summer growth, and is in consequence "open" in grain and light in weight, will not hold nails as tightly as a good average specimen on the behaviour of which the tables are based. Wet unseasoned timber, as it dries out, will tend to warp and shrink and will thereby reduce the strength of nailed joints by three-fourths when straight withdrawal from side grain is involved and by one-third where lateral resistance is concerned.
2. Joints involving lateral resistance of nails in end grain have 25% less strength than the values given in the table on page 154.
3. Joints should never be so designed that they involve the straight withdrawal of nails from end grain.
4. For calculating the resistance to lateral stresses or to withdrawal, only the length of the nail imbedded in the larger timber should be taken into account (see diagram).
5. For nails that pass completely through both timbers, and are clenched, 25% should be added to the permissible lateral load, when the nails are driven without boring. If driven through bored holes, clinching gives no added strength.

The above tables and formulae are for two pieces of wood connected together, with nails not clenched. If three pieces are connected together with nails that are clenched with the ends lying across the grain, research has shown that the joint is increased by about 50%, if the nails are not driven through bored holes. If driven through bored holes, only a 10% increase is given. Generally, if nails are driven into pre-bored timber, joints are stronger by about 14%.

From the values given in the tables, it will be seen, that nailed joints are most suitable for light loads—which, however, include the normal loads on structures designed as domestic dwellings. Moreover light loads do not always mean small structures: the Belfast truss, made entirely from thin boards, is well known, and can be made for large spans.

TIMBER CONNECTORS

Timber connectors are really stress plates, which spread the load on a joint over a greater area of the timbers connected than does a bolt. Approximately 80% of the strength of the timber can thus be utilised, whereas with plain bolted joints, the efficiency may be as low as 15%.

The safe loads in the tables below can be compared with those for the lateral withdrawal of nails in a foregoing table, when it will be seen how many nails will be needed to provide a joint as strong as one made with a bolt and a timber connector. These connectors are coming more and more into use in structural work. Washers larger than the commercial size should be used: $1\frac{1}{2}$in square washers for 2in connectors, 2in for $2\frac{1}{2}$in connectors, and $2\frac{1}{2}$in for 3in connectors, placed under both bolt head and nut.

B.S. 1579 gives dimensions and sizes of heavy pressed steel and malleable cast-iron connectors, in addition to light round and square toothed connectors. The table below applies to a light, round, toothed connector. Heavy pressed and cast-iron connectors are used only in structures incorporating very heavy timbers. For design data, see CP 112.

ALLOWABLE LOADS FOR ONE "BULLDOG" CONNECTOR UNIT

Dia of con-nector	Thickness of member		Load parallel to grain (lbs)		Load perpendicular to grain (lbs)	
	Connector on one side of member	Connector on both sides of member (on same bolt)	Group I	Group II	Group I	Group II
2"	$\frac{3}{4}$"	$1\frac{1}{2}$"	1,150	830	880	640
	1"	2"	1,225	905	925	670
	$1\frac{1}{2}$"	3"	1,280	1,010	1,065	750
	2" and over	4" and over	1,280	1,010	1,135	820
$2\frac{1}{2}$"	1"	2"	1,385	1,035	1,085	800
	$1\frac{1}{2}$"	3"	1,440	1,140	1,225	880
	2" and over	4" and over	1,440	1,140	1,295	950
3"	1"	2"	1,550	1,165	1,250	930
	$1\frac{1}{2}$"	3"	1,605	1,270	1,390	1,010
	2" and over	4" and over	1,605	1,270	1,460	1,080
$1\frac{1}{4}$"	—	—	Test results not yet available on this size			
$3\frac{3}{4}$" single sided	1"	2"	2,365	1,775	1,885	1,415
	$1\frac{1}{2}$"	3"	2,450	1,935	2,095	1,535
	2" and over	4" and over	2,450	1,935	2,205	1,650

METRIC VERSION—LOAD IN KILOGRAMS

Depth of beam (mm)	Connector on one side of member	Connector on both sides of member (on same bolt)	Group I	Group II	Group I	Group II
50	19	38	679	377	398	312
	25	50	544	411	419	303
	38	76	579	579	458	339
	50	101	579	579	573	371
62	25	50	627	468	544	363
	38	76	652	473	866	369
	50	101	652	473	583	431
75	25	50	701	468	525	422
	38	76	726	527	633	457
	50	101	726	527	651	489
81	25	50	1,061	793	563	422
	38	76	1,149	877	947	699
	50	101	1,149	877	998	747

MINIMUM TIMBER DIMENSIONS FOR "BULLDOG" TIMBER CONNECTORS

Diameter of connector	$1\frac{1}{2}$	2	$2\frac{1}{2}$	3	$3\frac{3}{4}$	inches
	38	50	68	75	94	mm
Minimum nominal size of member for connector on one side	$\frac{3}{4} \times 2$	$\frac{3}{4} \times 2\frac{1}{2}$	$\frac{7}{8} \times 3$	$1 \times 3\frac{1}{2}$	$1 \times 4\frac{1}{2}$	inches
	19×50	19×68	22×75	25×94	25×113	mm
Minimum nominal size of member for connector on both sides	$1\frac{1}{2} \times 2$	$1\frac{1}{2} \times 2\frac{1}{2}$	$1\frac{3}{4} \times 3$	$2 \times 3\frac{1}{2}$	$2 \times 4\frac{1}{2}$	inches
	35×50	38×68	44×75	50×94	50×113	mm

(Basis for conversion: 25mm equals 1 in)

JOIST HANGERS

The illustration shows a type of joist hanger that can be built into walls instead of leaving pockets or, in alteration work, cutting out pockets to take joists. There is also a modification that can be hung over rolled steel or timber or concrete beams, and which provides a simple method of inserting floors or platforms in existing framed structures. The sizes available are shown in the table overleaf:

STRUCTURAL DATA

Breadth of joist		Depth of joist		
in	mm	in		mm
1	25	7, 8		175–200
1½	38	3– 8 in 1 in steps		75–200 in 25 mm steps
2	50	3–11	″	75–275 ″
2½	63	4–10	″	100–250 ″
3	75	3–12	″	75–300 ″
3½	88	7, 9		175, 225
4	100	4–12 in 1 in steps		100–300 in 25mm steps
4½	113	7		175
5	125	12		300

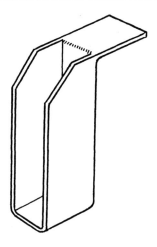

FRAMING ANCHORS

Framing anchors, long employed in the USA, are steel plates and angles that can be employed, with special nails, instead of framing joints in carpentry, and in place of the skew-nailed and other nailed joints that are so often used instead of true carpentry joints, such as the mortise and tenon in its many varieties. They have the advantage that, after many laboratory tests, the safe working loads of these anchors are known with tolerable certainty.

Design loads for many possible applications may be determined from the following data. The recommendations are for use with softwood comparable to Douglas Fir. In general, when two or more anchors are used in combination, the allowable load will be the sum of the individual loads.

RECOMMENDED SAFE WORKING VALUES

Timber Group	Load Direction (Figs. 1-3)											
	A		B		C		D		E		F	
	lb	kg	lb	kg	lb	kg	lb	kg	lb	kg	lb	kg
Dead load plus superimposed load	1	300	135	530	239	290	131	200	90	300	135	450	203
	2	240	109	424	192	184	83	160	72	240	109	360	163
Dead load plus superimposed load plus snow plus wind load	1	450	203	825	373	420	191	300	135	450	203	675	305
	2	360	163	660	300	266	119	240	303	360	163	540	144

(By courtesy of Messrs. Macandrews & Forbes, Ltd.)

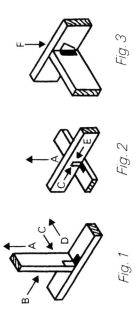

Fig. 1   *Fig. 2*   *Fig. 3*

STRUCTURAL DATA

CONCRETE LINTELS

A large range of concrete lintels for 4½in walls, to span openings for standard windows and doors, is given in B.S. 1239, from which the table below is abstracted.

Precast lintels to B.S. 1239, and pre-stressed concrete-cum-hollow brick lintels, of not more than 3in depth for moderate openings, can be purchased from concrete products manufacturers.

(4½in = 113mm 3in = 76mm)

LINTELS: DETAILS OF SIZES, ETC.

Clear span meters—feet	Depth inches—mm	Number	Reinforcing bars: size
0·696—2	5⅝ ⎫ 143	1	$\frac{5}{16}$in 0·312mm
0·914—3	5⅝ ⎭	2	¼in 0·250mm
1·219—4	8⅝ ⎫ 219	2	$\frac{5}{16}$in 0·312mm
1·524—5	8⅝ ⎭	2	⅜in 0·375mm
1·828—6	11⅝ ⎫ 295	2	⅜in 0·375mm
2·438—8	11¾ ⎭	3	⅝in 0·375mm

Reinforcement to have 1in of cover. Top or handling steel (three ¼in bars) required in 8ft spans.

(in = 25mm 8ft = 2.44m)

6. THE EXCAVATOR

Codes of Practice	165
Good Practice Guide	165
Excavating Data	168
Planking and Strutting	168
Embankments	169
Bearing Capacities of Foundations	169
Hardcore and Pitching	172
Trench Excavation	173

CODES OF PRACTICE

CP 101: 1972: Foundations and Substructures
Concrete in good ground is not to be leaner than 1—4—8 or 1—8 'all-in' ballast. In weak ground, where mass concrete footings are stressed in bending as well as in compression, concrete mix is not to be leaner than 1—3—6 or 1—6 'all-in' ballast. Excavation and filling should be done as soon as possible after clearing site. Filling should be placed in layers not more than 6 in/150 mm deep. Excavations to be cleared of water before placing concrete. In soils affected by exposure to the atmosphere, place 3 in/75 mm thickness of 'blinding' concrete immediately after excavation. For reinforced concrete foundations, place 3 in/75 mm of 'blinding' concrete before placing the foundation proper. Remove datum pegs and fill holes with concrete. Depth of foundations and type depend on type of soil and load to be taken. In some clays, known as 'shrinkable clays', footings must be at least 3 ft/900 mm below ground.

For further information consult CP 101.

GOOD PRACTICE GUIDE

Strip Foundations

If the foundations of a building are constructed as strip foundations of plain concrete situated centrally under the walls, good practice will be satisfied if—

(a) there is no made ground or wide variation in the type of subsoil within the loaded area and no weaker type of soil exists below the soil on which the foundations rest within such a depth as may impair the stability of the structure;

(b) the width of the foundations is not less than the width specified in the Table to this regulation in accordance with the related particulars specified in the Table;

(c) the concrete is composed of cement and fine and coarse aggregate conforming to BS882: Part 2: 1973 in the proportion of 50 kg of cement to not more than 0.1 m^3 of fine aggregate and 0.2 m^3 of coarse aggregate;

(d) the thickness of the concrete is not less than its projection from the base of the wall or footing and is in no case less than 150 mm;

(e) where the foundations are laid at more than one level, at each change of level the higher foundations extend over and unite with the lower foundations for a distance of not less than the thickness of the foundations and in no case less than 300 mm; and

(f) where there is a pier, buttress or chimney forming part of a wall, the foundations project beyond the pier, buttress or chimney on all sides to at least the same extent as they project beyond the wall.

Table: Minimum width of strip foundations

Type of subsoil	Condition of subsoil	Field test applicable	Minimum width in millimetres for total load in kilonewtons per lineal metre of loadbearing walling of not more than—					
			20 kN/m	30 kN/m	40 kN/m	50 kN/m	60 kN/m	70 kN/m
(1)	(2)	(3)	(4)	(5)	(6)	(7)	(8)	(9)
I Rock	Not inferior to sandstone, limestone or firm chalk	Requires at least a pneumatic or other mechanically operated pick for excavation	In each case equal to the width of wall					
II Gravel Sand	Compact Compact	Requires pick for excavation. Wooden peg 50 mm square in cross-section hard to drive beyond 150 mm	250	300	400	500	600	650
III Clay Sandy clay	Stiff Stiff	Cannot be moulded with the fingers and requires a pick or pneumatic or other mechanically operated spade for its removal	250	300	400	500	600	650
IV Clay Sandy clay	Firm Firm	Can be moulded by substantial pressure with the fingers and can be excavated with graft or spade	300	350	450	600	750	850
V Sand Silty sand Clayey sand	Loose Loose Loose	Can be excavated with a spade. Wooden peg 50 mm square in cross-section can be easily driven	400	600	Note: Foundations do not fall within the provisions of regulation D7 if the total load exceeds 30 kN/m			

THE EXCAVATOR

Table continued

Minimum width of strip foundations

Type of subsoil	Condition of subsoil	Field test applicable	Minimum width in millimetres for total load in kilonewtons per lineal metre of loadbearing walling of not more than–					
			20 kN/m	30 kN/m	40 kN/m	50 kN/m	60 kN/m	70 kN/m
(1)	(2)	(3)	(4)	(5)	(6)	(7)	(8)	(9)
VI Silt Clay Sandy clay Silty clay	Soft Soft Soft Soft	Fairly easily moulded in the fingers and readily excavated	450	650	Note: In relation to types VI and VII, foundations do not fall within the provisions of regulation D7 if the total load exceeds 30 kN/m			
VII Silt Clay Sandy clay Silty clay	Very soft Very soft Very soft Very soft	Natural sample in winter conditions exudes between fingers when squeezed in fist	600	850				

EXCAVATING DATA

TRANSPORT CAPACITIES

Type of Vehicle	cu. yds solid	m³ solid
Wheelbarrow	0·10	0·08
2 ton (2·03 t) lorry	1·50	1·15
3 ton (3·05 t) lorry	2·25	1·72
4 ton (4·06 t) lorry	2·90	2·22
5 ton (5·08 t) lorry	3·50	2·68
6 ton (6·10 t) lorry	4·50	3·44
2 cu. yd (1·53 m³) dumper	1·50	1·15
3 cu. yd (2·29 m³) dumper	2·25	1·72
6 cu. yd (4·59 m³) dumper	4·50	3·44
10 cu. yd (7·65 m³) dumper	7·50	5·73

INCREASE IN BULK AFTER EXCAVATION

1 cu. ft or 1 cu. yd or 1 m³	becomes after excavation	cu. ft, cu. yds or m³
Chalk	″	1·6
Clay	″	1·2
Clay (stiff)	″	1·5
Earth (ordinary)	″	1·25
Gravel	″	1·1
Rock	″	1·5
Sand	″	1·05

PLANKING AND STRUTTING

AVERAGE DEPTHS OF EXCAVATIONS OF VARIOUS SOILS WITHOUT USE OF TIMBERING

Type of Soil	Feet	Metres
Clay, stiff	10	3·05
Earth, ordinary	3	0·91
Gravel	2	0·61
Loam, drained	6	1·83
Sand, dry	1	0·30
Soil, compact	12	3·66

Note: Each case must be taken on its merits, and, as the limiting distances given above are approached, careful watch must be kept for the slightest signs of caving-in.

THE EXCAVATOR

TIMBER REQUIRED FOR VARIOUS TYPES OF PLANKING AND STRUTTING

SITUATION	Poling boards	Walings	Struts	Uprights	Shores	Sole plates
TRENCHES						
inches	$1\frac{1}{2} \times 7$	2×6	4×4	—	—	—
mm	38×175	50×150	100×100	—	—	—
BASEMENTS						
inches	$9\frac{1}{4} \times 1\frac{1}{2}$	6×8	$9 \times 1\frac{1}{2}$	9×6	9×9	11×3
mm	232×38	150×200	225×38	225×150	225×225	275×75
PIER HOLES	Poling boards	Walings	Angle posts	Hangers	Cross beams	—
inches	$9 \times 1\frac{1}{2}$	4×4	4×4	9×3	11×9	—
mm	225×38	100×100	100×100	225×75	275×225	—

Note 1: The timber sizes given for poling boards for trenches are for open or close poling boards. Basements and pier holes require close poling boards. (Conversion factor: 25 mm equals 1 in.)

Note 2: 14 g corrugated steel sheets may be used for poling boards and telescopic (screw) struts for struts, in trenches.

EMBANKMENTS

SAFE HEIGHTS OF EMBANKMENTS

Type of Soil	Height in Feet	Metres
Compact gravel or sand	120–100	36·5–30·5
Loamy sand	70– 55	21·3–16·7
Damp clay	30– 25	9·1– 7·6
Sandy clays and clay loams	30– 15	9·1– 4·5
Clay soil of river beds	20– 6	6·0– 1·8
Peat moss, marsh earth, consolidated mud and silt, soft pasty clay or marsh clay	7– 0	2·1– 0·0

SAFE SLOPES OF EARTH IN EMBANKMENTS

The usual slopes for cuttings and embankments range from 1 to 1 for firm earth, to 4 to 1. The most frequent is from 1 to 1 up to $1\frac{1}{2}$ to 1 for firm earth in cuttings, and $1\frac{1}{2}$ to 1 in embankments. Clays, from 3 to 1 for solid blue clays, to 12 to 1 for soft, mushy clays. Chalk in cuttings may be from nearly vertical to $\frac{1}{2}$ to 1, but 1 to 1 in embankments. In the ratios given above, the first figure is the length of the base and the second the vertical height, which is always 1.

BEARING CAPACITIES OF FOUNDATIONS

A table of typical bearing capacities; from British Standard Code of Practice **CP 101:172**: Foundations and substructures.

TABLE 1. TYPICAL BEARING CAPACITIES

NOTE. Due care should be paid to ensuring an adequate depth of the given soil, and in certain cases, in order to limit the amount of settlement, consideration may have to be given to restricting the allowable bearing pressure to a lower value than the bearing capacity.

Sand and gravel. Table 1 assumes that the width of the foundation is of the order of 1 m. For narrower foundations on sand and gravel, the permissible bearing capacity decreases as the width decreases. In such cases the permissible bearing capacity should be the value given in Table 1 multiplied by the width of the foundation in metres. In such soils the permissible bearing capacity can be increased by 12.5 kN/m^2 for every 0.3 m depth of the loaded area below the lowest ground surface immediately adjacent. If the ground water-level in sand or gravel is likely to be at a depth of less than the foundation width below the base of the foundation, then the submerged value given in Table 1 should be used.

Clay soils. For the types of building considered in this code, the width and depth of the foundations do not have an appreciable influence on the permissible bearing capacity on clay soils. If a clay is examined under dry summer conditions the probable deterioration under winter conditions should be borne in mind. Mud stones and clay shales may deteriorate very rapidly if exposed to the weather or to ground water.

Group		Types of rocks and soils	Bearing capacity kN/m^2	Remarks
I Rocks	1	Igneous and gneissic rocks in sound condition	10 000	These values are based on the assumption that the foundations are carried down to unweathered rock
	2	Massively-bedded limestones and hard sandstones	4 000	
	3	Schists and slates	3 000	
	4	Hard shales, mudstones and soft sandstones	2 000	
	5	Clay shales	1 000	
	6	Hard solid chalk	600	
	7	Thinly-bedded limestones and sandstones		To be assessed after inspection
	8	Heavily-shattered rocks and the softer chalks		

TABLE 1. TYPICAL BEARING CAPACITIES (cont.)

Group		Types of rocks and soils	Bearing capacity kN/m²		Remarks
			Dry	Submerged	
II Non-cohesive soils	9	Compact gravel or compact sand and gravel	> 600	> 300	Width of foundation not less than 1 m. Ground water level assumed to be a depth not less than the width of foundation below the base of the foundation
	10	Medium dense gravel or medium dense gravel and sand	200 to 600	100 to 300	
	11	Loose gravel or loose sand and gravel	< 200	< 100	
	12	Compact sand	> 300	> 150	
	13	Medium dense sand	100 to 300	50 to 150	
	14	Loose sand	< 100	< 50	
III	15	Very stiff boulder clays and hard clays	300 to 600		This group is susceptible to long-term consolidation settlement
	16	Stiff clays	150 to 300		
	17	Firm clays	75 to 150		
	18	Soft clays and silts	75		
	19	Very soft clays and silts	< 75		
IV	20	Peat			Foundations should be carried down through peat and organic soil to a reliable bearing stratum below
V	21	' Made ' ground			It should be investigated with extreme care

Mixed soils. Soils intermediate between the main types given in Table 1 may need to be assessed by test.

Peat and 'made' ground. If the bedding slopes appreciably, special measures may be necessary for these soils (see also 3.7.3 and 3.7.4).

HARDCORE AND PITCHING
APPROXIMATE COVERING CAPACITIES

Consolidated thickness laid in inches	Yards super to one ton covered							
	Stone Pitching (roughly thrown)	Stone Pitching (hand packed)	Hardcore rubbish (brick, concrete or stone)	Clinker Ashes	Broken Stone	Whinstone	Slag	Shale
1	—	—	—	33	28	24	32	24
2	—	—	—	$16\frac{1}{2}$	14	12	16	12
3	—	—	10	11	9	8	$10\frac{1}{2}$	8
4	—	—	$7\frac{1}{2}$	$8\frac{1}{4}$	7	6	8	6
5	—	—	6	$6\frac{1}{2}$	$5\frac{1}{2}$	5	$6\frac{1}{4}$	5
6	$4\frac{1}{4}$	4	5	$5\frac{1}{2}$	$4\frac{2}{3}$	4	$5\frac{1}{4}$	4
7	$3\frac{2}{3}$	$3\frac{1}{2}$	$4\frac{1}{3}$	—	—	—	—	—
8	$3\frac{1}{4}$	3	$3\frac{3}{4}$	—	—	—	—	—
9	$2\frac{7}{8}$	$2\frac{3}{4}$	$3\frac{1}{3}$	—	—	—	—	—
10	$2\frac{1}{2}$	$2\frac{1}{4}$	3	—	—	—	—	—
Basis in tons per cu. yd consolidated	1·4	1·45	1·2	1·09	1·28	1·5	1·125	1·5

METRES SQUARE PER METRIC TONNE COVERED

in	mm	Stone Pitching (roughly thrown)	Stone Pitching (hand packed)	Hardcore rubbish (brick concrete or stone)	Clinker Ashes	Broken Stone	Whinstone	Slag	Shale
1	25	—	—	—	27·5	23·4	20·0	26·7	20·0
2	50	—	—	—	13·7	11·7	10·0	13·3	10·0
3	75	—	—	8·3	9·2	7·5	6·6	8·7	6·6
4	100	—	—	6·2	6·8	5·8	5·2	6·6	5·2
5	125	—	—	5·0	5·4	4·5	4·1	5·2	4·1
6	150	3·5	3·3	4·1	4·5	4·3	3·3	4·3	3·3
7	175	2·8	2·9	3·7	—	—	—	—	—
8	200	2·7	2·5	3·1	—	—	—	—	—
9	225	2·4	2·3	2·9	—	—	—	—	—
10	250	1·8	1·8	2·7	—	—	—	—	—

(Based on 25 mm equals 1 inch, 1 ton equals 1 tonne)

APPROXIMATE WEIGHTS OF SOILS AND ROCKS

Material	Weight cwt per cu. yd	kg per m^3
Chalk	36	1 400
Clay	31	1 200
Granite	42	1 600
Gravel	30	1 150
Marl	26	1 000
Mud	25	950
Quartz	41	1 550
Sand	30	1 150
Sandstone	39	1 500
Shale	40	1 535
Slate	43	1 650
Trap	42	1 600

Note: 1,000 kilograms equals 1 tonne.

TRENCH EXCAVATION

For estimating the quantity of excavation in a trench tapering in depth (or width), use this formula:

The sum of the area of the two end sections *plus* four times the area of the middle section *multiplied* by one-sixth of the length.

$$\frac{(A + a + 4m)}{6}$$

ANGLE OF REPOSE OF VARIOUS SOILS

Material	Angle of repose in degrees from the horizontal
Mud	1 to 0
Gravel and Sand (natural mixture)	25 to 30
Rubble stone	45
Gravel	40
Ashes	40
Sand	
dry	30
moist	35
wet	25
Clay	
dry	30
moist	45
wet	15
Vegetable soil	
dry	30
moist	45
wet	20

WIDTHS REQUIRED FOR TRENCHES FOR VARIOUS DIAMETERS OF PIPES

Diameter of pipe—	4 in & under 101 mm & under	5–6 in 127–152 mm	7–9 in 177–228 mm	10–12 in 254–304 mm	13–15 in 330–381 mm	16–18 in 406–457 mm	19–24 in 482–609 mm
Depth up to 5ft 0 in (1·5 m)	1 ft 6 in 0·456 m	1 ft 9 in 0·496 m	2 ft 0 in 0·609 m	2 ft 3 in 0·649 m	2 ft 6 in 0·761 m	3 ft 0 in 0·914 m	3 ft 9 in 1·096 m
Over 5 ft 0 in (1·5 m)	2 ft 0 in 0·609 m	2 ft 3 in 0·649 m	2 ft 6 in 0·761 m	2 ft 9 in 0·801 m	3 ft 0 in 0·912 m	3 ft 6 in 1·06 m	4 ft 0 in 1·284 m

7. BUILDERS' PLANT AND MACHINERY

Conversion Table	...	...	...	...	176
Plant Economics	...	...	...	...	176
Machinery Data	...	...	...	...	178
Pump Performances	...	...	...	...	181
Steel Scaffolding	...	...	...	...	181

CONVERSION TABLE

HORSEPOWER AND KILOWATTS

1·34	1	0·75
2·68	2	1·49
4·02	3	2·24
5·36	4	2·98
6·71	5	3·73
8·05	6	4·48
9·39	7	5·22
10·73	8	5·96
12·07	9	6·71
13·41	10	7·45
26·82	20	14·91
40·23	30	22·37
53·64	40	29·83
67·05	50	37·21
134·10	100	74·59
268·20	200	149·41

(A rough conversion is 1 hp equals ¾ kw, or 3 kw equals 4 hp)

PLANT ECONOMICS

Today the builder has a bewildering variety of machines at his disposal for a large variety of operations. These are largely earth-moving machines and transport: the trades that have become largely mechanized are excavator, concretor and joiner. There are also a large variety of portable powered tools, from the well-known drill to intricate machines that can be used with profit on site work or in site workshops. Much time can be saved by the use of machines and this can lead to considerable savings in costs, if the plant is well maintained and well managed.

Machines for their own sake do not make sense. Machines that are not used must nonetheless be maintained and are a wasting asset; being a drain on capital without providing any return. In these days of credit restriction and high interest rates, the locking-up of capital in plant is something to be avoided if possible.

It is always as well, when thinking about buying a machine, to consider if it would not be more profitable to buy its end-product. For instance, ready-mixed concrete and ready-mixed mortar could be cheaper than investing in a large concrete mixer, and factory-made joinery can easily work out cheaper than setting up and maintaining a joiner's shop to make it for one's self. Specialist sub-contractors, too, can often quote prices that are lower than one's own costs. There is, too, the danger that the possession of machines will lead to their use on jobs for which they are not suited; and this results in inflated costs.

PLANT AND MACHINERY

Nevertheless, there is no doubt that plant pays. Many jobs on modern sites would be impossible or too costly without it. What is more, the use of plant is often forced on the builder by the operatives or by the architects. Labourers will no longer carry the hod (even if we could try to let them do it) and hoists or lifts must be installed. The architect may specify large, prefabricated panels and a tower crane is needed to handle them. Machines we must have; but the question is, shall they be bought or hired?

HIRING OR BUYING

Hiring has much to commend it. The machines are always in good condition. There is no worry about breakdowns—the hire firm attends to them. The latest type of machine can be had and one can always be certain of the right machine for the job, instead of having to make do with something from one's own limited range. The plant-hire industry is fiercely competitive, which keeps hire charges from being excessive. Good plant-hire firms have specialists who can give valuable advice as to which machine to use, and how best to use it on any particular site.

Against this, many builders, especially those in the medium and large categories, find it more profitable to buy their own plant. A survey of plant-users conducted by the Building Research Establishment, shows that most plant-users reckon on having each machine on site for 200 days a year and working for 150 days of that time. This may serve as a preliminary guide to intending purchasers. Unless a firm can judge from past experience that it will have something like this quantity of work for a machine, hiring is likely to be better than buying.

THE COST OF A MACHINE

Besides the cost of the machine itself (the capital cost) there are the annual recurring costs of:

1. insurance;
2. maintainance;
3. licence;
4. administrative costs.
    These generally amount to about 30 per cent per annum of the capital cost. To these must be added:
5. depreciation or replacement costs;
6. daily running costs;
7. labour costs; and
8. transport and erection costs.

For daily running costs, which is fuel etc., see page 176. For labour costs, see the estimating section of this book.

Transport and erection costs can be a considerable item. Most excavators and tractors need carrying on a low-loader. The erection of a tower crane is an engineering job that can take 50 hours for a small rail-mounted crane and several hundred hours for a very large tower crane.

When the above eight items have been costed out, the exact comparison between hiring and buying costs can be made, by totalling the items 1 to 6, dividing by 200 and comparing with the daily hire rate. If fuel and driver are included, add items 6 to 7. Item 8 is sometimes classed as an extra.

MACHINERY DATA

HORSEPOWER REQUIRED TO DRIVE BUILDERS' MACHINERY

Cranes		hp
Overhead cranes	(a) hoist	$12\frac{1}{2}$–20
12–15 tons	(b) travel	$7\frac{1}{2}$–20
Tower cranes, each movement separately motorised; per movement		3 – 7

Static Woodworking Machinery	
Circular saw, 15–36 in	$2\frac{1}{2}$–15
Crosscut, 24 in–36 in	3 –10
Universal woodworker, 5 in–7 in rip saw	3 – 5
Surfacer, 6–24 in width	2 – 5
Planer and thicknesser, up to 30 in × 9 in	4 –10
Spindle Moulder, up to 6 in	2 – 4
Mortising machine, up to 3 in × $1\frac{1}{4}$ in	2 – $7\frac{1}{2}$
Tenoning machine	5 – $7\frac{1}{2}$
Dovetailing machine	1
Router	2
Three-drum sander, 6 in width	20 –30
Disc sander, up to 48 in disc	6 – $7\frac{1}{2}$
Belt sander, up to 20 in belt width	5 – $7\frac{1}{2}$
Woodturning lathe	$1\frac{1}{2}$– $2\frac{1}{2}$

Note: The above horsepower ratings apply to individually-motorized machines powered by electric motors. For belt-driven machines, connected by shafting to one central power source, and for machines individually powered by petrol or diesel engines, other ratings obviously apply.

PLANT AND MACHINERY

FUEL CONSUMPTION OF MACHINES

Machine	Size	Petrol gals	Diesel gals
Mixer	5/3½	1	½
	7/5	1½	1
	10/7	2	1½
Hoist	(cwt)		
	5	2	1
	7/10	2¼	1¼
	12/15	2½	1½
	20	2¾	1¾
	30	—	2¼
Dumper	(cu. yd)		
	½/1	1¼	¾
	1½	—	1
	2	—	1¼
	3	—	1½
	4	—	1¾
Excavator	(cu. yd)		
	5/8	—	5¼
	3/4	—	7¼
	7/8	—	9¼
	1	—	11¼
Tractor	(bhp)		
	25	5½	2¾
	40	7½	5
	60	—	8¼
Shovel and Loader	(cu. yd)		
	5/8	7	2
	3/4	8	3
	1	10	5
Trencher	(16 to 60 in)	—	6
Pump (bore)	inches		
	3	2	1¾
	4	2½	2
	5	2½	2¼
	6	—	2½
Compressor	(cu. ft)		
	150	7	4¾
	200	8	5¾
	250	9¼	7
	315	10¾	8½
	510	—	12¾

AVERAGE OUTPUT IN CUBIC YARDS OF BUILDERS' MACHINES

Type of machine	Surface Soil	Soft clay	Heavy clay	Sand	Gravel	Soft rock
EXCAVATOR						
Skimmer						
$\frac{1}{4}$ yard bucket	11	9	7	11	8	5
$\frac{3}{8}$ yard bucket	18	14	12	19	13	9
$\frac{1}{2}$ yard bucket	24	19	16	26	18	14
$\frac{3}{4}$ yard bucket	42	33	28	46	31	23
Face Shovel						
$\frac{1}{4}$ yard bucket	14	11	9	15	10	7
$\frac{3}{8}$ yard bucket	23	18	16	25	17	12
$\frac{1}{2}$ yard bucket	32	26	22	34	24	17
$\frac{3}{4}$ yard bucket	55	44	37	59	41	30
Backactor						
$\frac{1}{4}$ yard bucket	10	8	7	10	7	5
$\frac{3}{8}$ yard bucket	16	13	11	17	12	9
$\frac{1}{2}$ yard bucket	22	17	14	23	16	13
$\frac{3}{4}$ yard bucket	37	29	25	39	27	20
Dragline						
$\frac{1}{4}$ yard bucket	14	12	10	15	10	6
$\frac{3}{8}$ yard bucket	23	19	16	26	17	11
$\frac{1}{2}$ yard bucket	33	25	21	35	24	16
$\frac{3}{4}$ yard bucket	55	44	36	60	41	28
TRENCHER						
John Allen 16/60						
Depth 1–3 ft	40	28	24	40	26	13
3–5 ft	48	34	28	48	31	16
5–7$\frac{1}{2}$ ft	60	42	36	60	39	16
7$\frac{1}{2}$–10 ft	55	38	33	55	36	13
10-12 ft	48	34	28	48	31	12
John Allen 12/21						
Depth 1–3 ft	25	18	15	25	17	8
3–5 ft	25	18	14	25	17	9
John Allen 140						
Depth 1–3 ft	66	53	33	66	43	20
3–5 ft	60	50	30	60	39	19
Cleveland 110 (USA)						
Depth 1–3 ft	20	16	13	22	16	11
3–5 ft	22	18	15	24	18	12
BULLDOZER						
D.4 50 ft push	36	30	27	39	34	—
100 ft push	22	18	16	24	22	—
200 ft push	13	10	9	13	12	—
D.6 50 ft push	45	37	33	49	42	—
100 ft push	30	24	21	33	28	—
200 ft push	16	14	12	18	15	—
D.8 50 ft push	100	82	72	110	94	—
100 ft push	70	57	50	75	65	—
200 ft push	35	28	25	37	32	—

PUMP PERFORMANCES

(IN GALLONS PER HOUR)

Bore	Type of Pump	Average gallons per hour
3 in	Single Diaphragm, engine-driven	3,600
4 in	Single Diaphragm, engine-driven	6,000
4 in	Double Diaphragm, engine-driven	12,000
5 in	Centrifugal Pump	24,000
6 in	Centrifugal Pump	36,000
8 in	Centrifugal Pump	48,000
3 in	Hand Pump	18,000

STEEL SCAFFOLDING

TUBES AND FITTINGS REQUIRED TO COVER 30 ft LONG × 30 ft HIGH OR 100 yds SUPER (9·1 m × 9·1 m, 83 m²)

	Tubing	Double couplers	Putlog couplers	Putlog heads	Base plates
	Feet	No off	No off	No off	No off
Bricklayers'—					
Putlog scaffolding	710	35	70	70	5
Independent scaffolding	1130	80	70	—	10
Plasterers' or Painters'—					
Putlog scaffolding	470	20	40	40	5
Independent scaffolding	800	50	80	—	10
Birdcage scaffolding (30 ft square × 30 ft high)	1860	144	40	—	16

8. THE CONCRETOR

The Design of Concrete Mixes	183
Maximum Aggregate Size	186
Materials Requirements	186
Steel Bars	189
Hooks and Passings in Reinforcement	189
Removal of Formwork	190
Ready-mixed Concrete	190
Prestressed Concrete	191

THE DESIGN OF CONCRETE MIXES

Within the last few years, a revolution in concrete making has taken place. The advent of high-strength concrete, and the general introduction of prestressed concrete, has led to greater control over the production of concrete on the site and in the factory. It has long been the complaint of engineers that concrete made in a laboratory could be up to ten times stronger than the same mix when made into concrete on a building site. "Quality control" of concrete is, however, expensive and can only be applied economically to large works: it does, however, produce a concrete of higher and more consistent strength from the same mix than can be achieved by ordinary methods, or will permit the use of a leaner mix to obtain the same strength as a concrete of uncontrolled quality. Very high strength concretes are required for prestressed concrete, and the achievement, under factory conditions and site conditions also, of such high and consistent strengths has led to the introduction, into civil engineers' specifications, of the concept of "guaranteed-strength concrete". The strength that the concrete is required to attain in 28 days is specified, and the contractor is allowed to use whatever mix will produce concrete of that required strength. Sometimes, a mix is suggested by the engineer, but the onus rests with the builder. He is free to adapt that mix, or use another. The secrets of such good concrete production are:

1. Batching by weight;
2. Strict control of water-cement ratio;
3. Correct grading of aggregate;
4. Care in placing and curing.

To achieve a high-strength concrete the mix should be designed in accordance with the instructions given in *Road Research Note No. 4*—"*The Design of Concrete Mixes* (H.M.S.O.). Readers are warned that this implies some familiarity with test procedure and the handling of a few simple scientific instruments.

The subject is too vast to summarize in these pages, and for the majority of concreting work would be too costly to put into practice. However, it has taken the place of the "absolute volume" method for large jobs.

Normal conversions, as given in the tables in previous sections, can be employed: but in some instances, the numbers will be so large as to be cumbersome and time-wasting when used in calculations. To avoid this as far as possible, some of the tables in this section have been worked to the following approximations: 50 kg = 1 cwt. 1 ton = 1 metric tonne. 1 cu. yd = $\frac{3}{4}$ cu. metre.

These approximations are sufficiently close for all practical purposes. A building site is not a cement-chemists' laboratory, and the values assumed will make no difference to the strength of the concrete.

For maximum strength concrete must be:
1. Adequately mixed;
2. Properly placed;
3. Consolidated by hand or vibration;
4. Adequately cured.

These are key operations: failure to observe any one of them can result in immense losses in final strength. Meticulous accuracy in proportioning can never be attained on site, unfortunately, but it is in no way as important as the four operations listed above.

APPROXIMATE AVERAGE WEIGHTS OF MATERIALS

Material	Voids	Loose weight		
		per cu.ft	per cu.yd	per m^3
	%	lb	tons	kg
Sand	39	100	1·2	1 661
Gravel $\frac{3}{8}$"–1"	45	90	1·1	1 441
Gravel 1$\frac{1}{2}$"–3"	42	97	1·2	1 553
Crushed stone	50	83	1·0	1 329
Crushed granite (under $\frac{3}{4}$")	50	84	1·0	1 345
Crushed granite ($\frac{3}{4}$" and over)	47	90	1·1	1 441
"All-in" ballast	32	112	1·3	1 794
Portland cement	54	90	1·1	1 441
Lime (lump)	73	45	·55	720
Lime (hydrated)	60	35	·42	560
Plaster (heavy)	57	62	·75	993
Plaster (light)	65	50	·60	880

MIXES FOR VARIOUS TYPES OF WORK

Mix	Class of work suitable for	Recommended Mix
A.	Roughest type of mass concrete, such as footings, road haunching over 12in thick, etc.	1 : 3 : 6
B.	Mass concrete of better class than A, such as bases for machinery, walls below ground, etc.	1 : 2$\frac{1}{2}$: 5

Mix	Class of work suitable for	Recommended Mix
C.	Most ordinary uses of concrete (usually specified 1 : 2 : 4), such as mass walls above ground, road slabs, etc., and general reinforced concrete work	1 : 2 : 4
D.	Watertight floors, pavements and walls, tanks, pits, steps, paths, surface of 2-course roads, reinforced concrete where extra strength is required.	1 : 1½ : 3
E.	Work of thin section, such as fence posts and small precast work.	1 : 1 : 2

Note: The above mixes are by proportions, by volume. The cement is given first, the fine aggregate next and the coarse aggregate last.

MIXES FOR VERY SMALL JOBS

Cement (loose) Buckets	Sand (damp) Buckets	Coarse Aggregate Buckets	Water (net) Buckets
A	4	6	¾
B	3¼	5	between ¾ and ½
C	2½	4	
D	2	3	
E	1¼	2	½

MIXES FOR A ONE BAG BATCH

MIX	Cement (bags) cwt	(kg)	Sand (damp) cu.ft	(m^3)	Coarse Agg. cu.ft	(m^3)	Water (net) gals	(litres)
A	1	50	5	0·14	7½	0·21	6	27
B	1	50	4	0·11	6¼	0·17	5½	24
C	1	50	3¼	0·09	5	0·14	*5	*22
D	1	50	2½	0·07	3¾	0·11	*4½	*20
E	1	50	1½	0·03	2½	0·06	4	18

*Increase slightly for reinforced concrete if not fluid enough for easy placing around reinforcement.

Note: If the proportions suggested do not give a 'workable mix' with the materials used vary slightly the quantities of sand and coarse aggregate, but do not add more water, except in the case noted above.

MAXIMUM AGGREGATE SIZE

For all reinforced concrete the coarse material should be graded from $\frac{3}{4}$in to $1\frac{1}{4}$in. For plain concrete the larger particles of coarse material should not be greater in diameter than one quarter of the thickness of the work. Maximum sizes usually suitable are: Mix A—$2\frac{1}{2}$in, Mix B—2in, Mix C—1in (roads $1\frac{1}{2}$in), Mix D—$\frac{3}{4}$in, Mix E—$\frac{1}{2}$in (or less if work is very thin). All sand should be graded from $\frac{3}{16}$in down with a good proportion of the larger particles.

PRECAUTIONS

Use clean water, clean sand and clean aggregate. Mix the materials thoroughly before adding the water and again afterwards. Place the concrete within 30 minutes of mixing; tamp and spade it well. Protect the concrete from sun and wind and keep it damp for 10 days.

The table is based on loose cement weighing 90lb per cu. ft; damp sand (30% bulked) 84lb per cu. ft; broken stone 90lb per cu. ft.

(*By courtesy of the Cement and Concrete Association*)

MATERIALS REQUIREMENTS

MATERIALS REQUIRED PER CU. YD OF CONCRETE

Based on Absolute Volume method, with cement weighing 90 lb per cu ft, sand 84 lb per cu. ft when damp and bulked 30 per cent gravel or shingle 109 lb per cu. ft, and broken stone 90 lb per cu. ft.

Type of Mix	Type of Coarse Aggregate	Cement lb	Sand (Damp) cu. yd	Sand (Damp) tons	Coarse Aggregate cu. yd	Coarse Aggregate tons	Bags of cement per cu. yd of combined aggregates
A	Shingle	335	0·54	0·55	0·83	1·09	3
	Broken stone	370	0·59	0·60	0·91	0·99	$3\frac{1}{2}$
B	Shingle	392	0·52	0·53	0·81	1·06	$3\frac{1}{2}$
	Broken stone	432	0·58	0·59	0·89	0·96	4
C	Shingle	481	0·15	0·52	0·79	1·04	4
	Broken stone	524	0·56	0·57	0·86	0·93	$4\frac{1}{2}$
D	Shingle	596	0·48	0·49	0·74	0·97	$5\frac{1}{2}$
	Broken stone	653	0·52	0·53	0·80	0·87	6
E	Shingle	813	0·43	0·44	0·67	0·88	$7\frac{1}{2}$
	Broken stone	880	0·47	0·48	0·72	0·78	8

MATERIALS REQUIRED PER CUBIC METER OF CONCRETE

Type of Mix	Type of coarse aggregate	Cement kg	Sand m³	Coarse Aggregate m³	No. of 50 kg bags of cement per m³ of combined aggregates
A	Shingle	203	0·54	0·83	4
	Broken stone	213	0·59	0·91	4¼
B	Shingle	237	0·52	0·81	4¾
	Broken stone	261	0·58	0·89	5¼
C	Shingle	291	0·55	0·79	6
	Broken stone	317	0·56	0·86	6¼
D	Shingle	361	0·48	0·74	7¼
	Broken stone	380	0·52	0·80	7¾
E	Shingle	491	0·43	0·67	10
	Broken stone	531	0·47	0·72	11

Note: A rough approximation of these mixes is obtained by using the number of bags of cement given in the last column, to 1 cu. yd of combined aggregates or "all-in" ballast. To use this table, multiply the cubic yardage of concrete required by the amount of cement in pounds, and the amount of coarse aggregate and cement in yards or tons.

APPROXIMATE STRENGTH AND WATER CONTENT
of concrete with a workable slump of 2-4in

Nominal Mix	Water/cement Ratio by weight	Compressive Strength (28 days)	WATER per cu yd of concrete	WATER per bag of cement	MN/m/²	Litres per 50 kg bag of cement
		lb/sq.in	gals	gals		
1 : 1 : 2	·40	5300	35⅔	4½	38·66	20·4
1 : 3	·44	5000	34	5	33·33	22·7
1 : 1¼ : 2½	·45	4850	33⅓	5	32·33	22·7
1 : 3½	·47	4700	32½	5¼	31·33	23·8
1 : 1½ : 3	·49	4500	31¾	5½	30·00	24·9
1 : 4	·50	4400	31	5½	29·33	24·9
1 : 2 : 3	·52	4250	30¾	5¾	28·33	26·1
1 : 1½ : 4	·54	4000	30	6		
1 : 1½ : 4½ **1 : 5** **1 : 2 : 4**	·57	3800	29½	6⅓	25·33	29·5
1 : 2½ : 4	·60	3500	28¾	6⅔	23·33	29·75
1 : 6	·63	3250	28	7	21·66	31·8

Nominal Mix	Water/cement Ratio by weight	Compressive Strength (28 days)	WATER per cu yd of concrete	WATER per bag of cement	MN/m^2	Litres per 50 kg bag of cement
		lb/sq.in	gals	gals		
1 : 2½ : 5	·65	3100	28	7⅓	20·66	32·3
1 : 3 : 5	·68	2800	27½	7⅔	18·33	34·8
1 : 7	·70	2650	27¼	7¾	17·66	35·2
1 : 3 : 6	·73	2400	26¾	8¼	16·00	37·5
1 : 9	·83	1500	26	9¼	10·00	42·0
1 : 3 : 9 ⎫ 1 : 10 ⎬	·89	1000	25½	10	6·66	45·4
1 : 4 : 8	·90	800	25⅓	10	5·33	45·4

Note 1: The compressive strength is related to the water/cement ratio. The water/cement ratio remains the same for metric as for imperial quantities.

Note 2: If strength is important and it is desired to gain greater workability, it is better to increase the proportions of cement and sand in the mix than to increase the water content.

Note 3: The nearest 'All-in' mix to a given three-part mix is that which gives the nearest equivalent water/cement ratio.

Note 4: The strengths of 'All-in' mixes should be assumed as being lower than shown, because of the variability of aggregate grading.

TABLE 1. PREFERRED TYPES OF STEEL FABRIC

BS reference	Mesh sizes (nominal pitch of wires)		Wire Sizes		Cross-sectional area per metre width		Nominal mass per square metre
	Main	Cross	Main	Cross	Main	Cross	
SQUARE MESH FABRIC							
	mm	mm	mm	mm	mm2	mm2	kg
A393	200	200	10	10	393	393	6.16
A252	200	200	8	8	252	252	3.95
A193	200	200	7	7	193	193	3.02
A142	200	200	6	6	142	142	2.22
A 98	200	200	5	5	98.0	98.0	1.54
STRUCTURAL MESH FABRIC							
B1131	100	200	12	8	1131	252	10.9
B 785	100	200	10	8	785	252	8.14
B 503	100	200	8	8	503	252	5.93
B 385	100	200	7	7	385	193	4.53
B 283	100	200	6	7	283	193	3.73
B 196	100	200	5	7	196	193	3.05
LONG MESH FABRIC							
C785	100	400	10	6	785	70.8	6.72
C503	100	400	8	5	503	49.0	4.34
C385	100	400	7	5	385	49.0	3.41
C283	100	400	6	5	283	49.0	2.61
WRAPPING FABRIC							
D98	200	200	5	5	98.0	98.0	1.54
D49	100	100	2.5	2.5	49.1	49.1	0.770

NOTE 1. Cross wires for all types of long mesh may be of plain hard drawn steel wire.
NOTE 2. 5 mm size is available in hard drawn wire only. The 6, 7, 8, 10 and 12 mm sizes are available in either hard drawn wire or in bar form. The 7 mm size is not a preferred size of cold worked bar.

THE CONCRETOR

TABLE 2. RECOMMENDED FABRIC FOR CARRIAGEWAYS

BS reference	Mesh sizes (nominal pitch of wires)		Wire sizes		Cross-sectional area per metre width		Nominal mass per square metre
	Main	Cross	Main	Cross	Main	Cross	
	mm	mm	mm	mm	mm2	mm2	kg
C636	80-130	400	8-10	6	636	70.8	5.55

STEEL BARS
SIZES AND PROPERTIES
(B.S. 4449M; 1969)

Size mm	Cross-sectional area mm^2	Weight per meter run kg
6	28·3	0·222
8	50·3	0·395
10	78·5	0·616
12	113·1	0·888
16	201·1	1·597
20	314·2	2·466
25	490·9	3·854
32	804·2	6·313
40	1256·6	9·864
50	1963·5	15·413

Note: B.S. 4461M "Cold worked steel bars for the reinforcement of concrete" gives identical data with above, except that the 50 mm size is not included.

WEIGHT IN POUNDS PER FOOT RUN OF ROUND REINFORCING BARS

Diam in inches	Weight in lbs per foot	Diam in inches	Weight in lbs per foot	Diam in inches	Weight in lbs per foot	Diam in inches	Weight in lbs per foot
$\tfrac{3}{16}$	·094	$\tfrac{9}{16}$	·845	$1\tfrac{1}{16}$	3·015	$1\tfrac{9}{16}$	6·52
$\tfrac{1}{4}$	·167	$\tfrac{5}{8}$	1·043	$1\tfrac{1}{8}$	3·380	$1\tfrac{5}{8}$	7·05
$\tfrac{9}{32}$	·211	$\tfrac{11}{16}$	1·262	$1\tfrac{3}{16}$	3·77	$1\tfrac{11}{16}$	7·6
$\tfrac{5}{16}$	·261	$\tfrac{3}{4}$	1·502	$1\tfrac{1}{4}$	4·172	$1\tfrac{3}{4}$	8·18
$\tfrac{11}{32}$	·316	$\tfrac{13}{16}$	1·763	$1\tfrac{5}{16}$	4·6	$1\tfrac{13}{16}$	8·77
$\tfrac{3}{8}$	·376	$\tfrac{7}{8}$	2·044	$1\tfrac{3}{8}$	5·05	$1\tfrac{7}{8}$	9·39
$\tfrac{7}{16}$	·511	$\tfrac{15}{16}$	2·347	$1\tfrac{7}{16}$	5·52	2	10·68
$\tfrac{1}{2}$	·668	1	2·670	$1\tfrac{1}{2}$	6·008	—	—

HOOKS AND PASSINGS IN REINFORCEMENT

The girth of a standard hook is 12d, where d = the diameter of the rod.

The length of overlap, or passing, to secure adequate bond in 1 : 2 : 4 concrete (bond stress 100lb per sq. in and steel stress 18,000lb per sq. in) is 59d. The minimum overlap in longitudinal steel in columns is 24d.

The necessary bond length (with or without hook) depends upon the mix of the concrete and the local stress on the steel. Minimum end length is 14d straight, or a hook equivalent to 12d.

B.S. 4466: Bending dimensions and scheduling of bars for the reinforcement of concrete: gives minimum hook and bend allowances as 9d, taken to the nearest 10 mm over or not less than 100 mm, to be added to the overall length of the bar as when bent. These dimensions are hooks with radius of 2d for mild steel and 3d for high-yield bars. The Standard gives illustrations of some 35 different shapes, with full details of bending dimensions, and a suggested form of bar schedule.

REMOVAL OF FORMWORK

Formwork in parts of building	Days before formwork should be struck			
	Ordinary concrete		Rapid-hardening concrete	
	Just above freezing	60° F (15·6 C)	Just above freezing	60° F (15·6 C)
Beams (except soffits) and walls	6	2	5	1
Beam soffits	14	7	10	4
Slabs	10	3	7	2

Props should be left under beam soffits and slabs, and be removed 1 week after. When temperature drops below freezing point, add the days of such temperature to the times given above.

READY-MIXED CONCRETE

Concrete can now be obtained, in any quantity from $1\frac{1}{3}$ cubic yards upwards, delivered ready-mixed to the building site. Any type of mix, containing any variety of cement or aggregate, can be now obtained from upwards of 100 firms in the larger cities in Great Britain. The cement and aggregate are loaded into mobile mixers at the factory; during the journey from factory to site the mix is agitated, water is added and freshly-mixed concrete is thus ready when the mixer-lorry reaches its destination. Cost is often greater than materials mixed on site, but for "rush" jobs, confined sites or contractors with small labour-forces, there is often a great saving in time, and therefore in cost. It should be remembered that this material can only be used profitably by very careful pre-planning, and working out, with the supplier, a schedule of arrivals of truck-mixers.

This pre-planning must also be carried out on the building site. Arrangements should be made to have sufficient labourers to cope with the placing of the concrete, as it arrives. Suppliers allow a 20 minute wait for their trucks, after which they charge waiting time. A truck-mixer is 8 ft wide, 11 ft high, 24 ft long and weighs 20 tons. Adequate access must be provided on the site including bridging and metallised ("Sommerfield track") or sleeper roads, if necessary. It is well to remember that the discharge chute allows concrete to be deposited 9 ft from the mixer.

Special mixes, suitable for pumping can be supplied. Concrete can be lifted 5 to 6 storeys by concrete pumps. If used in conjunction with ready-mixed concrete, spectacular results can be obtained. For example, over 150 cubic yards of pumped concrete have been placed in one day, using only two of the builder's own men. Most suppliers have technicians and engineers on their staff, who can advise on the type of mix and on problems that arise on the site also.

PRESTRESSED CONCRETE

Prestressed concrete is concrete compressed by an initial tension given to the reinforcement. Steel thus tensioned or stretched tends to return to its normal, unstretched length, just as does a piece of elastic. This tendency induces, by reason of the position of the reinforcement in the slab or beam, tension in the top half of the beam or slab and compression in the bottom. These stresses must be reversed before the beam or slab begins to act as a normal reinforced concrete unit: in practice, the load is never enough to reverse the stresses completely.

Prestressing makes better use of the strength of the concrete than does normal reinforcement. Concrete, though strong in compression, is weak in tension and a normal concrete beam always tends to crack at the bottom because, when loaded, the reinforcing steel stretches as it takes up tension and thus allows the bottom of the beam to crack. Prestressing so compresses the concrete that, when under load, there will be no tension in it, so that the concrete will not crack. In short, the concrete will always be acting in compression, where it is strongest. This means that full advantage is taken of the strength of concrete in compression, so that a prestressed concrete beam will always be smaller in size than a reinforced concrete one, designed to take the same load. The weight of concrete and steel will be less and this reduction in the dead or self-load of a structure may well reduce the size of foundations and supports. Because they are smaller, prestressed components will weigh less than normally reinforced ones and this is a vital consideration in the handling of components in industrialised building and of large beams used in bridge construction.

Two methods of prestressing are currently employed; pre-tensioning and post-tensioning. For pre-tensioning, cold-drawn steel wires of very high tensile strength are used. These wires are stretched very tightly, by means of jacks, between two fixed abutments and concrete is then poured round them. When the concrete is set and hardened, the wires are released from the abutments. They immediately try to return to their original length, but are prevented from so doing by the grip or bond of the concrete on them, with the result that the concrete is powerfully compressed.

In post-tensioning, the concrete is cast first, either in one piece or in several units. Holes are formed for the insertion of high-tensile steel wire cables or rods of special alloy steel. When the concrete is set or, if in units, when these are assembled to form a beam, the cables or rods are inserted, stretched by jacks and then anchored in this stretched condition to the concrete. The steel then acts to compress the concrete or concrete units.

9. THE MASON

Waterproofing Stone and Other Walling	194
British Standards	194
Classification of Stone	194
Masonry Data	196
The Laying of Masonry	196

WATERPROOFING STONE AND OTHER WALLING

The application of waterproofing compounds to stonework or to old brickwork, especially if pointed or laid in lime mortar, should be done only after consultation with an expert technician, either an architect well versed in the subject, or a building scientist or consultant. Care must be taken to see that all work is pointed and that the walling is sound, before applying waterproofing liquids. The best of these preparations have a life of approximately five years, when the treatment must be renewed. They can sometimes be used with effect to waterproof porous window-sills and other projections, but the treatment should be renewed every three years. Much was expected of liquids containing silicone compounds. These are very successful for about three to six months, after which their efficacy is much reduced, and then is generally no better than the best of the older type of liquid waterproofers.

Damp walls can only be waterproofed by eliminating the source of moisture, which often entails inserting a dampcourse in the wall itself, or flashing all projections where water lodges with lead. Crumbling stone, due to being set on its wrong bed, or to being worked after the "quarry sap" has dried, cannot easily be waterproofed.

BRITISH STANDARDS

B.S. 1240: Natural stone lintels. Gives breadths and depths, for stone of three ranges of hardness, for openings up to 8ft.

B.S. 3798: Coping units (of clayware, unreinforced cast concrete, unreinforced cast stone, natural stone and slate) specifies material, workmanship, functional requirements and dimensions.

B.S. 4374: Sills of clayware, cast concrete, cast stone, slate and natural stone. (See table overleaf.)

Note: The Standards are designed for housing and similar work, using standard doorframes and windows. The finish is specified as sawn or rubbed.

CLASSIFICATION OF STONE

Natural stones can be broadly classified as:

1. Granites and other igneous rocks;
2. Limestones, including marbles;
3. Sandstones.

The only valid guide to the suitability of stones in classes 2 and 3 is by inspection of buildings constructed with such stone. Every quarrymaster will be glad to provide a list of buildings in which his stone has been used. Granites and other like rocks are suitable for plinth courses. Stone for

TABLE 1. MINIMUM DIMENSIONS

Material	*AA**		*A*		*B*		*C*		*D*		*E*		*F*	
	in	mm	in	mm	in	mm	in	mm	in	mm	in	mm	in	mm
Clayware	2⅛	54	2	51	½	13	½	13	½	13	5/16	8	½	13
Cast concrete	1⅞	48	1¾	44	⅝	16	½	13	½	13	5/16	8	½	13
Cast stone	1⅞	48	1¾	44	⅝	16	½	13	½	13	5/16	8	½	13
Natural stone	1 13/16	46	1¾	44	½	13	½	13	½	13	5/16	8	½	13
Slate	13/16	21	¾	19	⅝	16	½	13	½	13	5/16	8	½†	13

* The dimensions given in column *AA* allow for the tolerances specified for each particular material and ensure the minimum finished dimension shown in Column *A*.

† This upstand may be made of other material approved by the purchaser.

parapets, cornices, string courses, sill and buttress weatherings should be chosen with care, and must withstand frost. CP 121: 201 gives a large number of excellent illustrations of stonework details, which illustrate its requirements in detail, but which cannot be reproduced here.

MASONRY DATA

Ashlar masonry requires about one-eighth its bulk of mortar in laying and rubble masonry about one-fifth its bulk.

About one-sixth waste is an average figure allowed in the laying of rubble masonry.

One ton or tonne of flints will face an average of 33 sq. ft 3.06 m^2 if used whole and 50 sq. ft 4.6 m^2 if split or knapped.

One ton of Kentish ragstone, as hammer-dressed random squared rubble or random coursed rubble, will cover about 3 yds super/2.6 m^2 (without backing).

A cu. yd/0.76 m^3 of walling (usually 18 in thick/450 mm) requires 32 cu. ft/0.93 m^3 undressed stone for rubble walling, 36 cu. ft/1.02 m^3 for coursed rubble walling and 30 cu. ft/0.85 m^3 or rough flints for flint walling.

The most satisfactory cramps and keys used in masonry work are made of copper, bronze, gunmetal or slate.

For external work, a good building stone should be dense, hard and comparatively non-absorbent.

As a general rule, a stone that absorbs more than 10 per cent of its own weight in twenty-four hours is undesirable, as most of the impurities in the atmosphere are soluble in water. They can therefore attack the stone and cause disintegration.

THE LAYING OF MASONRY

Laminated stones (*i.e.* ordinary stones other than granite) are liable to flake off along the lines of lamination and should therefore be laid, as a general rule, with the lines of lamination horizontal in ordinary walling, thus providing the greatest resistance, as the load to be carried will be at right angles to the lines of lamination.

In stone voussoirs for arches the lines of lamination are laid normal to the curve of the arch and at right angles to the face, again giving the greatest resistance to the line of thrust of the arch.

In cornices, mouldings and undercut work, however, it is more usual to edge-bed the stone, that is, to lay the stone with the lines of lamination vertical and at right angles to the face. In this way the spalling off of small areas on the face of the work is prevented, but this method exposes the ends of the lines of lamination on the upper and projecting surface, and to prevent the penetration of moisture such surfaces should always be covered with lead.

DETAILS OF STONES

Stone	Descriptions	Crushing Strength in tons per sq ft
Sandstones		
Robin Hood	Bluish-grey in colour—even grained, weathers well; suitable for sills, etc.	574
Bramley Full	Coarse grained—hard and of excellent quality for heavy work	552
Darley Dale	Very hard and close grained—light grey in colour and of fine weathering qualities. Also known as "Stancliffe".	517
Limestones		
Portland	*Basebed.* Pale cream in colour, moderately weatherproof	205
	Whitbed. Has a white or brown hue—harder than basebed. Takes a good arris and weathers exceptionally well	218
	Roachbed. Harder than other two but gives a broken appearance in mass	
Bath	General colour creamy buff, but in several varieties. Harder stone, suitable for most classes of building work	223 to 62
Ancaster	"*Free Bed*". Fine grained, working—cream coloured.	218
	"*Weather Bed*". Dark brown bluish-grey—a good general purpose stone	
Clipsham	A lime stone quarried in Rutland. The shelly variety is darker, and suitable for external work	272
Bolsover Moor	A magnesian limestone, yellowish-brown, quarried in Derbyshire	484
Granites		
Rubislaw	Generally very hard and durable and very suitable for plinths and lower courses of a building. Its hardness makes it costly to work. Will take a high polish.	1,290
Peterhead		1,208
Cornwall		1,060

Collections of building stone samples can be seen at the Science Museum, London, the Building Centre, London, and at Cambridge.

10. THE BRICKLAYER

Codes of Practice	199
British Standards	199
Brick Requirements	209
Quantities of Bricks and Mortar Used in Walling	210
Mortar for Pointing	212
Materials for Damp-proof Coursing	213

CODES OF PRACTICE

CP 101: 1972: Foundations and substructures for non-industrial buildings of not more than four storeys.

Information on site exploration, effects of weather, ground movement, bearing capacities and types of foundation.

CP 102: 1973: Protection of buildings against water from the ground.
  Part 1: prevention at or below ground floor level.
  Part 2: damp-proofing walls and floors at ground level by use of dpc materials to B.S. 743.

CP 111: 1970: Structural recommendations for loadbearing walls.

Covers walls of brickwork and blockwork. Requirements for loading, load dispersion, lateral support, effective height, length and thickness. (See tables, pages 197-199.)

CP 121: Part 1: 1973: Brick and block masonry.

Covers erection of brick and block walls. Gives data on fire resistance, thermal transmittance and sound insulation and makes recommendations on excluding rain, reducing cracking, durability, mixing mortars, jointing and pointing.

CP 122: Walls and partitions of blocks and slabs.
  Part 1: 1966: deals with hollow glass blocks forming walls and partitions.

CP 131: 1974: Chimneys and flues for domestic appliances burning solid fuel.

Design and construction of natural draught chimneys and flues built of brick.

B.S. 5628: 1979: Code of Practice for the structural use of masonry.
  Part 1: the use of blocks and brickwork as structural material (can be used as an alternative to CP 111).

BRITISH STANDARDS

B.S. 187: Calcium silicate (sandlime and flintlime) bricks.

Covers materials, sizes (standard and dimensionally co-ordinated) appearance and methods of sampling and testing.

REQUIREMENTS FOR CALCIUM SILICATE BRICKS (SANDLIME AND FLINTLIME) B.S. 187

Class	Use	Minimum mean compressive strength (wet) of 10 bricks N/mm²	Minimum predicted lower limit of compressive strength. N/mm²	Maximum* Average drying test shrinkage of 3 bricks. %	Appearance
7	Brickwork that will be heavily loaded, repeatedly exposed to frost while in a saturated condition or continuously wet (e.g. normal under-building)	48.5	40.5	0.04	All faces reasonably free from visible cracks and noticeable balls of clay, loam and lime. Bricks specified as for facing purposes to be of agreed colour and texture with the arrises of one face and end free from damage.
6		41.5	34.5		
5		34.5	28.0		
4		27.5	21.5		
3		20.5	15.5		
2	General use, external and internal.	14.0	10.0		

*No limit where the bricks are to be used under permanently damp conditions or for Class 2 commons.

(Published by Calcium Silicate Brick Association)

TABLE 1. STIFFENING COEFFICIENT FOR WALLS STIFFENED BY PIERS

Pier spacing (centre to centre) Pier width	Pier thickness Effective wall thickness (see Subclause 307a)		
	1	2	3
6	1·0	1·4	2·0
10	1·0	1·2	1·4
20	1·0	1·0	1·0

Linear interpolation between the values given in this table is permissible, but not extrapolation outside the limits given.

TABLE 2. SPACING OF TIES

Least leaf thickness (one or both)	Cavity width	Spacing of ties	
		Horizontally	Vertically
mm	mm	mm	mm
75	50– 75	450	450
90 or more	50– 75	900	450
90 or more	75–100	750	450
90 or more	100–150	450	450

The spacing may be varied provided that the number of ties per unit area is maintained.

TABLE 3b. BASIC COMPRESSIVE STRESSES FOR BLOCKWORK MEMBERS (AT AND AFTER THE STATED TIMES)

Description of mortar	Mix (parts by volume)			Hardening time after completion of work† (days)	Basic stress in MN/m^2 corresponding to units whose crushing strength (in MN/m^2)‡ is:								
	Cement	Lime	Sand		2·8	3·5	7·0	10·5	14·0	21·0	28·0	35·0	52·0
Cement	1	0–¼*	3	7	0·28	0·35	0·70	1·05	1·25	1·70	2·10	2·50	3·50
	1	½	4½	14	0·28	0·35	0·70	0·95	1·15	1·45	1·75	2·10	2·80
Cement-lime	1	1	6	14	0·28	0·35	0·70	0·95	1·10	1·35	1·60	1·90	2·50
Cement with plasticizers§	1	—	6										
Masonry cement‖	—	—	—										
Cement-lime	1	2	9	14	0·28	0·35	0·55	0·85	1·00	1·20	1·45	1·70	2·05
Cement with plasticizers§	1	—	8										
Masonry cement‖	—	—	—										
Cement-lime	1	3	12	14	0·21	0·23	0·49	0·70	0·80	1·00	1·20	1·40	1·70
Hydraulic lime	—	1	2	14	0·21	0·23	0·49	0·70	0·80	1·00	1·20	1·40	1·70
Non-hydraulic	—	1	3	28¶	0·21	0·23	0·42	0·55	0·60	0·70	0·75	0·85	1·05

* The inclusion of lime in cement mortars is optional.
† These periods should be increased by the full amount of any time during which the air temperature remains below 4·4 °C, plus half the amount of any time during which the temperature is between 4·4 °C and 10 °C.
‡ Linear interpolation is permissible for units whose crushing strengths are intermediate between those given in the table.
§ Plasticisers must be used according to manufacturers' instructions.
‖ Masonry cement mortars must be used according to manufacturers' instructions, and mix proportions of masonry cement to sand should be such as to give comparable mortar crushing strengths with the cement: lime: sand mix of the grade.
¶ A longer period should ensue where hardening conditions are not very favourable.

B.S. 493: Airbricks and gratings for wall ventilation.
  Part 1: 1967: imperial units.
  Part 2: 1970: metric units, types and dimensions, wall-hold, etc.

B.S. 743: 1970: Materials for damp-proof courses.
  Selection and laying of damp-proof courses; mortar for bedding or laying of courses.

TABLE 1. COMPOSITION OF BITUMEN DAMP-PROOF COURSES AND SHEETING

1	2	8
Ref.	Type of damp-proof course	Minimum weight of complete material excluding packaging
		kg/m²
A	Hessian base	3.8
B	Fibre base	3.3
C	Asbestos base	3.8
D	Hessian base and lead	4.4
E	Fibre base and lead	4.4
F	Asbestos base and lead	4.9
G	Bitumen sheeting	5.4

B.S. 1180: 1972: Concrete bricks and fixing bricks.
  Bricks from natural and lightweight aggregates and cement.

B.S. 1181: 1971: Clay flue linings and flue terminals.
  Materials, workmanship, design, construction, dimensions, and tests etc.

B.S. 1198: 1199: and 1200: 1976: Building sands from natural sources.
  B.S. 1200 covers sands for mortar for plain and reinforced brickwork.

Mortars for brickwork and blockwork.

Table 1 Choosing the mortar

Type of construction	*Time of construction	Mortar designation									
		For clay units					For calcium silicate concrete units				
		I	II	III	IV	V	I	II	III	IV	V
External walls outer leaf of cavity walls – above dpc	Winter			•					•		
	Summer			•						•	
External walls – below dpc	All		•	•					•		
Backing to external solid walls	Winter			•	•				•	•	
	Summer				•					•	
Internal walls and inner leaf of cavity walls – facing	Winter			•	•				•	•	
	Summer				•					•	
Internal walls and inner leaf of cavity walls – non facing	Winter			•	•				•	•	
	Summer				•	•				•	•
Copings and cills	All	•						•			
Parapets and domestic chimneys – rendered	Winter		•	•					•		
	Summer		•	•						•	
Parapets and domestic chimneys – not rendered	Winter	•							•		
	Summer		•	•					•		
Retaining walls	All	•						•			
External free-standing walls	All		•	•					•		
Manholes	All		•						•		

*as a general guide summer is April to October, winter is November to March

Table 2 Equivalent mortar mixes

Mortar designation	Cement: lime: sand	Cement: coarse stuff*	Masonry cement: sand	Cement: sand with plasticizer
I	1:0-¼:3	1:3	—	—
II	1:½:4-4½	1:4-4½	1:2½-3½	1:3-4
III	1:1:5-6	1:5-6	1:4-5	1:5-6
IV	1:2:8-9	1:8-9	1:5½-6½	1:7-8
V	1:3:10-12	1:10-12	1:6½-7	1:8

*Coarse stuff may be purchased as ready-mixed lime sand for mortar.
Where alternative proportions of sand are shown eg 4-4½ these relate to the grading of the sand. Use the lower limit if the sand is either coarse or uniformity fine - use the higher limit if the sand is well-graded.

P.D. 6472: 1974: Guide to specifying the quality of building mortars.

General considerations, forms of specification, storing, batching, mixing, use of mortar, choice of proportions, performance, specifications, etc.

B.S. 1758: 1966: Fireclay refractories (bricks and shapes).

Chemical composition, dimensional and warpage limits.

B.S. 2028: 1364: 1968: Precast concrete blocks.

Specifies precast solid, hollow and cellular blocks and gives intended uses.

TABLE 1. DIMENSIONS

1	2	3	4
Block	Length × Height		Thickness (Work size)
	Co-ordinating* size	Work† size	
	mm	mm	mm
Type A	400 × 100 400 × 200	390 × 90 390 × 190	75, 90, 100, 140 and 190
	450 × 225	440 × 215	75, 90, 100, 140, 190 and 215
	400 × 100 400 × 200	390 × 90 390 × 190	75, 90, 100, 140 and 190
Type B	450 × 200 450 × 225 450 × 300 600 × 200 600 × 225	440 × 190 440 × 215 440 × 290 590 × 190 590 × 215	75, 90, 100, 140, 190 and 215

Block	Length × Height		Thickness (Work size)
	Co-ordinating* size	Work† size	
	mm	mm	mm
Type C	400 × 200 450 × 200 450 × 225 450 × 300 600 × 200 600 × 225	390 × 190 440 × 190 440 × 215 440 × 290 590 × 190 590 × 215	60 and 75

NOTE 1. Blocks of work size 448 mm × 219 mm × 51, 64, 76, 102, 152 or 219 mm thick, and 397 mm × 194 mm × 75, 92, 102, 143 and 194 mm thick will be produced as long as they are required.

NOTE 2. If blocks of entirely non-standard dimensions or design are required the limits of size or the design shall be agreed. Such blocks shall then be deemed to comply with this standard provided they comply with the other requirements.

* Co-ordinating size. A size of the space, bounded by co-ordinating planes, allocated to a component, including the allowance for joints and tolerances.
† Work size. A size of a building component specified for its manufacture to which its actual size should conform within specified permissible deviations.

B.S. 3679: 1963: Acid resisting bricks and tiles.

Dimensions and limits of porosity.

2. SIZES

Standard sizes of finished bricks and tiles shall be as follows:

Bricks:

in	mm
8⅝ × 4⅛ × 2⅝	219 × 105 × 67
8⅝ × 4⅛ × 2⅞	219 × 105 × 73
8⅝ × 4⅛ × 2	219 × 105 × 51
8⅝ × 4⅛ × 1½	219 × 105 × 38

Tiles:

in	mm
8⅝ × 4⅛ × ⅞	219 × 105 × 22
6 × 6 × ⅞	152 × 152 × 22
6 × 6 × ½	152 × 152 × 13
6 × 6 × ¾	152 × 152 × 19
6 × 6 × 1	152 × 152 × 25
6 × 6 × 1¼	152 × 152 × 32

The manufacturer shall be permitted to correct dimensional faults by machining if necessary.

B.S. 3798: 1964: Coping units (of clayware, unreinforced cast concrete, unreinforced cast stone, natural stone and slate).

Requirements, materials, workmanship and dimensions.

B.S. 3921: 1974: Clay bricks and blocks.

Requirements for clay bricks and blocks (See tables overleaf.)

TABLE 1. STANDARD FORMAT (BRICKS)

Designation	Work size		
	Length	Width	Height
	mm	mm	mm
225 × 112.5 × 75	215	102.5	65

TABLE 2. STANDARD FORMATS (BLOCKS)

Designation	Work size		
	Length	Width	Height
	mm	mm	mm
200 × 62.5 × 225	290	62.5	215
300 × 75 × 225	290	75	215
300 × 100 × 225	290	100	215
300 × 150 × 225	290	150	215

In addition, half blocks, 140 mm long, and three-quarter blocks, 215 mm long, shall be available for bonding.

TABLE 3. LIMITS OF SIZE (24 BRICKS)

Work size (see Table 1)	Overall measurement of 24 bricks	Limits of size	
		Maximum	Minimum
mm	mm	mm	mm
215	5160	5235	5085
102.5	2460	2505	2415
65	1560	1605	1515

NOTE 1. This method of measurement is also recommended for non-standard bricks. The limits applied to the length, width and height sizes should then be directly proportional to those specified for the corresponding dimensions of the standard brick.
NOTE 2. In the past certain manufacturers have, by special arrangement, supplied bricks to closer limits than those quoted. Where for special reasons closer limits are required it is therefore suggested that this can best be done by agreement between the user and the manufacturer on the basis of the latter's routine control charts of brick sizes.
NOTE 3. See also Appendices A† and B‡.

D.D. 34: 1974: Clay bricks with modular dimensions.
Requirements for clay bricks with modular dimensions.

TABLE 1. MODULAR FORMATS

Designation	Work size		
	Length	Width	Height
	mm	mm	mm
300 × 100 × 100	288	90	90
200 × 100 × 100	190	90	90
300 × 100 × 75	288	90	65
200 × 100 × 75	190	90	65

TABLE 2. LIMITS OF SIZE (24 BRICKS)

Work size (see Table 1)	Overall measurement of 24 bricks	Limits of size	
		Maximum	Minimum
mm	mm	mm	mm
65	1560	1605	1515
90	2160	2205	2115
190	4560	4626	4494
288	6912	7012	6812

Notes It is apparent from the evidence of control charts made available to the committee that some bricks such as hand made and stock bricks which are sought after for their other properties have some difficulty in meeting the present dimensional specification consistently. Others could meet a closer specification, although not 100% of the time; manufacturers of these bricks have sometimes, in the past, supplied bricks to closer tolerance by special arrangement with the user.

The limits and acceptance clauses in B.S. 3921: 1974 have been based on notional tolerances for individual bricks of ± 6.4 mm on length and ± mm on width and thickness. In this Draft for Development the tolerances and standard deviations for the 65 mm and 90 mm work sizes have been assumed to be equal to those for the 65 mm and 102.5 mm work sizes in B.S. 3921: 1974. The tolerances and standard deviations for the 190 mm and 288 mm work sizes have been claculated from those for the 215 mm work size by assuming that the parameters concerned are proportional to work size. This assumption appears reasonable.

If in the light of future experience the assumption is found to be incorrect, the tolerances may then be altered. In the meantime manufacturers may still, as in the past, supply bricks to closer tolerances by special arrangement with the user.

B.S. 4729: 1971: Shapes and dimensions of special bricks.
Shapes and dimensions for a special range of bricks.

BRICK REQUIREMENTS
PER UNIT AREA OF HALF BRICK THICK WALL SURFACE

Joint thickness	Per square yard			Per square metre		
	6 mm	10 mm	13 mm	6 mm (0·24″)	10 mm (0·39″)	13 mm (0·51″)
BRICK FACE						
215 mm × 65 mm	53·3	49·5	47·0	63·7	59·3	56·2
290 mm × 65 mm	39·8	37·2	35·4	47·7	44·5	42·4
215 mm × 70 mm	49·8	46·5	44·2	59·5	55·6	52·9
290 mm × 90 mm	29·4	27·9	26·8	35·3	33·3	32·0
215 mm × 50 mm	67·6	61·9	58·2	80·8	74·1	69·6
190 mm × 90 mm	44·4	41·8	40·0	53·1	50	47·8
190 mm × 65 mm	60·1	55·7	52·8	71·9	66·7	63·3

(No allowance for cutting and wastage)

(*Source: The Brick Development Association*)

For walls of more than one thickness the number of bricks required can be obtained by multiplying the appropriate value in the above table, by the thickness multiple, *e.g.*:

Wall thickness 327 mm using 215 × 102.5 × 65 mm brick = $\frac{327}{102.5}$

= 3 approx. allowing for collar joint thickness. Therefore number of bricks required per m^2 = 59.5 × 3.

MORTAR REQUIRED FOR WALLS OF HALF OR WHOLE BRICK THICKNESS

Size of Brick (mm)	Joints (mm)	Cubic Metre of Mortar per Metre Super of Brickwork			
		Half Brick		Whole Brick	
		Wirecut	Frogged	Wirecut	Frogged
215 × 102·5 × 65	6	0·011	0·016	0·035	0·048
215 × 102·5 × 70	6	0·011	0·015	0·034	0·045
215 × 102·5 × 50	6	0·013	0·021	0·040	0·055
190 × 90 × 90	6	0·008	0·013	0·029	0·040
190 × 90 × 65	6	0·010	0·017	0·033	0·047
215 × 102·5 × 65	10	0·018	0·024	0·048	0·060
215 × 102·5 × 70	10	0·017	0·022	0·046	0·056
215 × 102·5 × 50	10	0·021	0·028	0·055	0·068

		Cubic Metre of Mortar per Metre Super of Brickwork			
Size of Brick (mm)	Joints (mm)	Half Brick		Whole Brick	
		Wirecut	Frogged	Wirecut	Frogged
190 × 90 × 90	10	0·013	0·017	0·039	0·048
190 × 90 × 65	10	0·016	0·022	0·045	0·057
215 × 102·5 × 65	13	0·022	0·027	0·057	0·067
215 × 102·5 × 70	13	0·021	0·026	0·056	0·066
215 × 102·5 × 50	13	0·026	0·033	0·040	0·054
190 × 90 × 90	13	0·016	0·021	0·046	0·055
190 × 90 × 65	13	0·020	0·025	0·052	0·063

APPROXIMATE WEIGHTS OF DRY AND WET BRICKWORK IN
1 : 3 CEMENT: SAND: MORTAR

Types of brick	DRY		SATURATED	
	lb/ft^3	kg/m^3	lb/ft^3	kg/m^3
Cellular Flettons	App. 80	1280	App. 90	1440
Most clay and sand-lime, common & facing bricks	105–121	1680–1940	124–134	1980–2150
Most clay engineering bricks	124–146	1980–2340	134–149	2150–2390
Concrete bricks	Weight varies over wide range according to whether bricks are made with normal or lightweight aggregates			

(No allowance made for wastage)
(*Source: The Brick Development Association*)

QUANTITIES OF BRICKS AND MORTAR USED IN WALLING

The following quantities have been arrived at by calculation assuming joints to be filled. Only two cases are given, one for a solid wirecut and the other for a single frog brick laid frog up. The frog volume is taken as 73.7 cm^3.

In the case of double frog bricks, the mortar quantity should be increased by about 5 per cent. It is difficult to estimate how much mortar enters perforations in the case of perforated bricks since this depends on the pattern and size of perforations.

Modular metric bricks of the 300 module are perforated although one manufacturer has produced a solid brick experimentally in the 300 × 100 × 75 format. In the 200 module, the bricks may be either solid or perforated. In the case of the modular metric range therefore, it has been

THE BRICKLAYER

assumed that as much mortar would enter the perforations of the 300 format as would be present in a single frog brick of standard size (*e.g.* about 75 cm³). For the 200 format perforated brick, the mortar entering the perforations has been assumed to be 50 cm³.

No allowance has been made for wastage, nor are the figures applicable to deep frog or calculon or cellular bricks.

1. QUANTITY OF MORTAR PER 1000 BRICKS (m³)

1.1. Standard bricks	Solid wirecut	Single frog
102·5 mm wall	0·30	0·37
215 mm wall	0·38	0·46
327·5 mm wall	0·41	0·48
440 mm wall	0·42	0·50

1.2. Modular metric bricks	Solid	Perforated
1.2.1. 90 mm wall		
300 × 100 × 100 format	—	0·44
300 × 100 × 75 format	0·34	0·41
200 × 100 × 100 format	0·26	0·31
200 × 100 × 75 format	0·24	0·29
1.2.2. 190 mm wall		
200 × 100 × 100 format	0·36	0·41
200 × 100 × 75 format	0·31	0·36

2. QUANTITIES OF BRICKS AND MORTAR PER m² OF WALLING

2.1. Standard bricks	Bricks	Mortar – m³	
		Solid wirecut	Single frog
102·5 mm wall	59·26	0·018	0·022
215 mm wall	118·52	0·045	0·054
327·5 mm wall	117·78	0·073	0·086
440 mm wall	237·04	0·101	0·118

2.2. Modular metric bricks	Bricks	Mortar – m³	
		Solid	Perforated
2.2.1. 90 mm wall			
300 × 100 × 100 format	33·33	—	0·015
300 × 100 × 75 format	44·44	0·015	0·018
200 × 100 × 100 format	50·00	0·013	0·016
200 × 100 × 75 format	66·67	0·016	0·019
2.2.2. 190 mm wall			
200 × 100 × 100 format	100·00	0·036	0·041
200 × 100 × 75 format	133·33	0·042	0·048

2.2	Modular metric bricks	Bricks	Mortar m³	
			Solid	Perforated
2.2.3.	290 mm wall			
	200 × 100 × 100 format	150·00	0·059	0·067
	200 × 100 × 75 format	200·00	0·068	0·078

3. QUANTITY OF CONSTITUENTS PER m³ WET MORTAR

Nominal mix properties	Cement (tonne)	Hydrated lime (tonne)	Sand (m³)
1 : 3	0·50	—	1·04
1 : 4	0·40	—	1·12
1 : 1 : 6	0·26	0·12	1·10
1 : 2 : 9	0·19	0·17	1·13
1 : 3 : 12	0·13	0·19	1·14

(*Source: The Brick Development Association*)

MORTAR FOR POINTING

One yard super of brick facings and mortar for pointing equals

64 bricks and $\frac{1}{5}$ cu. ft of mortar in Flemish bond.
72 bricks and $\frac{1}{4}$ cu. ft of mortar in English bond.
48 bricks and $\frac{1}{6}$ cu. ft of mortar in stretcher bond (as in hollow walls).

One yard super of half-brick walling equals
48 bricks and $\frac{3}{4}$ cu. ft mortar.

One yard super of brick-nogging equals
48 bricks and $\frac{3}{4}$ cu. ft of mortar if laid flat.
32 bricks and $\frac{1}{2}$ cu. ft of mortar if laid on edge.

METRIC

Metric equivalents using the 225 × 112.5 × 75 mm brick, which is not exactly divisible into square metre units, would be:

79 bricks and .0056 m³ mortar using Flemish bond.
89 bricks and .007 m³ mortar using English bond.
59 bricks and .0037 m³ mortar using stretcher bond.

One square metre of half-brick walling equals
59 bricks and .021 m³ mortar.

One square metre of brick nogging equals
59 bricks and .021 m³ mortar laid flat.
40 bricks and .014 m³ mortar laid on edge.

One load of mortar equals 1 cu. yd or 27 cu. ft or 1 m³.

(*Source: The Brick Development Association*)

TABLE 1. MATERIALS FOR DAMP–PROOF COURSES

Material	Minimum Weight kg/m2	Thickness mm	Joint treatment to prevent water moving: upward	downward	Structural Performance	Durability	Other considerations
GROUP A: FLEXIBLE Lead to BS1173	Code No. 4	—	Lapped at least 100 mm preferably 150 mm	welted	Will not extrude under pressures met in normal construction.	Very suitable for high-rise construction. Corrodes when built in with freshly laid lime or cement mortar.	Corrosion can be prevented by applying bitumen or heavy consistency bitumen paint to the corrosion producing surface and both sides of the lead dpc.
Copper to BS2670, Grade A annealed condition.	Approx. 2.28*	0.25	Lapped at least 100 mm preferably 150 mm	welted	Will not extrude under pressures met in normal construction.	Highly resistant to corrosion.	Where soluble salts are known to be present (e.g. sea salt in sand) similar treatment to Lead is recommended. Copper may stain external surfaces.
Bitumen to BS743: A — hessian base B — fibre base C — asbestos base D — hessian base and lead E — fibre base and lead F — asbestos base and lead G — sheeting with hessian base	3.8 3.3 3.8 4.4 4.4 4.9 5.4	— — — — — — —	Lapped at lease 100 mm	Lapped & sealed	Likely to extrude under heavy pressure but this is unlikely to affect water resistance.	The hessian or felt may decay but this does not affect efficiency if the bitumen remains undisturbed. Types D, E & F are most suitable for buildings that are extended to have a very long life or where there is risk of movement.	Materials should be unrolled with care. In cold weather, the rolls should be warmed before use.
Generally to BS743: Type A based but with aluminium, copper, or plastic core.	—	—	As for other bitumen based materials				

Material	Weight kg/m2	Minimum Thickness mm	Joint treatment to prevent water moving: upward	Joint treatment to prevent water moving: downward	Structural Performance	Durability	Other considerations
Polythene, black, low density—to BS743	Approx. 0.5	0.46	Lapped for distance at least equal to width of dpc	welted	No evidence of deterioration in contact with other building materials.	Will not extrude under pressures met in normal construction.	Welted joints must be held in compression. When used as cavity trays exposed on elevation these materials may need bedding in mastic for the full thickness of the outer leaf of walling, to prevent penetration by driving rain.
AGRÉMENT CERTIFICATES:							
*No. 73/159—Black polythene sheet	1.67	1.14	Lapped at least 100 mm	welted	Will not extrude under load up to point of compressive failure of the wall.	Tough, resilient and remains flexible over temperature range likely to occur in service when used in orthodox manner.	At temperature below 0°C liable to impact damage.
*No. 73/173—Black pitch/polymer sheet	1.5	1.3	Lapped at least 100 mm	Lapped & sealed	as above	Unlikely to be impaired by any of the movements normally occuring up to point of failure of the wall itself.	Accommodates considerable lateral movement without its performance becoming impaired and recovers after removal of the load. Standard range of cloaks available for changes in directions and levels. "Specials" can also be produced.
*No. 73/174—Black pitch/polymer sheet	1.45	1.19	Lapped at least 100 mm	Lapped & sealed	as above	as above	Has exceptional elasticity and recovery characteristics. Difficult to form certain details especially those involving bending through two angles. Where such details are necessary care must be taken to provide satisfactory seals.

Material	Weight kg/m2	Minimum Thickness mm ♦	Joint treatment to prevent water moving: upward	Joint treatment to prevent water moving: downward	Structural Performance	Durability	Other considerations
**No. 73/197 — Black low density polythene sheet, both faces knurled	0.87	0.84	Lapped for distance at least equal to width of dpc—min. 100 mm	welted	Will not extrude under compressive loads when in a wall.	Tough, resilient and remains flexible over the temperature range likely to occur under service conditions.	The knurled surfaces increase friction between dpc and wet mortar and increase resistance to sliding with wet mortar. Must not be laid on dry mortar.
**No. 73/201 — Asbestos based sheet, with coarse sand finish on each side	2.35	2.0	Lapped at least 100 mm	Lapped & sealed	Resistance to extrusion under load better than traditional bitumen felt dpc	Resistance to damage due to normal structural movement similar to traditional bitumen felt dpc. Sanded surfaces increase bond between dpc and mortar, and increased resistance to both shear and tensile loading when compared to conventional bitument felt dpc.	Particularly suitable as a flexible dpc in piers, garden or parapet walls, domelstic chimneys or like structures where increased resistance to displacement of construction at dpc level is required. Bed in wet mortar only.
GROUP B: SEMI-RIGID							
Mastic asphalt to BS1097 or BS1418	—	—	No joint problems		Liable to extrude under pressures above 65 KN/m2. Material of hardness appropriate to conditions should be used.	No deterioration.	To provide key for mortar below next course of brickwork, grit should be beaten into asphalt immediately after application and left proud of surface. Alternatively the surface should be scored whilst still warm.
GROUP C: RIGID							
Brick to BS3921 maximum water absorption 4.5%	—	—	—	Not suitable	Has tensile bending strength equal to that of main wall.	No deterioration.	Use two courses, laid to break joint, bedded in 1:3 Portland cement: sand.

Material	Minimum Weight kg/m2	Minimum Thickness mm	Joint treatment to prevent water moving: upward / downward	Structural Performance	Durability	Other considerations
Slates at least 230 mm long; should pass wetting and drying test and acid-immersion test of BS3798	—	5.0	— / Not suitable, although they may sometimes be used to resist lateral movement of water		No deterioration.	as above

*Where copper is to form in one piece a dpc and a projecting drip of flashing it should weigh at least 4.12 kg/m2.

**Proprietary material.

(Source: BDA Practical Note 'Damp-proofing courses and flashings with brickwork and blockwork'. Published by the Brick Development Association.)

11. THE CARPENTER AND JOINER

Conversion Tables	218
Basic Metric Timber Sizes	220
Marking Out	221
Calculating Timber Costs	222
Codes of Practice	223
British Standards:	224
Doors	226
Windows	229
Flooring	229
Fencing	230
Wood Trim	234
Boards and Sheets	235
Glulam	239
Fastenings	239
Locks, Hinges and Hardware	247
Builders' Hardware	256
Adhesives	256
Wood Joists — Floor, Ceiling and Roof	257
Stress Graded Softwood	257

CONVERSION TABLES (FOR TIMBER ONLY)

Inches	Millimetres	Feet	Metres
1	25	1	0·300
2	50	2	0·600
3	75	3	0·900
4	100	4	1·200
5	125	5	1·500
6	150	6	1·800
7	175	7	2·100
8	200	8	2·400
9	225	9	2·700
10	250	10	3·000
11	275	11	3·300
12	300	12	3·600
13	325	13	3·900
14	350	14	4·200
15	375	15	4·500
16	400	16	4·800
17	425	17	5·100
18	450	18	5·400
19	475	19	5·700
20	500	20	6·000
21	525	21	6·300
22	550	22	6·600
23	575	23	6·900
24	600	24	7·200

(Basics: 25mm = 1 in. 0·300m = 1 ft.)

STANDARDS (TIMBER) TO CUBIC METERS AND CUBIC METRES TO STANDARDS (TIMBER)*

Cu. metres		Standards
5	1	0·2
9	2	0·4
14	3	0·6
19	4	0·9
23	5	1·1
28	6	1·3
33	7	1·5
37	8	1·7
42	9	1·9
47	10	2·1
93	20	4·3
140	30	6·4
187	40	8·6
234	50	10·7
280	60	12·8
327	70	15·0
374	80	17·1
421	90	19·3
467	100	21·4

*Note: The above are approximate values. The cubic metres have been rounded off to the nearest whole number. Factors for use on calculating machines are:
1 cu. metre—35·3148 cu. ft—0·21403 std.
1 cu. ft—0·028317 cu. metre   1 std—4·67227 cu. metres.

BASIC METRIC TIMBER SIZES WITH IMPERIAL EQUIVALENTS AND RUNNING FEET AND METERS PER CUBIC METRE

SOFTWOOD

Thickness in	Thickness mm	3 / 75	4 / 100	4⅝ / 115	5 / 125	6 / 150	7 / 175	8 / 200	9 / 225	10 / 250	12 / 300	
5/8	16	2467	1851		1480	1234						ft
		751	563		551	376						metres
3/4	19	2256	1692		1393	1128						ft
		687	514		424	343						metres
7/8	22	1905	1428		1152	952						ft
		583	434		351	282						metres
1	25	1694	1296		1038	864	726	648	564	519	432	ft
		516	394		316	263	221	197	171	156	131	metres
1¼	32	1272	925	887	740	636	545	462	424	370	318	ft
		387	282	270	255	193	166	140	129	112	96	metres
1½	38	1128	846	720	677	564	483	423	379	338	282	ft
		344	257	219	206	162	147	128	115	103	85	metres
1¾	44	952	735	627	588	476	408	367	314	294	238	ft
		290	224	191	179	145	124	111	95	89	72	metres
2	50	864	648	551	518	432	370	324	288	259	216	ft
		263	197	167	157	131	112	98	84	79	65	metres
2½	63		475		380	318	272	247	212			ft
			144		115	96	82	75	64			metres
3	75		432		345	282	240	216	188	172	141	ft
			131		105	85	73	65	57	52	42	metres
4	100		324			211		162		129	105	ft
			98			64		49		39	32	metres
6	150					141		108			71	ft
						42		32			21	metres
8	200							81				ft
								24				metres
10	250									47		ft
										14		metres
12	300										36	ft
											10	metres

Note: The sizes in the table opposite are basic sizes, as imported, and can of course be re-sawn to other dimensions and/or surfaced as required. The sizes given above are calculated on the basis of 25 mm to 1 inch. As imported, some timbers will be over these sizes, being calculated on the basis of 25.4mm to 1 inch.

BASIC METRIC TIMBER SIZES

HARDWOOD

Thickness, in mm

19, 25, 32, 38, 50, 63, 75, 100, rising by 25mm stages.

150mm and up, by increments of 10mm "Strips" and "narrows".

50mm and up, by increments of 10mm.

Length, in metres

1.80m, rising by increments of 10cm "Shorts" (regularly imported of certain species only). Falling from 1.70m by stages of 10cm.

PRICES PER STANDARD AND PER CUBIC METRE

per standard £	per m³ £	per standard £	per m³ £
1	0·21	20	4·35
2	0·43	30	6·52
3	0·65	40	8·70
4	0·87	50	10·88
5	1·08	100	21·76
10	2·17	200	43·52

PLYWOOD: COMPARATIVE THICKNESS

Table showing standard thicknesses in millimeters of plywood and block boards and the equivalent in parts of an inch.

Milli-metres	Inches	Milli-metres	Inches	Milli-metres	Inches	Milli-metres	Inches
·8	$\frac{1}{32}$ full	5	$\frac{3}{16}$ full	15	$\frac{5}{8}$ bare	25	1
1·5	$\frac{1}{16}$	6	$\frac{1}{4}$	16	$\frac{5}{8}$ full	26	1 full
2·5	$\frac{3}{32}$ full	9	$\frac{3}{8}$	18	$\frac{3}{4}$	32	$1\frac{1}{4}$ full
3·0	$\frac{1}{8}$ bare	10	$\frac{3}{8}$ full	19	$\frac{3}{4}$ full	38	$1\frac{1}{2}$ full
4·0	$\frac{3}{16}$ bare	12	$\frac{1}{2}$ bare	22	$\frac{7}{8}$ full	—	—
4·5	$\frac{3}{16}$	12·5	$\frac{1}{2}$	24	1 bare	—	—

MARKING OUT

Full-size working drawings or 'rods' are not difficult to read if plans, elevations, sections and projections are not super-imposed one on top of the other. This is very often done with half-inch and such details supplied by architects, but this practice can lead to costly error, unless the craftsman is conversant with the methods employed by the technician, draughtsman or foreman who sets out the 'rods'.

It is an invariable rule when reading a rod that the face side is to the front (or bottom) and the top is to the left. Paper rods, as used by shopfitters, are often dyeline prints. They should always be dimensioned. Never 'offer up' any piece of work to a paper rod, because it is often inaccurate, due to expansion of the paper in the printing process.

Joints are not always fully or accurately shown; it is safe to take established dimensions or proportions. Dovetails are traditionally $\frac{1}{6}$ of the thickness of the material at the root and $\frac{1}{2}$ that thickness at the broadest part. Tenons are usually $\frac{1}{3}$ the thickness of the stuff, and double tenons are $\frac{1}{5}$ of the thickness.

SUPERFICIAL AREAS PER STANDARD

Thickness in	mm	sq ft	m²
$\frac{1}{2}$	13	3690	342·7
$\frac{5}{8}$	16	3168	285·9
$\frac{3}{4}$	19	2640	245·0
$\frac{7}{8}$	22	2263	210·4
1	25	1980	183·7
$1\frac{1}{4}$	32	1584	147·1
$1\frac{1}{2}$	38	1320	122·9
2	50	990	92·0
$2\frac{1}{2}$	63	792	73·6
3	75	660	49·3

METRES RUN PER CUBE METRE: TIMBER IN SIZES AVAILABLE EX STOCK

Thickness	Width mm	in	m run per m³
$12\frac{1}{2}$mm	38	$1\frac{1}{2}$	2064
($\frac{1}{2}$in)	75	3	1032
	100	4	788
19mm	25	1	2056
($\frac{3}{4}$in)	38	$1\frac{1}{2}$	1372
	50	2	1028
	100	4	514
	150	6	343

Thickness	Width mm	in	m run per m³
25mm	25	1	1336
(1in)	38	1½	1032
	50	2	688
	75	3	516
	100	4	394
	150	6	258
	225	9	172
38mm	38	1½	688
(1½in)	50	2	516
	75	3	344
	100	4	257
	155	6¼ full	190
50mm	50	2	394
(2in)	75	3	263
	100	4	197
	150	6	131
	175	7	112
	225	9	84
75mm	75	3	170
(3in)	100	4	131
	150	6	85
	255	9	57
100mm (4in)	100	4	98

CALCULATING TIMBER COSTS

In the metric system, the thickness and the width of scantlings are measured in millimetres while the length is given in metres.

The basis of all metric timber calculations, the cube metre, contains one million (one thousand times one thousand) pieces of timber, each 1 mm thick, 1 mm wide and 1 m long.

If the price is given in m³ (cubic metres), the cost of any quantity of any particular sectional area (*i.e.* thickness times width) proceed as follows:

1. Look up (page 218), the numbr of running metres per m³ of the sectional area of the timber. (If the sectional area is not given, multiply the thickness by the width and divide this into 1 million.) This will give the number of running metres per m³.
2. Divide the price per m³ (reduced to pence) by the number of running metres, as found in Stage 1. This gives the price per m run in pence.
3. Multiply this price per running metre by the number of running metres in the quantity being considered. This gives the price in pence: divide by 100 to obtain the price in pounds sterling.

If the price is given in 10 or 100 m run, reduce to price per m run and proceed as stage 3 above.

If the price is given per standard, reduce to m^3 by using table on page 218 and proceed as stages 1, 2 and 3 above.

If the price is given per cubic foot, divide by 35·3 and proceed as per stages 1, 2 and 3 above.

CODES OF PRACTICE

CP 112: The structural use of timber.

Part 2: 1971: Metric units: deals with the design of structures in timber and plywood. Covers selection of timber and plywood species and grades, giving tables of strengths and elastic moduli for visual and stress grades, the use of adhesives and mechanical fasteners the design of members and joints workmanship, testing and maintenance.

Part 3: 1973: Trussed rafters for roof of dwellings: materials, functional requirements, manufacture, design, testing, permissible spans, handling and erection of Fink and Fan type trussed rafters.

B.S. 5395: 1977: Code of practice for stairs.

Design and construction of stairs in different materials for all types of buildings.

CP 151: Doors and windows including frames and linings.

Part 1: 1957: Wooden doors: deals with wooden doors of all types and with different methods of hanging doors and fixing their frames and linings. Guidance on economy, strength, rigidity and dimensional stability.

CP 152: 1972: Glazing and fixing of glass for buildings.

Numerical values in SI units. Account has been taken of a better understanding of the strength of float glass (supersedes B.S. 973).

CP 201: Flooring of wood and wood products.

Part 1: 1967: imperial units.

Part 2: 1972: metric units: recommends bases.

CP 209: Care and maintenance of floor surfaces.

Part 1: 1963: Wooden flooring: deals with methods of providing polished finish to wooden flooring using

    (i) waxed finishes without prior sealing,

    (ii) treatment with seals, and

    (iii) treatment with oils.

BRITISH STANDARDS

TIMBER, QUALITY, GRADES AND DIMENSIONS

B.S. 1186: Quality of timber and workmanship in joinery.

Part 1: 1971: Quality of timber: requirements, classification, species of timber for various purposes, etc.

Part 2: 1971: Quality of workmanship: fixed joints of framed, edge-to-edge and staircase types; joints permitting movement for plywood panels, solid panels, bead butt, bead flush panels and joints for matchboard surfaces; moving parts such as doors and sashes, sliding drawers and flaps; gluing of joints, surface finish; laminating and finger jointing.

B.S. 565: 1972: Glossary of terms relating to timber and woodwork.

Terms covering forestry; physical structure; seasoning and conditioning; defects, blemishes and imperfections; sizes and quantities; plywood, battenboard, blockboard and laminboard; fibre, chipboard and other particle boards. Carpentry, joinery and joints used.

B.S. 881, 589: 1974: Nomenclature of commercial timbers including sources of supply.

Tabulates recognised hardwoods and softwoods, botanical species, commercial names, sources of supply and average densities.

B.S. 4471: Dimensions of softwood.

Part 1: 1969: basic sections, range of sawn softwood sizes and table of reductions to finished sizes.

TABLE 1. BASIC SIZES OF SAWN SOFTWOOD (CROSS-SECTIONAL SIZES)
(All dimensions are in millimetres)

Thickness	Width								
	75	100	125	150	175	200	225	250	300
16	x	x	x	x					
19	x	x	x	x					
22	x	x	x	x					
25	x	x	x	x	x	x	x	x	x
32	x	x	x	x	x	x	x	x	x
36	x	x	x	x					
38	x	x	x	x	x	x	x		
40*	x	x	x	x	x	x	x		
44	x	x	x	x	x	x	x	x	x
50	x	x	x	x	x	x	x	x	x
63		x	x	x	x	x	x		
75		x	x	x	x	x	x	x	x
100		x		x		x		x	x
150				x		x			x
200						x			
250								x	
300									x

* For 40 mm thickness, designers and users should check availability.

NOTE. The smaller sizes contained within the dotted lines are normally but not exclusively of European origin. The larger sizes outside the dotted lines are normally but not exclusively of North and South American origin.

TABLE 3. REDUCTIONS FROM BASIC SIZE TO FINISHED SIZE BY PROCESSING OF TWO OPPOSED FACES
(All dimensions are in millimetres)

Purpose	Reduction from basic size to finished size				
	15 to and including 22	for sawn sizes of width or thickness			
		Over 22 to and including 35	Over 35 to and including 100	Over 100 to and including 150	Over 150
(1) Constructional timber surfaced	3	3	3	5	6
(2) *Floorings	3	4	4	6	6
(3) *Matchings and interlocking boards (4) Planed all round	4	4	4	6	6
(5) Trim	5	5	7	7	9
(6) Joinery and cabinet work	7	7	9	11	13

* The reduction of width is overall the extreme size and is exclusive of any reduction of the face by the machining of a tongue or lap joint

Part 2: 1971: Small resawn sections: dimensions of small resawn sections for floor fillets, fixing grounds, noggins, and other uses.

TABLE 1. CROSS-SECTIONAL SIZES
(All dimensions are in millimetres)

Finished widths

Finished thickness	22	30	36	44	48
6	x		x		
14			x		x
17	x		x		x
22	x	x	x	x	x
30			x	x	x
36			x		x
44					x
48					x

B.S. 5450: 1977: Specification for sizes of hardwoods and methods of measurement.

Range of basic sizes of sawn hardwood at 15 per cent moisture content. Measurement of moisture content and sizes.

STAIRS

B.S. 585: 1972: Wood stairs.

Interior stairs and close strings for use in houses. Quality, design and construction.

DOORS

B.S. 459: Doors.

Part 1: 1954: Panelled and glazed wood doors: design, dimensions and construction of dowelled and morticed-and-tenoned panelled and glazed wood doors for internal, external and garage.

Part 2: 1962: Flush doors: internal and external, dimensions, timber, facings, plywood, hardboard, adhesives, lippings, provision for glazing, hinges, locks, etc.

Part 3: 1951: fire-check flush doors and wood and metal frames (half-hour and one-hour types), for internal and external use to provide effective barriers to the passage of fire for the time stated. Materials and construction.

Part 4: 1965: Matchboarded doors: ledged and braced, framed ledged and braced; timber, adhesives, construction and sizes.

B.S. 1567: 1953: Wood door frames and linings.

For external doors opening inwards and outwards, and for internal door frames and linings.

B.S. 4787: Internal and external wood doorsets, door leaves and frames.

Part 1: 1972: Dimensional requirements: sizes for dimensionally co-ordinated internal and external wood doorsests, door leaves and frames for traditional and industrialised forms of construction.

TABLE 1. SIZES OF INTERNAL DOORSETS AND THEIR COMPONENT PARTS

Dimension Description	Size in mm	Permissible deviation in mm
Co-ordinating dimension: height of door leaf height sets	2100	
Co-ordinating dimension: height of ceiling height set	2300 2350 2400 2700 3000	
Co-ordinating dimension: length of all doorsets (S. single leaf set, D. double leaf set)	600 S 700 S 800 S and D 900 S and D 1000 S and D 1200 D 1500 D 1800 D 2100 D	
Thickness	This dimension is not of great importance in the doorset, but consideration should be given to its relationship to standardised sizes of sawn timber and to partition thicknesses.	
Work size: height of door leaf height set	2090	2.0

Dimension / Description	Size in mm	Permissible deviation in mm
Work size: height of ceiling height set	2285 2335 2385 2685 2985	2.5
Work size; length of all doorsets (S. single leaf set, D. double leaf set)	590 S 690 S 790 S and D 890 S and D 990 S and D 1190 D 1490 D 1790 D 2090 D	2.5
Door leaf height for all doorsets	2040	1.5
Length of door leaf used in single leaf sets (F. flush leaf, P. panel leaf)	526 F 626 F 726 F and P 826 F and P 926 F and P	1.5
Length of one leaf of an equal double leaf doorset (F. flush leaf, P. panel leaf) (Dimensions of pairs of doors assume square meeting stiles with 2 mm clearance between leaves. If rebated meeting stiles are required, then leaf length will be 6 mm longer than stated to allow for a 12 mm rebate)	362 F 412 F 462 F 562 F and P 712 F and P 862 F and P 1012 F and P	1.5
Door leaf thickness	40	+0 −2.0
Doorstop width (solid or otherwise)	13	2.0
Rebate depth	42	0.5
Threshold (sill) height (see also (3) in Appendix A) of all doorsets	15	1.0

TABLE 2. SIZES OF EXTERNAL DOORSETS AND THEIR COMPONENT PARTS

Dimension / Description	Size in mm	Permissible deviation in mm
Co-ordinating dimension: height of door leaf height sets	2100 (if required, may be provided with packing at head to give 25 mm positive boundary condition at sill)	
Co-ordinating dimension: height of ceiling height set	2300, 2400, 2700, 3000 (if required may be provided with packing at head to give 25 mm positive boundary condition at sill) 2350 (intended for protected positions and recommended for use with zero boundary condition)	

Dimension Description	Size in mm	Permissible deviation in mm
Co-ordinating dimension: length of all doorsets (S. single leaf set, D. double leaf set)	900 S 1000 S 1200 D 1500 D 1800 D 2100 D	
Thickness	This dimension is not of great importance in the doorset, but consideration should be given to its relationship to standardized sizes of sawn timber and to partition thicknesses	
Worksize: height of door leaf height set	Zero boundary condition: 2090 25 mm boundary condition (see A^*): 2155	2.0
Work size: height of ceiling height set	Zero boundary condition: 2290, 2340, 2390, 2690, 2990 25 mm boundary condition (see G^*): 2315, 2415, 2715, 3015	2.5
Work size: length of all door sets (S. single leaf set, D. double leaf set)	890 S 990 S 1190 D 1490 D 1790 D 2090 D	2.5
Door leaf height for all doorsets	2000	1.5
Length of door leaf in single leaf sets (F. flush leaf, P. panel leaf)	807 907	1.5
Length of one leaf of a double leaf doorset (F. flush leaf, P. panel leaf) (Dimensions of pairs of doors assume square meeting stiles with 2 mm clearance between leaves If rebated meeting stiles are required, then leaf length will be 9 mm longer than stated, to allow for an 18 mm rebate)	552 F and P 702 F and P 852 F and P 1002 F and P	1.5
Door leaf thickness	40 44	1.5
Doorstop width (solid or otherwise)	18	1.0
Rebate depth	42 46	0.5
Threshold (sill) height (see also (3) in Appendix A) for all doorsets	45 20 †	1.0

* *Co-ordinating dimension: height of ceiling height set.* † Intended for protected positions, with packing at head.

B.S. 5277: 1976: Doors: Measurement of defects of general flatness of door leaves.

B.S. 5278: 1976: Doors: Measurement of dimensions and defects of squareness of door leaves.

B.S. 5278 and B.S. 5277 determine methods to use to measure twist and bending, etc.

WINDOWS

B.S. 644: Wood windows.

Part 1: 1951: Wood casement windows: types, sizes, construction with requirements for normal and easy-clean hinges.

TABLE 2. REBATE SIZES OF WINDOW UNITS

	Widths		Heights (For key to these reference letters, see Fig 2) A	B	C	D	E	F	G	H
	in	in	in	in	in	in	in	in	in	in
Frames Rebate sizes	13⅝	21⅝	10⅝	14⅛	26⅛	32⅛	38⅛	44⅛	44⅛	50⅛
Casement and ventlights Rebate sizes, no-bar types	11¼	19¼	8¼	(11¼)*a*	23¼	(29¼)*a*	35¼	41¼	41¼	47¼
Rebate sizes, bar types	11¼	19¼	8¼	11¾	23¾	29¾	35¾	41¾	41¾	47¾
Glazing rebates, bar types	11¼	2 × 9½	8¼	11¾	2 × 11¾	{1 × 8¾ 2 × 10¼} *b*	3 × 11¾	{1 × 9 5/16 3 × 10 5/16} *b*	4 × 10¼	4 × 11¾

NOTES. a. In the standard range this size of opening is not fitted with a no-bar casement.
b. This modification is necessary where the opening occurs below a ventlight or over a sub-light and is adjacent to a full-height opening, so as to ensure that the glazing bars are suitably aligned. The smaller pane is that immediately under the ventlight, or over the sub-light.

Part 2: 1958: Wood double-hung sash windows: types, sizes, construction of weight and spring balanced windows.

Part 3: 1951: Wood double-hung sash windows (Scottish type).

B.S. 990: Steel windows generally for domestic and similar buildings.

Part 1: 1967: Imperial units.

Part 2: 1972: Metric units: co-ordinating sizes, work sizes, tolerances and deviations.

B.S. 1285: 1963: Wood surrounds for steel windows and doors.

Sizes of surrounds complying with B.S. 990.

B.S. 4873: 1972: Aluminium alloy windows.

Specifies materials, construction, finishes, hardware and performance.

FLOORING

B.S. 1187: 1959: Wood blocks for floors.

Hardwood and softwood blocks for laying on level bases.

B.S. 1297: 1970: Grading and sizing of softwood flooring.

Tongued and grooved flooring; species, grading moisture content, sapwood, decay and dimensions, etc.

TABLE 1. TRANSVERSE DIMENSIONS

	mm	mm	mm	mm
Finished thicknesses	16	19	21	28
Finished widths of face	65	90	113	137

TABLE 2. TONGUE DIMENSIONS

		mm	mm	mm	mm
Finished thickness	A	16	19	21	28
Finished tongue thickness	B	4.5	6	6	6
Finished tongue top width	C	7	7	7	7
Face of board to top of tongue	D	7	7	8	12

FENCING

B.S. 1722: Fences:
Part 4: 1972: Cleft chestnut pale fences.

TABLE 1. CLEFT CHESTNUT PALE FENCES: GENERAL CHARACTERISTICS

Fence types		Height of top of pales	Applicability	Number of lines of wiring	Spacing between pales	Maximum distance between intermediate concrete posts	Maximum distance between intermediate wooden posts
With concrete posts	With wooden posts	m			mm	m	m
CC 90	CW 90	0.90	Housing		75		
CC 105A	CW 105A		Housing and inner fences of parks, etc.	2	75	3.0	2.50
		1.05					
CC 105B	CW 105B		Miscellaneous		100		
CC 120	CW 120	1.20	General purposes		75	2.75	2.25
CC 135	CW 135	1.35	Road boundaries				
CC 150	CW 150	1.50	Outer fences of parks and other boundaries	3	50	2.25	2.0
CC 180	CW 180	1.80	Factories and security				

NOTE. If barbed wire is to be fixed above the fencing, the length of the posts shall be increased by 150 mm for each line of barbed wire.

TABLE 2. CONCRETE POSTS FOR CLEFT CHESTNUT PALE FENCING

Fence types	No. of holes for lines of wiring	Intermediate posts		Straining posts		Struts	
		Length	Base dimension*	Length	Section	Length	Section
		m	mm	m	mm	m	mm
CC 90	2	1.50	100 × 100	1.50		1.50	
CC 105A	2	1.65	100 × 100	1.80		1.65	100 × 75
CC 105B	3	1.65		1.95	125 × 125	1.82	
CC 120	3	1.87		2.10		1.90	
CC 135	3	1.95	125 × 125	2.25		2.20	100 × 83
CC 150	3	2.25		2.63		2.59	
CC 180	3	2.63					

NOTE. If barbed wire is to be fixed above the fencing, the length of the posts shall be increased by 150 mm for each line of barbed wire.
* A demoulding draw allowance of 3 mm on each of two sides is permissible in addition to permissible deviations specified in 2.6.1.

THE CARPENTER AND JOINER

TABLE 3. WOODEN POSTS FOR CLEFT CHESTNUT PALE FENCING

Fence types	Intermediate posts		Straining posts		Struts	
	Length	Girth at mid-length	Length	Girth at top (round)	Length	Girth at top (round or cleft)
	m	mm	m	mm	m	mm
CW 90	1.52		1.52		1.52	
CW 105A	1.67		1.67		1.67	190 to 230
CW 105B	1.67	190 to 230	1.67	230 to 250	1.67	
CW 120	1.82		1.82		1.82	
CW 135	1.95		2.10		1.95	
CW 150	2.10	230 to 250	2.25	250 to 290	2.10	230 to 250
CW 180	2.43		2.58	280 to 350	2.43	250 to 290

NOTE. If barbed wire is to be fixed above the fencing the lengths of the posts shall be increased by 150 mm for each line of barbed wire.

Part 5: 1972: Close boarded fences: including oak pale fences. See table opposite.

TABLE 4. PRESERVATIVE TREATMENTS ACCORDING TO SPECIES

Species	Use	Treatment
Oak (European and English), sweet chestnut, European larch, yew, western red cedar	Rails and posts	Treatment only if specified by the purchaser....
Spruce, Douglas fir and hemlock	Rails Posts	*Appropriate treatments are specified*
Hardwoods or softwoods other than spruce and Douglas fir	Rails Posts	
Imported hardwoods	Posts	

Part 6: 1972: Wooden pallisade fences.

TABLE 1. WOODEN PALISADE FENCES

Type references		Height of top of pales	Applicability	No. of arris rails	Concrete posts	Wooden posts	
With concrete posts	With wooden posts				Length	Length	Section
		m			m	m	mm
WPC 100	WPW 100A	1.0	Housing	2	1,525		100 × 100
WPC 100	WPW 100B	1.0	Parks inner fences	2		1.60	100 × 125
WPC 120	WPW 120	1.20	General purposes	2	1,725	1.80	100 × 125
WPC 140	WPW 140	1.40	Roads and railways	3	1,925	2.00	100 × 125
WPC 160	WPW 160	1.60	Housing flank fences and parks	3	2,275	2.425	100 × 125
WPC 180	WPW 180	1.80	Parks, commercial and public buildings	3*	2,475	2.65	100 × 150

* For size of arris rails see 2.5.

TABLE 1. CLOSE BOARDED FENCES

Type reference							Height of top of filling	Applicability	No. of rails	Concrete posts length		Wooden posts	
Oak pale fences		With wooden Posts	Fences in other timbers							Recessed PCR or BCR	Morticed PCM or BCM	Length	Section
With concrete posts			With concrete posts		With wooden posts								
Recessed*	Morticed		Recessed*	Morticed									
						m			m	m	mm	
PCR 100	PCM 100	PW 100A PW 100B	BCR 100	BCM 100	BW 100A BW 100B	1.0 1.0	Housing Parks inner fences	2	1.525	1.65	1.60	100×100 100×125
PCR 120	PCM 120	PW 120	BCR 120	BCM 120	BW 120	1.20	General purposes	2	1.725	1.90	1.80	100×125
PCR 140	PCM 140	PW 140	BCR 140	BCM 140	BW 140	1.40	Roads and railways	3	1.925	2.08*	2.00	100×125
PCR 160 PCR 180A	PCM 160 PCM 180A	PW 160 PW 180A	BCR 160 BCR 180A	BCM 160 BCM 180A	BW 160 BW 180A	1.60 1.80	Housing, flank fences	3 3	2.275 2.475	2.45 2.65	2.35 2.60	100×125 100×125
PCR 180B	PCM 180B	PW 180B	BCR 180B	BCM 180B	BW 180B	1.80	Parks, railways, commercial and public buildings	3	2.475	2.70	2.70	100×150

* For use without cappings.
For use with cappings or gravel boards or both.
For size of rails, see 2.5.
For barbed wire, see 3.9.
If barbed wire is not specified, the length shall be 2.65 m.

TABLE 4. PRESERVATIVE TREATMENTS ACCORDING TO SPECIES

Species	Use	Treatment
Oak (European and English), sweet chestnut, European larch, yew, western red cedar	Rails and posts	Treatment only if specified by the purchaser....
Spruce, Douglas fir and hemlock	Rails Posts	*Appropriate treatments are specified*
Hardwoods or softwoods other than spruce and Douglas fir	Rails Posts	
Imported hardwoods....	Posts	

Part 7: 1972: Wooden post and rail fences.

TABLE 1. WOODEN POST AND RAIL FENCING

Type reference	Height to top of top rail	Main posts length	Prick posts length	Clear spacing between rails top to bottom
	m	m	m	mm
Morticed type				
MPR 11/3 (3 rail)	1·1	1·8	1·6	325—275—240 (to ground)
MPR 11/4 (4 rail)	1·1	1·8	1·6	225—200—175—150 (to ground)
MPR 13/4 (4 rail)	1·3	2·1	1·8	250—250—225—225 (to ground)
Nailed type				
SPR 11/3 (3 rail)	1·1	1·8	—	325—275—240 (to ground)
SPR 11/4 (4 rail)	1·1	1·8	—	225—200—175—150 (to ground)
SPR 13/4 (4 rail)	1·3	2·1	—	250—250—225—225 (to ground)

Part 11: 1972: Woven wood fences.

All parts of this British Standard (B.S. 1722) cover construction workmanship, types, height, selection and preservation of materials.

TABLE 1. DIMENSIONS OF PANELS

Type reference	Height
	m
WW 60	0.6
WW 90	0.9
WW 120	1.2
WW 150	1.5
WW 180	1.8

TABLE 2. LENGTHS OF POSTS

Type reference	Height of fence	Length of posts
	m	m
WW 60	0.6	1.2
WW 90	0.9	1.6
WW 120	1.2	1.8
WW 150	1.5	2.1
WW 180	1.8	2.5

B.S. 4092: Domestic front entrance gates.

Part 2: 1966: Wooden gates: sizes, design requirements, construction, finished sizes of components, quality of timber, workmanship and hanging.

TABLE 1. MINIMUM FINISHED SIZES OF COMPONENTS

	in	mm
Hanging stiles for pairs of gates	3¾ × 1¾	95 × 45
Hanging stiles for single gates	2¾ × 1¾	70 × 45
Shutting stiles	2¾ × 1¾	70 × 45
Rails behind infilling	2¾ × 1	70 × 25
Other rails	2¾ × 1¾	70 × 45
Brace	2¾ × 1	70 × 25
Thickness of infilling	9⁄16	14

3.2 Widths

The standard widths of gates, measured overall of the stiles, shall be:

Single gates: 2 ft 8 in (810 mm)
3 ft 4 in (1020 mm)
Pairs of gates: 7 ft 0 in (2130 mm)
7 ft 8 in (2340 mm)
8 ft 8 in (2640 mm)

These widths shall be subject to a tolerance of $+0 -\frac{1}{4}$ in (6 mm).

WOOD TRIM

B.S. 584: 1967: Wood trim (softwood).

Quality, design and dimensions of architraves, skirtings, picture rails, cover fillets, quadrants, half-round beads and scotias.

BOARDS AND SHEETS

B.S. 1142: Fibre building boards.

Part 1: 1971: Methods for testing.

Part 2: 1971: Medium board and hardboard: types defined, coding scheme for testing, performance levels.

3. REQUIREMENTS FOR SPECIAL PURPOSE BOARDS

3.1 General

As their classification implies, fibre building boards within the scope of this section have been manufactured or modified after manufacture to fulfil a specific use. The physical requirements of these boards shall be as agreed between the manufacturer or his representative and the purchaser.

3.2 Type FR Flame retardant hardboards

3.2.1 Surface spread of flame classification. The boards shall be tested in accordance with the requirements of Clause 2 of BS 476 : Part 7 ‡: 1971, and they shall be classified into one of the following groups:

(1) Boards giving the same flame spread classification on both faces in the ' Surface spread of flame test '.

Class 1A, both faces comply with Class 1
Class 2A, both faces comply with Class 2
Class 3A, both faces comply with Class 3
Class 4A, both faces comply with Class 4
} in Table 1 of BS 476 : Part 7‡ : 1971

(2) Boards giving a higher flame spread classification on one face than on the other in the ' Surface spread of flame test '.

Class 1B, one face complies with Class 1
Class 2B, one face complies with Class 2
Class 3B, one face complies with Class 3
Class 4B, one face complies with Class 4
} in Table 1 of BS 476 : Part 7* : 1971

With these Type ' B ' boards, the face having the superior performance shall be the unmarked face.

NOTE. Boards which have been treated or processed to improve their flame retardant properties may require the application of special type sealers or primers prior to decoration. Possible reduction of original flame retardant properties after applying surface treatments should be considered.

TABLE 1. TEST LEVELS FOR MEDIUM BOARD

Type	Specified manufacturing thickness mm	Minimum mean bending strength N/mm² (MPa)	Changes after water immersion		Effect of relative humidity change from 33% to 90%	
			Maximum mean absorption by weight %	Maximum mean thickness swelling %	Maximum mean increase in length and width %	Maximum mean increase in thickness %
HME	equal to or greater than 6.4 but equal to or less than 10	20				
	greater than 10 but equal to or less than 13	17	20	10	0.25	7
	greater than 13 but equal to or less than 16	15				
	greater than 16	13				
HMN	equal to or greater than 6.4 but equal to or less than 10	15				
	greater than 10 but equal to or less than 13	12	40	20	0.30	10
	greater than 13 but equal to or less than 16	10				
	greater than 16	8				
LME	6.4	14	25	13		
	greater than 6.4 but equal to or less than 10	11	20	10	0.30	5
	greater than 10	9	15	8		
LMN	6.4	12	35	18		
	greater than 6.4 but equal to or less than 10	10	30	15	0.40	8
	greater than 10	8	25	12		

Part 3: 1972: Insulating board (softboard) and tiles: types defined and performance levels.

3. REQUIREMENTS FOR SPECIAL PURPOSE BOARDS OR TILES

3.1 General

Fibre building boards within the scope of this section shall be those manufactured or modified after manufacture for a specific use. The permissible deviations in squareness, size and straightness of boards or tiles shall comply with the definitions given in 1.3 and qualified in Clause 2.

3.2 Flame retardant insulating boards and tiles (type FL.R.)

The physical requirements of these materials shall be agreed between the supplier and the purchaser.

3.2.1 Surface spread of flame classification. FL.R. boards or tiles shall be tested, after applying any decorative effects or treatments during manufacture, in accordance with the requirements of BS 476: Part 7‡, and they shall be classified as one of the following types:

(1) Boards giving the same classification (from Table 1 of BS 476: Part 7‡) on both faces in the surface spread of flame test.

Type 1A Both faces are within the limits of Class 1
Type 2A Both faces are within the limits of Class 2
Type 3A Both faces are within the limits of Class 3

(2) Boards giving a higher classification (from Table 1 of BS 476: Part 7‡) on one face than the other in the surface spread of flame test.

Type 1B2 One face falls within Class 1 and the other within Class 2
Type 1B3 One face falls within Class 1 and the other within Class 3
Type 1B4 One face falls within Class 1 and the other within Class 4
Type 2B3 One face falls within Class 2 and the other within Class 3
Type 2B4 One face falls within Class 2 and the other within Class 4
Type 3B4 One face falls within Class 3 and the other within Class 4

The face of these Type B boards having the superior performance shall be the unmarked face.

NOTE. Boards which have been treated or processed to improve their flame retardant properties may require the application of special sealers or primers prior to decoration. Possible reduction of original flame retardant properties after applying surface treatments should be considered.

B.S. 1455: 1972: Plywood manufactured from tropical hardwoods.

Describes grades and qualities.

B.S. 2604: Resin-bonded wood chipboard.

Part 2: 1970: Metric units: basic levels of quality and minimum acceptance limits.

TABLE 1. BASIC LEVELS OF QUALITY

1	2	3	
Property	Clause in BS 1811: Part 2: 1969*	Mean ($\bar{X}$)	
	12 mm thickness		
Density, kg/m³	7	480	
Bending strength, N/mm²§	8	13.8	
Tensile strength perpendicular to the plane of the board, N/mm²§		0.34	
Surface soundness, N	11	1100	
Edge screw holding, N	12	360	
Thickness swelling (1 hour) (%)	13	12	
	19.1		
	19 mm thickness		
Density, kg/m³	7	480	
Bending strength, N/mm²§	8	13.8	
Tensile strength perpendicular to the plane of the board, N/mm²§		0.34	
Surface soundness, N	11	1100	
Edge screw holding, N	12	360	
Thickness swelling (1 hour) (%)	13	12	
	19.1		

* BS 1811, 'Methods of test for wood chipboards and other particle boards', Part 2, 'Metric units'.
† BS 476, 'Fire tests on building materials and structures', Part 7, 'Surface spread of flame test for materials'.
‡ i.e., for swelling the mean value shall not be greater; for the other properties the mean values shall not be less.
§ *In the original standard these values are expressed in MN/m². Numerically they are identical.*

TABLE 3. CONSUMERS' MINIMUM ACCEPTANCE LIMITS

1	2	3
Property	Clause in BS 1811: Part 2: 1969†	Limit for sample means for all thicknesses complying with the standard
Density, kg/m³	7	460
Bending strength, N/mm²‡	8	13.1
Tensile strength perpendicular to the plane of the board, N/mm²‡		0.31
Surface soundness, N	11	1020
Edge screw holding, N	12	310
Thickness swelling (1 hour) (%)	13	13
	19.1	

B.S. 3444: 1972: Blockboard and laminboard.

Defines grades and qualities.

B.S. 3794: 1973: Decorative laminated plastics sheet.

Describes four classes of aminoplastic phenolic laminated sheet.

B.S. 4965: 1974: Decorative laminated plastics sheet veneered boards or panels.

Specifies requirements for eight types made from various core materials veneered on one or both sides with cured aminoplastic faced phenol laminated sheet.

GLULAM

B.S. 4169: 1970: Glued-laminated timber structural members.

Covers structural components manufactured from separate pieces of timber arranged in laminations parallel to the axis of the member, individual pieces being assembled with the grain approximately parallel and glued together to form a member which functions as a single structural unit.

Design requirements are given in CP 112: The structural use of timber.

FASTENINGS—NAILS, SCREWS AND CONNECTORS

B.S. 1202: Nails.

    Part 1: 1974: Steel nails.

TABLE 1. DIMENSIONS AND APPROXIMATE COUNT OF ROUND PLAIN HEAD NAILS

Length L	Shank diameter D	Approx. no. of nails per kg
mm	mm	
200	8.00	13
180	6.70	22
150	6.00	29
150	5.60	35
125	5.60	42
125	5.00	53
115	5.00	57
100	5.00	66
100	4.50	77

Length L	Shank diameter D	Approx. no. of nails per kg
mm	mm	
100	4.00	88
100	3.75	110
90	4.50	88
90	4.00	106
90	3.75	123
90	3.35	152
75	4.00	121
75	3.75	154
75	3.35	194
75	3.00	236
65	3.75	175
65	3.35	230
65	3.00	275
65	2.65	350
60	3.35	255
60	3.00	310
60	2.65	385
50	3.35	290
50	3.00	340
50	2.65	440
50	2.36	550
45	2.65	510
45	2.36	640
45	2.00	840
40	2.65	575
40	2.36	750
40	2.00	970
30	2.36	840
30	2.00	1170
30	1.80	1410
25	2.00	1430
25	1.80	1720
25	1.60	2120
20	1.60	2710
20	1.40	3750
15	1.40	4400

TABLE 2. DIMENSIONS AND APPROXIMATE
COUNT OF ROUND LOST HEAD NAILS

Length L mm	Shank diameter D mm	Approx. no. of nails per kg
75	3.75	160
65	3.35	240
65	3.00	270
60	3.35	270
60	3.00	330
50	3.00	360
50	2.65	420
40	2.36	760
30	2.00	1190
25	1.00	6100
20	1.00	8030
15	1.00	9400

TABLE 7. DIMENSIONS AND APPROXIMATE
COUNT OF PANEL PINS

Length L mm	Shank diameter D mm	Approx. no. of nails per kg
75	2.65	290
65	2.65	345
50	2.00	770
40	1.60	1590
30	1.60	1900
25	1.60	2340
25	1.40	3090
20	1.60	3140
20	1.40	3970
20	1.25	5290
15	1.25	6400
15	1.00	8800

TABLE 12. DIMENSIONS AND APPROXIMATE COUNT OF OVAL BRAD HEAD NAILS

Length L	Shank dimensions $D \times d$	Approx. no. of nails per kg
mm	mm	
150	7.10 × 5.00	31
125	6.70 × 4.50	44
100	6.00 × 4.00	64
90	5.60 × 3.75	90
75	5.00 × 3.35	125
65	4.00 × 2.65	230
60	3.75 × 2.36	340
50	3.35 × 2.00	470
45	3.35 × 2.00	655
40	2.65 × 1.60	940
30	2.65 × 1.60	1480
25	2.00 × 1.25	2530
20	2.00 × 1.25	4500

TABLE 24. DIMENSIONS AND APPROXIMATE COUNT OF CUT CLASP NAILS

Length L	Shank dimension D	Approx. no. of nails per kg
mm	mm	
200	6.00	11
175	5.60	13
150	5.60	19
125	5.00	30
100	4.00	48
90	3.75	66
75	3.35	103
65	3.00	171
60	2.65	202
50	2.65	286
40	2.00	616
30	1.80	858
25	1.60	1384

TABLE 25. DIMENSIONS AND APPROXIMATE COUNT OF CUT FLOOR BRADS

Length L	Shank dimension D	Approx. no. of nails per kg
mm	mm	
75	3.35	100
65	3.35	154
60	3.00	198
50	2.65	264
45	2.36	330
40	2.36	396

Part 2: 1974: Copper nails.

TABLE 4. DIMENSIONS AND APPROXIMATE COUNT OF FLAT COUNTERSUNK AND ROSE-HEAD SQUARE SHANK BOAT NAILS (ROUND OR DIAMOND POINT)

Length L	Shank dimension D	Approx. no. of nails per kg
mm	mm	
150	6.00	22
150	5.00	33
125	5.60	31
125	4.50	44
100	5.00	46
100	4.00	66
90	4.00	81
75	4.00	88
75	3.75	108
75	3.35	132
65	4.00	103
65	3.35	163
65	3.00	198
65	2.65	246
50	3.35	193
50	3.00	255
50	2.65	308
45	3.35	220

Length L	Shank diameter D	Approx. no. of nails per kg
mm	mm	
45	2.65	330
40	2.65	412
40	2.36	494
40	2.00	688
30	2.65	490
30	2.36	659
30	2.00	862
25	2.36	774
25	2.00	939
25	1.80	1336
20	2.00	1237
20	1.80	1740

Part 3: 1974: Aluminium nails.

TABLE 1. DIMENSIONS AND APPROXIMATE COUNT OF ROUND PLAIN HEAD NAILS

Length L	Shank diameter D	Approx. no. of nails per kg
mm	mm	
115	5.00	159
100	5.00	184
100	4.50	215
90	4.50	246
90	4.00	296
75	4.00	338
75	3.75	431
75	3.35	543
65	3.75	490
65	3.35	644
65	3.00	770
60	3.35	714
60	3.00	868
50	3.35	812
50	3.00	952

Length L	Shank diameter D	Approx. no. of nails per kg
mm	mm	
50	2.65	1232
45	2.65	1428
45	2.36	1792
40	2.65	1610
40	2.36	2100
40	2.00	2716
30	2.00	3276
30	1.80	2948
25	2.00	4004
25	1.80	4816
25	1.60	5936
20	1.60	7588

TABLE 2. DIMENSIONS AND APPROXIMATE COUNT OF ROUND LOST HEAD NAILS

Length L	Shank diameter D	Approx. no. of nails per kg
mm	mm	
75	3.75	448
65	3.35	672
60	3.35	756
50	3.35	860
50	3.00	1008
40	2.64	1390
40	2.36	2128

B.S. 1210: Wood screws.

Materials, dimensions, tolerances and other criteria for screws of steel, stainless steel, brass, aluminium, silicon bronze and nickel copper alloy.

TABLE 1. SLOTTED COUNTERSUNK HEAD WOOD SCREWS

1	2	5	6
Screw gauge	Size of screw and diameter of unthreaded shank D Nom. in	Number of threads per inch	Diameter of head V^* in
0	0.060	30	0.126
1	0.070	28	0.147
2	0.082	26	0.174
3	0.094	24	0.199
4	0.108	22	0.230
5	0.122	20	0.261
6	0.136	18	0.291
7	0.150	16	0.323
8	0.164	14	0.353
9	0.178	12	0.384
10	0.192	12	0.414
12	0.220	10	0.476
14	0.248	9	0.538
16	0.276	8	0.599
18	0.304	7½	0.660
20	0.332	7	0.721
24	0.388	6	0.845
28	0.444	5½	0.967
32	0.500	5	1.088

* The dimensions for 'V' are the theoretical diameters of head to sharp corners and are given for design purposes only.

TABLE 1M. METRIC EQUIVALENTS OF TABLE 1

NOTE. The metric conversions are approximate; the figures in British units in Table 1 are to be regarded as the standard.

1	2	5	6
Screw gauge	Size of screw and diameter of unthreaded shank D Nom. mm	Number of threads per cm	Diameter of head V^* mm
0	1.52	12	3.20
1	1.78	11	3.73
2	2.08	10	4.42
3	2.39	9½	5.05
4	2.74	8½	5.84
5	3.10	8	6.63
6	3.45	7	7.39
7	3.81	6½	8.20
8	4.17	5½	8.97
9	4.52	4½	9.75
10	4.88	4½	10.52
12	5.59	4	12.09
14	6.30	3½	13.67
16	7.01	3	15.21
18	7.72	3	16.76
20	8.43	2¾	18.31
24	9.86	2½	21.46
28	11.28	2	24.56
32	12.70	2	27.64

* The dimensions for 'V' are the theoretical diameters of head to sharp corners and are given for design purposes only.

THE CARPENTER AND JOINER

TABLE 2. STANDARD SIZES OF STEEL SLOTTED COUNTERSUNK HEAD WOOD SCREWS*

Length		Screw gauge														
in	mm	0	1	2	3	4	5	6	7	8	9	10	12	14	16	18	20
3/16	4.8		1														
¼	6.4	0	1	2	3	4	5										
5/16	7.9		1	2	3	4	5	6									
3/8	9.5	0	1	2	3	4	5	6	7	8							
7/16	11.1					4		6									
½	12.7	0	1	2	3	4	5	6	7	**8**	9	10					
5/8	15.9		1	2	3	**4**	5	**6**	7	**8**	9	**10**	12				
¾	19.1			2	3	**4**	5	**6**	7	**8**	9	**10**	**12**	14			
7/8	22.2				3	4	5	6	7	8	9	10	12	14			
1	25.4			2	3	**4**	5	**6**	7	**8**	9	**10**	**12**	**14**	16		
1¼	31.8					4	5	**6**	7	**8**	9	**10**	**12**	**14**	**16**	18	
1½	38.1					4	5	**6**	7	**8**	9	**10**	**12**	**14**	**16**	**18**	20
1¾	44.5					4	5	6	7	**8**	9	**10**	**12**	**14**	**16**	18	
2	50.8					4	5	**6**	7	**8**	9	**10**	**12**	**14**	**16**	**18**	20
2¼	57.2							6	7	8	9	10	12	**14**	*16		
2½	63.5							6	7	**8**	9	**10**	**12**	**14**	**16**	18	20
2¾	69.9							6		8		10	12	14	16		
3	76.2							6	7	**8**	9	**10**	**12**	**14**	**16**	**18**	20
3¼	82.6											10	12	14			
3½	88.9									8		10	**12**	**14**	16	18	
4	101.6									8		10	**12**	**14**	**16**	18	20
4½	114.3											10	12	14	16		20
5	127.0											10	**12**	**14**	**16**	18	20
6	152.4												12	14	16	18	

* Preferred standard sizes, in bold type, should be used whenever possible, as they are the sizes normally stocked.

Supplementary standard sizes, in light type, are less likely to be obtained from local stockists than the preferred standard sizes shown in bold type.

B.S. 1579: 1960: Connectors for timber.

Split rings, shear plates, round and square-toothed plates, appropriate bolts and special washers.

LOCKS, HINGES AND HARDWARE

B.S. 455: 1957: Schedule of sizes of locks and latches for doors in buildings.

TABLE 8. SLOTTED ROUND HEAD WOOD SCREWS

1	2	5	6
Size of screw and diameter of unthreaded shank		Number of threads per inch	Diameter of head
Screw gauge	D Nom.		V*
	in		in
0	0.060	30	0.126
1	0.070	28	0.147
2	0.082	26	0.174
3	0.094	24	0.199
4	0.108	22	0.230
5	0.122	20	0.261
6	0.136	18	0.291
7	0.150	16	0.323
8	0.164	14	0.353
9	0.178	12	0.384
10	0.192	12	0.414
12	0.220	10	0.476
14	0.248	9	0.538
16	0.276	8	0.599
18	0.304	7½	0.660
20	0.332	7	0.721
24	0.388	6	0.845
28	0.444	5½	0.967
32	0.500	5	1.088

* The dimensions for 'V' are the theoretical diameters of head to sharp corners and are given for design purposes only.

TABLE 1M. METRIC EQUIVALENTS OF TABLE 1

NOTE. The metric conversions are approximate; the figures in British units in Table 1 are to be regarded as the standard.

1	2	5	6
Size of screw and diameter of unthreaded shank		Number of threads per cm	Diameter of head
Screw gauge	D Nom.		V*
	mm		mm
0	1.52	12	3.20
1	1.78	11	3.73
2	2.08	10	4.42
3	2.39	9½	5.05
4	2.74	8½	5.84
5	3.10	8	6.63
6	3.45	7	7.39
7	3.81	6½	8.20
8	4.17	5½	8.97
9	4.52	4½	9.75
10	4.88	4½	10.52
12	5.59	4	12.09
14	6.30	3½	13.67
16	7.01	3	15.21
18	7.72	3	16.76
20	8.43	2½	18.31
24	9.86	2½	21.46
28	11.28	2	24.56
32	12.70	2	27.64

* The dimensions for 'V' are the theoretical diameters of head to sharp corners and are given for design purposes only.

TABLE 20. RECESSED COUNTERSUNK HEAD WOOD SCREWS (FOR STANDARD SIZES, SEE TABLES 21 AND 22)

1	2	5	6	9
Size of screw and diameter of unthreaded shank		Number of threads per inch	Diameter of head	Depth of head C†
Screw gauge	D Nom.		V*	Max.
	in		in	in
3	0.094	24	0.199	0.055
4	0.108	22	0.230	0.064
5	0.122	20	0.261	0.073
6	0.136	18	0.291	0.082
7	0.150	16	0.323	0.091
8	0.164	14	0.353	0.100
9	0.178	12	0.384	0.109
10	0.192	12	0.414	0.117
12	0.220	10	0.476	0.135
14	0.248	9	0.538	0.153
16	0.276	8	0.599	0.170

* The dimensions for 'V' are the theoretical diameters of head to sharp corners and are given for design purposes only.
† The maximum depth of head 'C' on a nominal size shank is given for the convenience of users only.

TABLE 20M. METRIC EQUIVALENTS OF TABLE 20
NOTE. The metric conversions are approximate; the figures in British units in Table 20 are to be regarded as the standard.

1	2	5	6	9
Size of screw and diameter of unthreaded shank		Number of threads per cm	Diameter of head	Depth of head C†
Screw gauge	D Nom.		V*	Max.
	mm		mm	mm
3	2.39	9½	5.05	1.40
4	2.74	8½	5.84	1.63
5	3.10	8	6.63	1.85
6	3.45	7	7.39	2.08
7	3.81	6½	8.20	2.31
8	4.17	5½	8.97	2.54
9	4.52	4½	9.75	2.77
10	4.88	4½	10.52	2.97
12	5.59	4	12.09	3.43
14	6.30	3½	13.67	3.89
16	7.01	3	15.21	4.32

* The dimensions for 'V' are the theoretical diameters of head to sharp corners and are given for design purposes only.
† The maximum depth of head 'C' on a nominal size shank is given for the convenience of users only.

B.S. 1227: Hinges.

Part 1A: 1967: Hinges for general building purposes: classified according to metal. Dimensions, weights, and diagrammatic figures.

TABLE 1. DIMENSIONS .... OF STEEL BROAD BUTT HINGES

1	2	3		6	7
BS type reference No.	Length of joint	Open width over flaps		Holes countersunk for screw gauge No.	Number of screw holes in each hinge
	in	in			
		min.	max.	min.	
1A/101	2 ± 0.008	1$^{15}/_{16}$	2	7	4
1A/102	2½ ± 0.008	2$^{7}/_{16}$	2½	7	6
1A/103	3 ± 0.008	2$^{15}/_{16}$	3	8	6
1A/104	3½ ± 0.008	3$^{7}/_{16}$	3½	8	6
1A/105	4 ± 0.008	3$^{15}/_{16}$	4	10	8
Metric equivalents					
	mm	mm			
		min.	max.	min.	
1A/101	50.8 ± 0.19	49.2	50.8	7	4
1A/102	63.5 ± 0.19	61.9	63.5	7	6
1A/103	76.2 ± 0.19	74.6	76.2	8	6
1A/104	88.9 ± 0.19	87.3	88.9	8	6
1A/105	101.6 ± 0.19	100.0	101.6	10	8

Standard finish: Bright, self colour. Other finishes available if specially ordered

NOTE. It is recommended that loose headed pins should be designated by adding 'L' to the BS type reference number in Column 1, e.g. '1A/101L'.

TABLE 2. DIMENSIONS .... OF STRONG STEEL BUTT HINGES

1	2	3			6	7
BS type reference No.	Length of joint	Open width over flaps			Holes countersunk for screw gauge No.	Number of screw holes in each hinge
	in	in				
		min.	max.		min.	
1A/201	1½ ± 0.008	1 13/32	1 15/32	...	7	4
1A/202	2 ± 0.008	1 27/32	1 29/32		8	4
1A/203	2½ ± 0.008	1 31/32	2 1/32		8	6
1A/204	3 ± 0.008	2 3/8	2 7/16		10	6
1A/205	3½ ± 0.008	2 11/16	2 3/4		10	6
1A/206	4 ± 0.008	2 7/8	2 15/16		10	8
1A/207	5 ± 0.008	3 7/16	3 1/2		12	8
1A/208	6 ± 0.008	3 9/16	3 5/8		12	8

Metric equivalents

	mm	mm				
		min.	max.		min.	
1A/201	38.1 ± 0.19	35.7	37.3	...	7	4
1A/202	50.8 ± 0.19	46.8	48.4		8	4
1A/203	63.5 ± 0.19	50.0	51.6		8	6
1A/204	76.2 ± 0.19	60.3	61.9		10	6
1A/205	88.9 ± 0.19	68.3	69.9		10	6
1A/206	101.6 ± 0.19	73.0	74.6		10	8
1A/207	127.0 ± 0.19	87.3	88.9		12	8
1A/208	152.4 ± 0.19	90.5	92.1		12	8

Standard finish: Bright, self colour. Other finishes available if specially ordered.

TABLE 3. DIMENSIONS .... OF NARROW STEEL BUTT HINGES

1	2	3		6	7
BS type reference No.	Length of joint	Open width over flaps		Holes countersunk for screw gauge No.	Number of screw holes in each hinge
	in	in			
		min.	max.	min.	
1A/301	1 + .005 − .009	29/32	31/32	4	4
1A/302	1¼ + .005 − .012	1 3/32	1 5/32	4	4
1A/303	1½ + .005 − .014	1 11/32	1 13/32	5	4
1A/304	2 + .005 − .014	1½	1 9/16	6	4
1A/305	2½ + .008 − .014	1 21/32	1 23/32	6	6
1A/306	3 + .008 − .020	1 29/32	2	7	6
1A/307	3½ + .008 − .022	2 9/32	2⅜	8	6
1A/308	4 + .008 − .026	2 23/32	2⅞	8	8
1A/309	5 + .008 − .026	3 7/16	3 9/16	10	8
1A/310	6 + .008 − .026	3⅞	4	10	8
Metric equivalents					
	mm	mm			
		min.	max.	min.	
1A/301	25.4 + 0.13 − 0.23	23.0	24.6	4	4
1A/302	31.8 + 0.13 − 0.31	27.8	29.4	4	4
1A/303	38.1 + 0.13 − 0.36	34.1	35.7	5	4
1A/304	50.8 + 0.19 − 0.36	38.1	39.7	6	4
1A/305	63.5 + 0.19 − 0.36	42.1	43.7	6	6
1A/306	76.2 + 0.19 − 0.51	48.4	50.8	7	6
1A/307	88.9 + 0.19 − 0.56	57.9	60.3	8	6
1A/308	101.6 + 0.19 − 0.66	70.0	73.0	8	8
1A/309	127.0 + 0.19 − 0.66	81.0	84.1	10	8
1A/310	152.4 + 0.19 − 0.66	98.4	101.6	10	8

Standard finish: Bright, self colour. Other finishes available if specially ordered

NOTE. It is recommended that loose headed pins should be designated by adding ' L ' to the BS type reference, e.g. ' 1A/301 L '.

TABLE 6. DIMENSIONS .... OF STEEL RISING BUTT HINGES

1	2	3			6	7
BS type reference No.	Length of joint	Open width over flaps			Holes countersunk for screw gauge No.	Number of screw holes in each hinge
	in	in				
		min.		max.	min.	
1A/601	3 +0.008 / −0.020	22³⁰⁄₃₂		22²⁹⁄₃₂	8	6
1A/602	3½ +0.008 / −0.022	22²⁷⁄₃₂		22²⁹⁄₃₂	10	8
1A/603	4 +0.008 / −0.026	23¹⁷⁄₃₂		3¹⁄₃₂	10	8

Metric equivalents

	mm	mm			min.	
		min.		max.		
1A/601	76.2 +0.19 / −0.51	69.1		70.6	8	6
1A/602	88.9 +0.19 / −0.56	72.2		73.8	10	8
1A/603	101.6 +0.19 / −0.66	75.4		77.0	10	8

Standard finish: Bright, self colour. Other finishes available if specially ordered

NOTE. These hinges are handed and it is essential that purchasers indicate, when ordering, whether left-hand or right-hand hinges are required.

TABLE 18. DIMENSIONS .... OF CAST IRON BUTT HINGES

1	2	3		6	7	
BS type reference No.	Length of joint	Open width over flaps	...	Holes countersunk for screw gauge No.	Number of screw holes in each hinge	...
	in	in		min.		
1A/1801	2½ ± ¹⁄₁₆	1¾ ± ¹⁄₁₆		10	6	
1A/1802	3 ± ¹⁄₁₆	2 ± ¹⁄₁₆		10	6	
1A/1803	4 ± ¹⁄₁₆	2⅜ ± ¹⁄₁₆		12	8	
Metric equivalents						
	mm	mm		min.		
1A/1801	63.5 ± 1.59	44.5 ± 1.59		10	6	
1A/1802	76.2 ± 1.59	50.8 ± 1.59		10	6	
1A/1803	101.6 ± 1.59	60.3 ± 1.59		12	8	

TABLE 19. DIMENSIONS .... OF CAST IRON RISING BUTT HINGES

1	2	3		8	9
BS type reference No.	Length of joint	Open width over flaps	....	Holes countersunk for screw gauge No.	Number of screw holes in each hinge
	in	in		min.	
1A/1901	3 ± 1/16	2 5/16 ± 1/16		12	6
1A/1902	4 ± 1/16	2 3/4 ± 1/16		12	6
Metric equivalents					
	mm	mm	....	min.	
1A/1901	76.2 ± 1.59	58.7 ± 1.59		12	6
1A/1902	101.6 ± 1.59	69.9 ± 1.59		12	6

NOTE. These hinges are handed and it is essential that purchasers indicate, when ordering, whether left-hand or right-hand hinges are required.

BUILDERS' HARDWARE

B.S. 1331: 1954: Builders' hardware for housing.

1. SCOPE

.... materials, main dimensions, minimum weights and fixing accessories for ....

Schedule One	**Door and gate equipment**
	Knobs
	Lever handles
	Lock sets
	Pull handles
	Finger plates
	Thumb latches
	Cabin hooks
	Gate latches
Schedule Two	**Window equipment**
	Casement stays
	Casement fasteners
	Cremorne bolts
	Sash pulleys
	Sash lifts
	Sash fasteners
	Sash pivots
Schedule Three	**Fanlight equipment**
	Catches
	Stays
Schedule Four	**Cupboard equipment**
	Thread escutcheon
	Knobs
	Turns
	Catches
	Buttons
	Bales catches
	Ball catches
	Hat and coat hooks
	Wardrobe hooks
Schedule Five	**Drawer equipment**
	Pulls

ADHESIVES

B.S. 745: 1969: Animal glue for wood (joiner's glue) (dry glue; jelly or liquid glue).

Specifies dry glue supplied in cakes, pieces, granules, pearl, cubes or powder, and jelly or liquid glue.

B.S. 1204: Synthetic resin adhesives (phenolic and aminoplastic) for wood.

Part 1: 1964: Gap-filling adhesives: three groups, according to weathering and a fourth group for interior use.

Part 2: 1965: Close-contact adhesives: four groups in terms of weathering.

B.S. 1444: 1970: Cold-setting casein adhesive powders for wood.

Selection of samples, instructions for use, etc.

B.S. 4071: 1966: Polyvinyl acetate (PVA) emulsion adhesives for wood.

Close contact adhesives for interior use.

B.S. 4643: 1970: Glossary of terms relating to joints and jointing in building.

Defines terms for jointing products, joint dimensions and functions.

WOOD JOISTS—FLOOR, CEILING AND ROOF

STRESS GRADED SOFTWOOD

The following excerpts are taken from TRADA Wood Information Sheet 7, *Guide to stress graded softwood*. Copies of this, and other publications in the series are available from the Timber Research and Development Association.

HOW TO COMPLY WITH THE BUILDING REGULATIONS 1976

There are several methods deemed to satisfy the main body of the regulations, the most usual of which are listed below:

1. Use the permissible span tables in Schedule 6 of the Building Regulations 1976. Table to Rule 2 of Schedule 6 gives the species, origin and grade of timber to which the span tables in Schedule 6 relate. The span tables are divided into two sets and the Table to Rule 2 lists the species and grades which are appropriate for use in accordance with each set of span tables.
2. Design to CP 112: 1971 (including amendment slips 1, 2 and 3) using timber graded and marked GS, MGS, M50, SS, MSS, M75 in accordance with B.S. 4978: 1973.
3. Design to CP 112: 1971 (including amendment slips 1, 2 and 3) using timber visually graded in Canada to the NLGA 1978 rules. Amendment slip No. 2 to CP 112 also introduced grade stresses for light framing and studs.

4. Design trussed rafters in accordance with CP 112: Part 3: 1973: Trussed rafters for roofs of dwellings: using the grades as defined in B.S. 4978 or CP 112: Part 2: 1971.

Notes In addition to the methods listed above, a structure can be prototype tested in accordance with CP 112; 1971.

It should be noted that any species or grade, the stresses for which are not less than those given for the Schedule 6 species and grades, can be safely used to the permissible spans given in that Schedule. Stress values should be compared in CP 112; 1971 (including amendment slips 1, 2 and 3).

The so-called 'numbered grades' defined in Appendix A of CP 112: 1971 are being phased out in favour of B.S. 4978: 1973: grades. However these grades (40, 50, 65, 75 or composite 40/50 grade) will continue to be *deemed-to-satisfy* in the Building Regulations until CP 112: 1971 is amended.

Table 1: Floor Joists

GS, MGS, M50, M75 or No.2 grade timber

Size of joist (in mm)	Dead load (in kg/m²) supported by joist, excluding the mass of the joist											
	Not more than 25			More than 25 but not more than 50			More than 50 but not more than 125		
	Spacing of joists (in mm)											
	400	450	600	400	450	600	400	450	600
	Maximum span of joist (in m)											
38 × 75	1.05	0.95	0.72	0.99	0.90	0.69	0.87	0.79	0.62
38 × 100	1.77	1.60	1.23	1.63	1.48	1.16	1.36	1.24	1.00
38 × 125	2.53	2.35	1.84	2.33	2.12	1.69	1.88	1.73	1.40
38 × 150	3.02	2.85	2.48	2.83	2.67	2.26	2.41	2.23	1.83
38 × 175	3.51	3.32	2.89	3.29	3.11	2.71	2.82	2.66	2.27
38 × 200	4.00	3.78	3.30	3.75	3.55	3.09	3.21	3.03	2.64
38 × 225	4.49	4.24	3.70	4.21	3.98	3.47	3.61	3.41	2.96
44 × 75	1.20	1.08	0.83	1.13	1.02	0.79	0.98	0.89	0.70
44 × 100	2.01	1.82	1.41	1.83	1.67	1.31	1.51	1.39	1.12
44 × 125	2.71	2.56	2.09	2.54	2.38	1.90	2.08	1.92	1.56
44 × 150	3.24	3.06	2.67	3.04	2.87	2.50	2.60	2.45	2.03
44 × 175	3.77	3.56	3.10	3.53	3.34	2.91	3.02	2.86	2.48
44 × 200	4.29	4.06	3.54	4.02	3.80	3.31	3.45	3.26	2.83
44 × 225	4.81	4.55	3.97	4.51	4.27	3.72	3.87	3.66	3.18

Table 1 continued: Floor joists

GS, MGS, M50, M75 or No.2 grade timber

Size of joist (in mm)	Dead load (in kg/m²) supported by joist, excluding the mass of the joist									
	Not more than 25			More than 25 but not more than 50			More than 50 but not more than 125		
	Spacing of joists (in mm)								
	400	450	600	400	450	600	400	450	600
	Maximum span of joist (in m)								
50 × 75	1.35	1.22	0.93	1.26	1.14	0.89	1.08	0.99	0.78
50 × 100	2.22	2.03	1.58	2.03	1.85	1.46	1.66	1.53	1.23
50 × 125	2.84	2.72	2.33	2.70	2.55	2.10	2.27	2.09	1.71
50 × 150	3.40	3.26	2.84	3.23	3.05	2.66	2.76	2.61	2.21
50 × 175	3.95	3.78	3.30	3.75	3.55	3.09	3.22	3.04	2.64
50 × 200	4.51	4.31	3.76	4.27	4.04	3.52	3.67	3.46	3.01
50 × 225	5.06	4.83	4.22	4.79	4.53	3.95	4.11	3.89	3.39
63 × 150	3.66	3.52	3.17	3.50	3.38	2.97	3.09	2.92	2.54
63 × 175	4.25	4.10	3.68	4.07	3.93	3.45	3.59	3.40	2.96
63 × 200	4.84	4.67	4.20	4.64	4.48	3.93	4.09	3.87	3.37
63 × 225	5.43	5.24	4.70	5.21	5.02	4.41	4.59	4.34	3.78
75 × 200	5.10	4.93	4.51	4.90	4.72	4.27	4.43	4.20	3.67
75 × 225	5.72	5.52	5.06	5.49	5.30	4.79	4.97	4.71	4.11

Table 1 continued: Floor joists

SS or MSS grade timber

Size of joist (in mm)	Dead load (in kg/m²) supported by joist, excluding the mass of the joist									
	Not more than 25			More than 25 but not more than 50			More than 50 but not more than 125		
	Spacing of joists (in mm)									
	400	450	600	400	450	600	400	450	600
	Maximum span of joist (in m)									
38 × 75	1.41	1.32	1.01	1.35	1.23	0.96	1.15	1.05	0.84
38 × 100	2.11	2.00	1.71	2.00	1.90	1.57	1.77	1.63	1.32
38 × 125	2.74	2.63	2.39	2.62	2.52	2.25	2.34	2.23	1.83
38 × 150	3.28	3.15	2.87	3.14	3.02	2.75	2.82	2.72	2.35
38 × 175	3.81	3.67	3.35	3.65	3.52	3.20	3.29	3.17	2.76
38 × 200	4.35	4.19	3.82	4.16	4.01	3.66	3.76	3.62	3.16
38 × 225	4.88	4.70	4.29	4.68	4.51	4.11	4.22	4.06	3.54
44 × 75	1.51	1.43	1.16	1.45	1.37	1.09	1.29	1.18	0.94
44 × 100	2.25	2.13	1.87	2.13	2.02	1.77	1.89	1.80	1.47
44 × 125	2.87	2.76	2.52	2.74	2.64	2.40	2.47	2.37	2.02
44 × 150	3.43	3.31	3.01	3.29	3.16	2.88	2.96	2.85	2.55
44 × 175	3.99	3.85	3.51	3.83	3.68	3.36	3.45	3.32	2.97
44 × 200	4.55	4.39	4.00	4.36	4.20	3.83	3.94	3.79	3.39
44 × 225	5.11	4.93	4.50	4.90	4.72	4.31	4.42	4.26	3.81

Table 1 continued: Floor joists

SS or MSS grade timber

Size of joist (in mm)	Dead load (in kg/m²) supported by joist, excluding the mass of the joist									
	Not more than 25			More than 25 but not more than 50			More than 50 but not more than 125		
	Spacing of joists (in mm)								
	400	450	600	400	450	600	400	450	600
	Maximum span of joist (in m)								
50 × 75	1.60	1.51	1.30	1.53	1.45	1.21	1.39	1.30	1.04
50 × 100	2.38	2.26	1.98	2.25	2.14	1.89	1.99	1.90	1.61
50 × 125	2.99	2.88	2.62	2.86	2.75	2.51	2.58	2.48	2.21
50 × 150	3.57	3.44	3.14	3.42	3.30	3.01	3.09	2.97	2.71
50 × 175	4.16	4.01	3.66	3.98	3.84	3.50	3.60	3.46	3.15
50 × 200	4.74	4.57	4.17	4.54	4.38	3.99	4.10	3.95	3.60
50 × 225	5.32	5.13	4.68	5.10	4.91	4.49	4.61	4.44	4.05
63 × 150	3.84	3.70	3.38	3.68	3.55	3.24	3.32	3.20	2.92
63 × 175	4.47	4.31	3.94	4.28	4.13	3.77	3.87	3.73	3.40
63 × 200	5.09	4.91	4.49	4.88	4.71	4.30	4.41	4.25	3.88
63 × 225	5.71	5.51	5.04	5.48	5.28	4.83	4.95	4.77	4.36
75 × 200	5.37	5.18	4.74	5.15	4.97	4.54	4.66	4.49	4.10
75 × 225	6.02	5.81	5.32	5.78	5.57	5.10	5.23	5.04	4.61

Table 2: Ceiling joists

GS, MGS, M50, M75 or No.2 grade timber

Size of joist (in mm)	Dead load (in kg/m²) supported by joist, excluding the mass of the joist					
	Not more than 25			More than 25 but not more than 50		
	Spacing of joists (in mm)					
	400	450	600	400	450	600
	Maximum span of joist (in m)					
38 × 75	1.88	1.80	1.57	1.71	1.61	1.40
38 × 100	2.50	2.39	2.08	2.27	2.14	1.86
38 × 125	3.11	2.97	2.59	2.82	2.67	2.32
38 × 150	3.72	3.55	3.10	3.37	3.19	2.78
38 × 175	4.32	4.12	3.60	3.92	3.71	3.23
38 × 200	4.92	4.69	4.10	4.46	4.22	3.68
38 × 225	5.51	5.25	4.60	4.99	4.73	4.13
44 × 75	1.97	1.90	1.68	1.83	1.73	1.51
44 × 100	2.61	2.52	2.23	2.43	2.30	2.00
44 × 125	3.25	3.14	2.78	3.03	2.86	2.49
44 × 150	3.89	3.75	3.32	3.62	3.42	2.98
44 × 175	4.52	4.36	3.86	4.20	3.97	3.47
44 × 200	5.14	4.96	4.39	4.77	4.52	3.95
44 × 225	5.76	5.56	4.92	5.34	5.06	4.43

Table 2 continued: Ceiling joists

GS, MGS, M50, M75 or No.2 grade timber

Size of joist (in mm)	Dead load (in kg/m²) supported by joist, excluding the mass of the joist					
	Not more than 25			More than 25 but not more than 50		
	Spacing of joists (in mm)					
	400	450	600	400	450	600
	Maximum span of joist (in m)					
50 × 75	2.05	1.98	1.79	1.91	1.84	1.60
50 × 100	2.72	2.62	2.38	2.53	2.44	2.13
50 × 125	3.39	3.26	2.95	3.15	3.04	2.65
50 × 150	4.04	3.90	3.53	3.77	3.63	3.17
50 × 175	4.70	4.53	4.10	4.38	4.22	3.68
50 × 200	5.34	5.16	4.66	4.99	4.79	4.19
50 × 225	5.98	5.78	5.22	5.59	5.37	4.70

Table 3: Ceiling joists

SS or MSS grade timber

Size of joist (in mm)	Dead load (in kg/m²) supported by joist, excluding the mass of the joist					
	Not more than 25			More than 25 but not more than 50		
	Spacing of joists (in mm)					
	400	450	600	400	450	600
	Maximum span of joist (in m)					
38 × 75	1.98	1.90	1.73	1.84	1.77	1.61
38 × 100	2.62	2.53	2.30	2.44	2.35	2.14
38 × 125	3.27	3.15	2.87	3.04	2.93	2.67
38 × 150	3.91	3.77	3.44	3.64	3.51	3.20
38 × 175	4.54	4.38	4.00	4.23	4.08	3.72
38 × 200	5.17	4.99	4.56	4.82	4.65	4.25
38 × 225	5.80	5.59	5.12	5.41	5.22	4.77
44 × 75	2.07	1.99	1.82	1.93	1.85	1.69
44 × 100	2.75	2.65	2.42	2.56	2.46	2.24
44 × 125	3.42	3.30	3.01	3.19	3.07	2.80
44 × 150	4.09	3.94	3.60	3.81	3.67	3.35
44 × 175	4.75	4.58	4.19	4.43	4.27	3.90
44 × 200	5.41	5.22	4.78	5.05	4.87	4.45
44 × 225	6.06	5.85	5.36	5.66	5.46	4.99

Table 3 continued: Ceiling joists

SS or MSS grade timber

Size of joist (in mm)	Dead load (in kg.m⁻¹) supported by joist, excluding the mass of the joist					
	Not more than 25			More than 25 but not more than 50		
	Spacing of joists (in mm)					
	400	450	600	400	450	600
	Maximum span of joist (in m)					
50 × 75	2.16	2.08	1.89	2.01	1.93	1.76
50 × 100	2.86	2.76	2.52	2.66	2.57	2.34
50 × 125	3.56	3.43	3.14	3.32	3.20	2.92
50 × 150	4.25	4.10	3.75	3.97	3.82	3.49
50 × 175	4.94	4.77	4.36	4.61	4.45	4.06
50 × 200	5.62	5.42	4.97	5.25	5.06	4.63
50 × 225	6.29	6.08	5.57	5.88	5.68	5.20

Table 4: Joists for flat roofs with access not limited to the purposes of maintenance or repair

GS, MGS, M50, M75 or No.2 grade timber

Size of joist (in mm)	Dead load (in kg/m²) supported by joist, excluding the mass of the joist								
	Not more than 25			More than 25 but not more than 75			More than 75 but not more than 100		
	Spacing of joists (in mm)								
	400	450	600	400	450	600	400	450	600
	Maximum span of joist (in m)								
38 × 75	1.29	1.16	0.89	1.14	1.03	0.81	1.09	0.99	0.78
38 × 100	1.96	1.86	1.51	1.79	1.65	1.31	1.70	1.55	1.24
38 × 125	2.60	2.50	2.22	2.38	2.27	1.87	2.28	2.16	1.75
38 × 150	3.11	3.00	2.73	2.87	2.76	2.45	2.77	2.67	2.28
38 × 175	3.63	3.49	3.18	3.34	3.22	2.85	3.23	3.11	2.71
38 × 200	4.14	3.98	3.63	3.81	3.67	3.26	3.68	3.55	3.09
38 × 225	4.64	4.47	4.08	4.28	4.12	3.66	4.14	3.98	3.47
44 × 75	1.40	1.33	1.02	1.29	1.17	0.92	1.22	1.11	0.88
44 × 100	2.10	1.99	1.72	1.91	1.81	1.48	1.83	1.73	1.39
44 × 125	2.73	2.63	2.38	2.51	2.41	2.09	2.42	2.31	1.95
44 × 150	3.26	3.14	2.87	3.01	2.90	2.63	2.91	2.80	2.50
44 × 175	3.80	3.66	3.34	3.50	3.37	3.06	3.38	3.26	2.91
44 × 200	4.33	4.17	3.81	4.00	3.85	3.50	3.86	3.72	3.32
44 × 225	4.86	4.68	4.28	4.49	4.32	3.93	4.34	4.18	3.73

Table 4 continued: Joists for flat roofs with access not limited to the purposes of maintenance or repair

GS, MGS, M50, M75 or No.2 grade timber

Size of joist (in mm)	Dead load (in kg/m²) supported by joist, excluding the mass of the joist									
	Not more than 25			More than 25 but not more than 75			More than 75 but not more than 100		
	Spacing of joists (in mm)								
	400	450	600	400	450	600	400	450	600
	Maximum span of joist (in m)								
50 × 75	1.49	1.41	1.15	1.38	1.30	1.02	1.34	1.23	0.98
50 × 100	2.22	2.11	1.85	2.01	1.91	1.63	1.94	1.84	1.54
50 × 125	2.84	2.74	2.49	2.62	2.52	2.26	2.53	2.43	2.14
50 × 150	3.40	3.27	2.99	3.13	3.02	2.75	3.03	2.92	2.66
50 × 175	3.95	3.81	3.48	3.65	3.51	3.20	3.53	3.40	3.09
50 × 200	4.51	4.34	3.97	4.16	4.01	3.66	4.02	3.87	3.53
50 × 225	5.06	4.88	4.45	4.67	4.50	4.11	4.52	4.35	3.96
63 × 150	3.66	3.52	3.22	3.37	3.25	2.96	3.26	3.14	2.86
63 × 175	4.25	4.10	3.74	3.93	3.78	3.45	3.79	3.66	3.33
63 × 200	4.84	4.67	4.27	4.47	4.31	3.94	4.33	4.17	3.80
63 × 225	5.43	5.24	4.79	5.02	4.84	4.42	4.86	4.68	4.27
75 × 200	5.10	4.93	4.51	4.72	4.55	4.16	4.57	4.40	4.02
75 × 225	5.72	5.52	5.06	5.30	5.11	4.67	5.12	4.94	4.52

Table 6: Joists for flat roofs with access only for the purposes of maintenance or repair

SS or MSS grade timber

Size of joist (in mm)	Dead load (in kg/m²) supported by joist, excluding the mass of the joist									
	Not more than 25			More than 25 but not more than 75			More than 75 but not more than 100		
	Spacing of joists (in mm)								
	400	450	600	400	450	600	400	450	600
	Maximum span of joist (in m)								
38 × 75	1.98	1.90	1.73	1.73	1.66	1.51	1.64	1.58	1.44
38 × 100	2.62	2.53	2.30	2.30	2.21	2.02	2.19	2.10	1.92
38 × 125	3.27	3.15	2.87	2.87	2.76	2.52	2.73	2.62	2.39
38 × 150	3.91	3.77	3.44	3.43	3.31	3.01	3.27	3.14	2.86
38 × 175	4.54	4.38	4.00	3.99	3.85	3.51	3.80	3.66	3.34
38 × 200	5.17	4.99	4.56	4.55	4.39	4.00	4.34	4.18	3.81
38 × 225	5.80	5.59	5.12	5.11	4.93	4.50	4.87	4.69	4.28
44 × 75	2.07	1.99	1.82	1.81	1.75	1.59	1.72	1.66	1.51
44 × 100	2.75	2.65	2.42	2.41	2.32	2.11	2.29	2.21	2.01
44 × 125	3.42	3.30	3.01	3.01	2.89	2.64	2.86	2.75	2.51
44 × 150	4.09	3.94	3.60	3.60	3.46	3.16	3.42	3.30	3.00
44 × 175	4.75	4.58	4.19	4.18	4.03	3.68	3.98	3.84	3.50
44 × 200	5.41	5.22	4.78	4.77	4.60	4.20	4.54	4.38	3.99
44 × 225	6.06	5.85	5.36	5.35	5.16	4.71	5.10	4.91	4.48

Table 6 continued: Joists for flat roofs with access only for the purposes of maintenance or repair

SS or MSS grade timber

Size of joist (in mm)	Dead load (in kg/m²) supported by joist, excluding the mass of the joist									
	Not more than 25			More than 25 but not more than 75			More than 75 but not more than 100		
	Spacing of joists (in mm)								
	400	450	600	400	450	600	400	450	600
	Maximum span of joist (in m)								
50 × 75	2.16	2.08	1.89	1.89	1.82	1.66	1.80	1.73	1.57
50 × 100	2.86	2.76	2.52	2.51	2.42	2.20	2.39	2.30	2.09
50 × 125	3.56	3.43	3.14	3.13	3.01	2.75	2.98	2.87	2.61
50 × 150	4.25	4.10	3.75	3.74	3.61	3.29	3.56	3.43	3.13
50 × 175	4.94	4.77	4.36	4.35	4.20	3.83	4.15	4.00	3.65
50 × 200	5.62	5.42	4.97	4.96	4.78	4.37	4.73	4.56	4.16
50 × 225	6.29	6.08	5.57	5.56	5.37	4.91	5.30	5.11	4.67

Table 7: Common or Jack rafters for roofs having a pitch more than 10° but not more than 22½° with access only for the purposes of maintenance or repair

GS, MGS, M50, M75 or No.2 grade timber

Size of rafter (in mm)	Dead load (in kg/m²) supported by rafter, excluding the mass of the rafter								
	Not more than 50			More than 50 but not more than 75			More than 75 but not more than 100		
	Spacing of rafters (in mm)								
	400	450	600	400	450	600	400	450	600
	Maximum span of rafter (in m)								
38 × 100	2.42	2.28	1.97	2.20	2.08	1.79	2.03	1.92	1.65
38 × 125	3.01	2.84	2.46	2.74	2.59	2.23	2.53	2.39	2.06
38 × 150	3.60	3.39	2.94	3.28	3.09	2.67	3.03	2.86	2.46
44 × 75	1.96	1.85	1.60	1.79	1.68	1.45	1.65	1.55	1.34
44 × 100	2.60	2.45	2.12	2.37	2.24	1.93	2.19	2.06	1.78
44 × 125	3.23	3.05	2.65	2.95	2.78	2.41	2.73	2.57	2.22
44 × 150	3.86	3.65	3.16	3.53	3.33	2.88	3.26	3.08	2.66
50 × 75	2.09	1.97	1.71	1.91	1.80	1.55	1.76	1.66	1.43
50 × 100	2.77	2.61	2.27	2.53	2.38	2.06	2.34	2.20	1.90
50 × 125	3.44	3.25	2.82	3.14	2.97	2.57	2.91	2.74	2.37
50 × 150	4.10	3.88	3.37	3.75	3.54	3.07	3.48	3.28	2.84

Table 8: Common or jack rafters for roofs having a pitch more than 22½° but not more than 30° with access only for the purposes of maintenance or repair

GS, MGS, M50, M75 or No.2 grade timber

Size of rafter (in mm)	Dead load (in kg/m²) supported by rafter, excluding the mass of the rafter									
	Not more than 50			More than 50 but not more than 75			More than 75 but not more than 100		
	Spacing of rafters (in mm)									
	400	450	600	400	450	600	400	450	600
	Maximum span of rafter (in m)									
38 × 100	2.68	2.53	2.19	2.44	2.30	1.99	2.25	2.12	1.84
38 × 125	3.33	3.15	2.73	3.04	2.87	2.48	2.80	2.65	2.29
38 × 150	3.98	3.76	3.27	3.63	3.43	2.97	3.35	3.16	2.74
44 × 75	2.17	2.05	1.78	1.97	1.86	1.61	1.82	1.72	1.49
44 × 100	2.88	2.72	2.36	2.62	2.47	2.14	2.42	2.28	1.98
44 × 125	3.58	3.38	2.94	3.26	3.08	2.67	3.01	2.84	2.47
44 × 150	4.27	4.04	3.51	3.89	3.68	3.20	3.60	3.40	2.95
50 × 75	2.31	2.18	1.89	2.10	1.98	1.72	1.94	1.83	1.59
50 × 100	3.06	2.89	2.51	2.79	2.63	2.29	2.58	2.43	2.11
50 × 125	3.80	3.59	3.13	3.47	3.28	2.85	3.21	3.03	2.63
50 × 150	4.53	4.29	3.74	4.14	3.91	3.40	3.83	3.62	3.14

Table 9: Softwood floor boards (tongued and grooved)

Finished thickness of board (in mm)	Maximum span of board (in mm)
(1)	(2)
16	505
19	600
21	635
28	790

12. TIMBER PESTS AND PRESERVATION

Wood Preservation	275
Timber Pests	275
Prevention	278
Preservatives — Their Choice and Application ...	278

WOOD PRESERVATION

To combat the spread of the house longhorn beetle, special treatment of softwood timber used in building is required in the following areas; the Urban District Councils of Chertsey, Esher, Farnborough, Farnham, Frimley and Camberley, Staines, Sunbury, Walton and Weybridge and Woking, and in the Rural District Councils of Easthampstead, Guildford, Hartley, Wintney and Hambledon. The following treatments are regarded as good practice.

1. Impregnation under pressure with copper-chrome solution to B.S. 3452 or with copper-chrome-arsenate solution, or
2. Total immersion for ten minutes in:
    (a) chlorinated phenols, metallic napthenates or chlorinated naphthelenes in a solution, or
    (b) a coal-tar oil.

Few, if any, building contractors have the plant to carry out these treatments. Readers intending to build or to tender for building in these areas are advised to contact one of the firms specialising in wood preservation.

TIMBER PESTS

Timber is a natural material: it was man's first building material and in many of its properties is superior to all man-made materials, such as reinforced concrete or steel, which are used in load-bearing frames.

When we use it in building, it is dead and is subject to all manner of attack by insects, birds and fungi. In nature these serve a useful purpose, by replenishing the soil through breaking-down dead timber into what the gardeners call humus.

In houses, insects and plants that prey on dead timber are unmitigated nuisances. They are expensive nuisances also: their activities cost the country millions of pounds every year. It is safe to say that every houseowner has had to spend a considerable sum on repairs necessitated by the activities of these pests. Some have become wholly parasitic on timber structures and, like the "dry rot" fungus, can no longer be found under natural conditions.

DRY ROT

Attack by fungi is today far more widespread than ever before, in spite of most houses built since 1945 having solid ground floors. The principal fungus, which is responsible for 80 per cent of damage by fungal attack is *Merulius Lachrymans*. Largely confined to timber ground floors, it can infest a whole house, including its roof timbers, if it be left shut up and untenanted for a year or so. It produces its own moisture, soaking the timber around it and so preparing the way for further spread. Instances are known of its travelling 15ft along a steel joist in search of further timber. Walls, 3ft to 5ft thick, have become honeycombed with it. It cannot live in a current of air until it produces a fruiting body that liberates 3 million spores or seeds, each so fine as to be properly visible only through a microscope. Thus the attack goes on, out of sight, until the damage has reached most serious proportions. Special surveying techniques have been devised to detect attacks in their early stages, but these can only be applied by trained experts.

When an attack is discovered:

1. Remove and burn all infected timber.
2. Remove all loose rubble, shavings and such (this particularly applies to attacks in ground floors).
3. Cut off all affected timbers to 2ft beyond the point of visible attack.
4. Brush down all walls.
5. Extract all dust by vacuum, or by brushing.
6. Apply a reliable and appropriate fungicide to walls and timber (both new and old).

Do not attempt to sterilize walls with a blowlamp. Unless the wall is brought up to 120°-130° F the fungus will still live. If brought up to this temperature the building will probably be set on fire in the process. Spend the money on a good water-soluble fungicide and use it liberally on the walling. Similarly, use a good fungicide liberally on all timbers of the original construction and on all timber used in replacement. "Liberally" means using the fungicide like water; drenching the walling and timber.

Everyone in building has seen the dry-rot fungus; if in doubt send a sample to the Timber Research and Development Association for a free identification. There are other fungi, such as the *Poria* or mine fungus, and *Coniophera Cerebella*, the cellar fungus. But, in the average attack, do not be pedantic about identification. Treat them all as if they were the dry-rot fungus and all will be well. Only in large attacks by the cellar fungus can this advice be varied with safety, and this should only be done under expert guidance.

Generally, the average medium-sized or small builder should be able to cope with minor attacks of fungal decay, such as a decayed ground floor. For larger attacks, or for attacks involving historic or valuable woodwork, the work of rectification should be placed in the hands of members of the British Wood Preserving Association. They have immense experience, special equipment and trained workmen. Moreover, many of these firms will guarantee to eradicate the trouble—which no general builder is in a position to do—and will back their guarantees with insurance policies.

WOODWORM

If fungal attack is most likely to be found at ground floor level, infestation by beetles is most common in upper storeys and in roofs. Here, well out of sight in dark roof spaces or beneath linoleum, the attack goes on, year after year, until a chance visit or a little pile of dust on the floor reveals the danger.

Beetles have a complex life-history. They begin as eggs, laid in cracks in timber, or old wormholes: then they become grubs or "worms", changing to a sort of hibernation (called pupation), in which they undergo fundamental changes, emerging finally as beetles, to lay eggs and start the life cycle all over again. The first and last two states are very short, almost the whole existence being spent as a grub or "worm". It is in this state, lasting from three to seven years and sometimes longer, according to the type of beetle, that the insect feeds on timber, tunnelling through it and reducing it to a hollow shell, or, at least, greatly impairing its strength.

Common in Great Britain are two beetles. One is the furniture beetle, *Anobium punctatum*, which is responsible for between 80 and 85 per cent of all damage. The death-watch beetle (*Xestobium rufovillosum*) is responsible for only 10 per cent of the damage, contrary to popular supposition.

Everyone knows the small round holes made by the furniture beetle. Those of the death-watch beetle are considerably larger, being $\frac{1}{8}$in in diameter. Even larger are those of the house longhorn beetle, which are oval and $\frac{3}{16}$in across the major axis. As yet, this beetle is not common in England.

An active attack is indicated by flying beetles in late spring and early summer, and by piles of bore dust found below timbers being attacked. The furniture beetle excretes a sandy sort of dust, but the death-watch beetle makes little oval, bun-shaped pellets, as well as bore dust.

Treatment is much the same for all types of beetle attack:
1. Remove all bore dust and loosened particles of wood dust by scraping, and then by vacuum cleaner.
2. Apply a suitable insecticide, during the emergence and breeding season (April to June). For the furniture beetle, one application is enough, but for all others two should be given.

3. Renew this application every year for three years in the case of the furniture beetle, five to seven years in the case of death-watch beetle, and up to nine years for the house longhorn beetle.

Here again, the medium-sized and small builder should confine his efforts to small outbreaks: for anything serious that threatens the structural stability of a floor or roof, or involves panelling or joinery, a specialist firm should be engaged.

PREVENTION

It is more than ever necessary to consider the prevention of decay and insect attack. Since the war, the danger to timber structures has increased. The house longhorn beetle, which can destroy a roof in five years, is now established at Camberley, in Surrey, and has been recently found in Middlesex—on its way to London. The termite, or white ant, the scourge of tropical climates, is now firmly established in Germany and on the Atlantic seaboard of France. At least one outbreak has been reported in Britain. It is but common sense to take precautions against these actual and threatened dangers by applying preservative to the new timber we build into our structures. Impregnation of timber under pressure with a suitable preservative gives complete protection against fungal and insect attack, The cost is a cheap form of insurance against having to spend at least twenty times that sum.

The following notes on the choice of a preservative, both for new and old work, are a very brief introduction to the subject.

PRESERVATIVES—THEIR CHOICE AND APPLICATION

Preservatives are classified in three types:

1. Tar oil;
2. Organic solvent, and
3. Water solution type.

They are applied to timber in three ways:

1. By pressure impregnation;
2. By the open-tank method, and
3. By brushing or spraying.

Pressure impregnation. The timber is impregnated with the preservative in closed tanks or autoclaves under pressure alone (the "full-cell" process) or under pressure, followed by vacuum which draws out the superfluous preservative left in the cells of the timber after impregnation (the "empty-cell" process). The latter makes the material easier to handle when an oil preservative has been used, and easier to reseason when a water-solution preservative has been employed. Pressure impregnation can only be executed by firms possessing the necessary special equipment.

The open-tank method. In the "hot and cold" process, the timber is immersed in a tank of preservative, which is afterwards heated, and is left there until the preservative has cooled. This gives a far greater depth of penetration than simple immersion and subsequent drainage, which is the "cold dip" process. Both these processes can be undertaken by building contractors, but the "hot and cold" method will only be economic when large amounts of timber are to be treated.

Brushing and spraying. While this method affords the least protection, it is often the only possible treatment, as in repair work or when very small amounts of timber are to be treated. The preservative should be applied liberally and should coat all surfaces and joints. It should not be applied to wet or frozen timber. In many instances, hands and clothing must be protected against contamination by the preservative, and in others ample ventilation must be provided, since some preservatives give off toxic or irritant vapours.

Choice of a preservative. Some water-solution preservatives, such as mercuric chloride, attack ironwork. Tar oil and organic solvent types stain plaster and paint and have a solvent effect on linoleum. Tar oil preservatives cause damage to carpets both by contact and by the fumes they give off. In confined situations, and when labour unused to the work is employed, caustic solutions and preservatives that give off toxic or irritant gases should be avoided.

13. THE ROOFER

Codes of Practice	281
British Standards	281
Conversion Table	285
Slates	285
Asbestos Slates	288
Cedar Shingles	288
Aluminium Roofing Sheets	290
Tiles	292
Minimum Pitches	294

CODES OF PRACTICE

CP 142: Slating and tiling.
  Part 1: Imperial units: materials, concrete and clay in two sections, recommends lap and pitch, and batten sizes.
  Part 2: Metric units: superseded by B.S. 5534: Part 1.
B.S. 5534: Code of Practice for slating and tiling.
  Part 1: 1978: Design: brought into line the wind code (CP 3).
CP 143: Sheet roof and wall coverings.
  Part 1: 1958: aluminium corrugated and troughed.
  Part 2: 1961: galvanised corrugated steel.
  Part 3: 1960: lead (imperial units).
  Part 5: 1964: zinc.
  Part 6: 1962: corrugated asbestos-cement.
  Part 8: 1970: semi-rigid asbestos bitumen sheets.
  Part 10: 1973: galvanised corrugated steel (metric units).
  Part 11: 1970: lead (metric units).
  Part 15: 1973: aluminium (metric units).
  Part 16: 1974: semi-rigid asbestos bitumen sheet (metric units).
CP 144: Roof coverings.
  Part 1: 1968: built-up bitumen felt (imperial units).
  Part 2: 1966: mastic asphalt (imperial units).
  Part 3: 1970: built-up bitumen felt (metric units).
CP 199: Roof deckings.
  Part 1: 1973: asbestos-cement; design of roof decks with units complying with B.S. 3717; information on thermal insulation, and fire hazard, etc.

BRITISH STANDARDS

B.S. 402: Clay plain roofing tiles and fittings.
  Part 1: 1945: imperial units: plain tiles, eaves and top course tiles, tile-and-a-half tiles, various hip, valley and ridge tiles, external and internal angle tiles.
  Part 2: 1970: Metric units.

TABLE 1. SIZE OF CLAY PLAIN TILES

	Co-ordinating sizes		Maximum manufacturing size		Minimum manufacturing size		Work size	
	length	width	length	width	length	width	length	width
	mm	mm	mm	mm	mm	mm	mm	mm
Standard roof tile	—	—	268	168	262	162	265	165
Standard tile for vertical tiling	—	150	268	150	262	145	265	148
	—	200	268	200	262	196	265	198

B.S. 473 and B.S. 550: Concrete roofing tiles and fittings.
Part 1: 1967: imperial units.
Part 2: 1971: metric units.

TABLE 1. GROUP A. DOUBLE-LAP (NON-INTERLOCKING CONCRETE TILES AND CONCRETE SLATES

Maximum limit of manufacturing size		Minimum limit of manufacturing size		Overall work size	
Length mm	Width mm	Length mm	Width mm	Length mm	Width mm
270	168	264	162	267	165
462	335	452	325	457	330

TABLE 2. GROUP B. SINGLE-LAP (INTERLOCKING) CONCRETE TILES AND CONCRETE SLATES

Maximum limit of manufacturing size		Minimum limit of manufacturing size		Overall work size	
Length mm	Width mm	Length mm	Width mm	Length mm	Width mm
386	234	376	224	381	229
418	335	408	325	413	330
425	335	415	325	420	330
435	385	425	375	430	380

B.S. 680: Roofing slates.
Part 1: 1944: Imperial units.
Part 2: 1971: Metric units; with dimensions of slates in work sizes, maximum and minimum manufacturing sizes.

TABLE 1. STANDARD LENGTHS AND WIDTHS OF SLATES
All dimensions are in millimetres.

Maximum manufacturing size		Minimum manufacturing size		Work size	
Length	Width	Length	Width	Length	Width
615	340	605	350	610	355
615	310	605	300	610	305
565	310	555	300	560	305

THE ROOFER

Maximum manufacturing size		Minimum manufacturing size		Work size	
Length	Width	Length	Width	Length	Width
565	285	555	275	560	280
515	310	505	300	510	305
515	260	505	250	510	255
465	310	455	300	460	305
465	260	455	250	460	255
465	235	455	225	460	230
410	310	400	300	405	305
410	260	400	250	405	255
410	235	400	225	405	230
410	210	400	200	405	205
340	310	350	300	355	305
340	230	350	250	355	255
340	210	350	200	355	205
340	185	350	175	355	180
335	285	325	275	330	280
335	260	325	250	330	255
335	210	325	200	330	205
335	185	325	175	330	180
310	230	300	250	305	255
310	210	300	200	305	205
310	155	300	145	305	150
300	300	250	250	255	255
300	210	250	200	255	205
300	155	250	145	255	150

TABLE 2. RANGE OF LENGTHS FOR RANDOMS AND PEGGIES

All dimensions are in millimetres.

Length range	Length range	Length range	Length range
610 to 355	560 to 355	510 to 305	460 to 305
610 to 305	560 to 305	510 to 255	460 to 255
405 to 305	560 to 255	510 to 230	460 to 230
405 to 225	355 to 255	305 to 230	255 to 230
405 to 230	355 to 230	305 to 205	255 to 205
			255 to 150

B.S. 690: Asbestos-cement slates and sheets.
  Part 1: 1963: Imperial units: slates, corrugated sheets and semi-compressed flat sheets.
  Part 2: 1971: flat sheets and fully compressed.
  Part 3: 1973: corrugated sheets.
  Part 4: 1974: slates.
  Part 5: 1975: lining sheets and panels.
  Part 6: 1976: fittings for use with corrugated sheets.
B.S. 747: 1977: Specification for roofing felts.
  Felts for use in the British Isles and similar climates.

TABLE 1. WEIGHTS.... AND LENGTHS OF ROLLS OF CLASS 1 BITUMEN FELTS (FIBRE BASE)

Type of felt		Nominal weight* kg/10 m²	Nominal length of roll m
1A	Saturated bitumen	7.0	20
1B	Fine sanded surface bitumen	17.0	20
1C	Self-finished bitumen	9.0 13.0 18.0 23.0	10† 10† 10† 10†
1D	Coarse sand surfaced bitumen	20.0 30.0	10 10
1E	Mineral surfaced bitumen	36.0	10
1F	Reinforced bitumen	11.0	20

NOTE 1. Type 1B and Type 1C felts (13 kg/m²), when used in built-up roofing, are regarded as equivalent and interchangeable.
* *Exclusive of wrappings and accessories.*
† *Or multiples thereof.*

TABLE 2. WEIGHTS .... AND LENGTHS OF ROLLS OF CLASS 2 BITUMEN FELTS (ASBESTOS BASE)

Type of felt		Nominal weight* kg/10 m²	Nominal length of roll m
2A	Saturated bitumen asbestos	7.0	20
2B	Fine sand surfaced bitumen asbestos	16.0	20
2C	Self-finished bitumen asbestos	13.0 18.0 23.0	10† 10† 10†
2E	Mineral surfaced bitumen asbestos	36.0	10

NOTE. Type 2B and Type 2C felts (13 kg/10^2), when used in built-up roofing, are regarded as equivalent and interchangeable.
* *Exclusive of wrappings and accessories.*
† *Or multiples thereof.*

B.S. 849: 1939: Plain sheet zinc roofing.
  Construction and method of laying.
B.S. 1178: 1969: Milled lead sheet and strip for building purposes.
  Dimensions, thickness, tolerances on thickness, marking. In sheets width 2.40m, length up to 12m. Coded for thickness:
  Code 3, colour of marking; Green,   thickness 1.25mm.
  Code 4, colour of marking; Blue,   thickness 1.80mm.
  Code 5, colour of marking; Red,   thickness 2.24mm.
  Code 6, colour of marking, Black,   thickness 2.50mm.
  Code 7, colour of marking; White,   thickness 3.15mm.
  Code 8, colour of marking; Orange, thickness 3.55mm.

B.S. 2717: 1956: Glossary of terms applicable to roof coverings.
Main headings include: (1) general terms (2) asphalt (3) corrugated sheets (4) flexible sheets (5) patent glazing (6) shingles, slates and tiles (7) thatch.
B.S. 3083: 1959: Hot-dipped galvanised corrugated steel sheets.
Types of materials and weights of zinc coating.
B.S. 3717: 1972: Asbestos-cement decking.
Requirements for non-combustible decking of two types.
B.S. 5247: Code of practice for sheet roof and wall coverings.
Part 14: 1975: corrugated asbestos-cement.

CONVERSION TABLE
SQUARES AND SQUARE METRES

0·108	1	9·290
0·216	2	18·580
0·324	3	27·870
0·432	4	37·161
0·540	5	46·451
0·648	6	55·741
0·756	7	65·032
0·864	8	74·324
0·972	9	83·612
1·080	10	92·903
2·160	20	185·806
3·240	30	278·709
4·320	40	371·612
5·400	50	464·515

SLATES

A good slate should give a sharp metallic ring when struck and should not splinter under the slater's axe. It should be easily holed without fracture. If a clayey odour is emitted when the slate is breathed upon, it will not weather well.

THICKNESSES OF SLATES

First quality ... ... $\frac{3}{16}''$
Second quality ... ... $\frac{1}{4}''$

To determine the gauge of Slates the following formula can be used:

HEAD NAILING

$$\text{Gauge} = \frac{\text{Length of Slate} - 1'' - \text{Lap}}{2}$$

CENTRE NAILING

$$\text{Gauge} = \frac{\text{Length of Slate} - \text{Lap}}{2}$$

Centre Nailed

Name	Size In	Number of sq. yd covered by 1000: Lp $2\frac{1}{2}''$	Lp $3''$	Lp $3\frac{1}{2}''$	Lp $4''$	Number to cover one square (100 feet super): Lp $2\frac{1}{2}''$	Lp $3''$	Lp $3\frac{1}{2}''$	Lp $4''$	Approx. wgt. per 1,000 cwts.
Empresses	..26×16	$144\frac{1}{3}$	$140\frac{1}{3}$	$138\frac{3}{4}$	$135\frac{1}{2}$	77	79	80	82	$68\frac{3}{4}$
Princesses	..24×14	$115\frac{3}{4}$	$113\frac{1}{3}$	110	108	96	98	101	103	$55\frac{1}{2}$
Duchesses	..24×12	99	$96\frac{2}{3}$	94	$92\frac{2}{3}$	112	115	118	120	$47\frac{1}{2}$
Small Duchesses	22×12	$89\frac{2}{3}$	$87\frac{1}{3}$	$85\frac{1}{2}$	83	124	127	130	134	$43\frac{3}{4}$
Marchionesses	22×11	$82\frac{1}{3}$	$80\frac{1}{2}$	$78\frac{1}{4}$	76	135	138	142	146	40
Wide Countesses	20×12	$80\frac{1}{2}$	$78\frac{1}{4}$	76	74	138	142	146	150	40
Countesses	..20×10	$67\frac{1}{3}$	65	$63\frac{1}{2}$	$61\frac{3}{4}$	165	170	175	180	$31\frac{2}{3}$
Viscountesses	..18× 9	$53\frac{1}{2}$	52	$50\frac{1}{4}$	$48\frac{1}{2}$	207	214	221	229	27
Wide Ladies	..16×10	52	50	48	$46\frac{1}{4}$	214	222	231	240	$24\frac{1}{2}$
Ladies ..	..16× 8	41	40	$38\frac{1}{2}$	37	267	277	288	300	$19\frac{2}{3}$
Wide Headers	..14×12	53	$50\frac{3}{4}$	$48\frac{1}{2}$	$46\frac{1}{4}$	209	219	229	240	$26\frac{1}{4}$
Headers	..14×10	$44\frac{1}{4}$	$42\frac{1}{3}$	$40\frac{1}{3}$	$38\frac{1}{2}$	251	262	275	288	$21\frac{3}{4}$
Small Ladies	..14× 8	$35\frac{1}{3}$	$33\frac{1}{3}$	$32\frac{1}{3}$	$30\frac{3}{4}$	314	328	343	360	$17\frac{1}{2}$
Doubles	..13× 7	$28\frac{1}{3}$	$26\frac{2}{3}$	$25\frac{2}{3}$	$24\frac{1}{4}$	392	412	434	458	$14\frac{1}{2}$
Wide Doubles	..12×10	$36\frac{1}{2}$	$34\frac{3}{4}$	$32\frac{1}{3}$	$30\frac{3}{4}$	304	320	339	360	20

No allowance has been made for waste. Suggest 5 per cent

The traditional names for slates are going out of use and more and more slates are being quoted by size, instead of the traditional names. Suppliers are still quoting in Imperial measure only, but since some specifications are bound to be in metric, metric equivalents are also given.

SLATES (MODERN SIZES)

IMPERIAL Length (ins.	Widths (ins)	METRIC Length (mm)	Widths (mm)
24	14:12	600	350:300
22	12:11	550	300:280
20	12:10	500	300:250
18	12:10:9	450	300:250:230
16	12:10:9:8	400	300:250:230:200
14	12:10:8	350	300:250:200
13	10:8:7	330	250:200:180
12	8	300	200

THE ROOFER

RUN OF SLATING BATTENS REQUIRED PER SQUARE

Lengths of slate		\multicolumn{8}{c}{Lap centre nailed}							
in	mm	2½"	64 mm	3"	76 mm	3½"	89 mm	4"	102 mm
12	305	253	82·90	267	87·57	284	93·14	300	98·39
13	330	229	75·10	240	78·71	253	82·90	266	87·22
14	356	209	68·54	219	71·82	229	75·10	240	78·71
16	406	178	58·38	185	60·68	192	62·97	200	65·59
18	457	153	50·18	160	52·48	166	54·45	172	56·41
20	508	138	45·26	142	46·57	146	47·88	150	49·19
22	559	123	40·24	127	41·65	130	42·64	134	43·95
24	610	112	36·74	115	37·72	118	38·70	120	39·36
26	660	102	33·46	105	34·54	107	35·10	109	35·75

Add 10 per cent for waste

SLATE NAILS: NUMBER PER POUND/KILOGRAM

Type	1½ in.	1¾ in.	2 in	28·5 mm	34·4 mm	50·8 mm
Plain Wire	148	130	112	325	286	242
Galvd Wire	140	121	102	305	256	224
Copper	148	120	88	325	254	194
Zinc	190	133	91	415	292	200

Note: Two nails are required to each slate, but add 5 per cent for waste.

NAILS FOR BATTENS
REQUIREMENTS IN LB/KG PER SQUARE

Length of Slate in		\multicolumn{8}{c}{Lap Centre Nailed}							
in	mm	2½"	62 mm	3"	75 mm	3½"	87 mm	4"	100 mm
12	312	1·6	0·725	1·7	0·771	1·8	0·816	1·9	0·861
13	338	1·4	0·635	1·5	0·703	1·6	0·725	1·7	0·771
14	363	1·3	0·589	1·4	0·635	1·45	0·680	1·6	0·725
16	414	1·1	0·498	1·2	0·544	1·2	0·544	1·25	0·589
18	464	·9	0·408	1·0	0·453	1·05	0·476	1·1	0·498
20	515	·85	0·385	·9	0·408	·92	0·417	·95	0·430
22	566	·75	0·340	·8	0·362	·82	0·371	·85	0·385
24	616	·65	0·294	·7	0·317	·75	0·340	·75	0·340
26	667	·6	0·272	·65	0·294	·67	0·303	·7	0·317

Add 5 per cent for waste

ASBESTOS SLATES

TABLE FOR DIAGONAL PATTERN, STANDARD SIZE 15¾ × 15¾"

Lap				$2\frac{3}{4}''$	$3''$	$3\frac{1}{2}''$	$4''$
Gauge				$8\frac{7}{8}''$	$8\frac{1}{2}''$	$8\frac{1}{4}''$	$7\frac{5}{8}''$
Number per square				90	93	102	110
Feet run of battens				134	140	216	166
Rivets (number)				90	93	102	110
Nails (number)				180	186	204	220

Waste has been allowed for in the above. For rectangular pattern asbestos slates, see page 289. They are made in the following sizes: 24 × 12, 20 × 10, 16 × 8 and 12 × 6.
200 rivets weigh approx. 1lb.

CEDAR SHINGLES

Length ... 16in.
Width ... Random, varying from 4in to 12in.
Thickness ... 25in at butt and taper to top.
Weight ... 144lb per square when laid at 5in gauge.

GAUGE
All pitches down to 30 degrees, 5in. Under 30 degrees but not less than 20 degrees, $3\frac{3}{4}$in.
Unusually steep pitches, $5\frac{1}{2}$in or 6in.
Walls 6in, $6\frac{1}{2}$ or $7\frac{1}{2}$in (first two recommended).
(Cedar shingles are imported, not metricated).

COVERING CAPACITIES OF CEDAR SHINGLES

Gauge				$3\frac{3}{4}$	5	$5\frac{1}{2}$	6	$6\frac{1}{2}$	$7\frac{1}{2}$	in
Lap				9	6	5	4	3	2	in
Bundles per square				5·3	3·9	3·5	3·25	3	2·6	
One bundle covers				19·2	25·7	28·3	30·8	33·4	38·5	sq. ft

Note: One bundle contains tiles of random widths which should be laid in no special order but just as they come from the bundle.

THE ROOFER

LAP, NUMBER AND AVERAGE WEIGHT PER SQUARE OF PLAIN TILES AND AVERAGE WEIGHT PER THOUSAND

Tile Dimensions	Standard 10½" × 6½"				11" × 7"				10" × 6"				Unit
Gauge	4	3¾	3½	3¼	4	3¾	3½	3¼	3½	3¼	3	ins
Lap	2½	3	3¼	4	3	3¼	4	2½	3	3½	4	ins
Tiles per square	550	590	635	680	485	515	550	590	640	685	740	800	—
Weight per square													
Hand-made	12.7	13.6	14.7	15.7	13.0	13.8	14.7	15.8	12.6	13.5	14.5	15.7	cwts
Machine-made	11.6	12.5	13.4	14.4	11.9	12.6	13.4	14.5	—	—	—	—	cwts
Average weight													
Hand-made				23.0			26.0				19.6		cwts
Machine-made				21.1			23.8				—		cwts

Tile Dimensions	Standard 165 × 165 mm				11" × 7"				10" × 6"				Unit
Gauge	102	95	89	83	115	102	95	89	89	83	76	mm
Lap	63	76	83	102	63	76	89	102	76	89	102	162	hand
Tiles per 10m²	592	635	604	732	522	554	592	635	609	737	796	862	kg
Weight per 10m²													
Hand-made	645	692	747	798	660	700	747	803	640	686	736	798	mach.
Machine-made	595	635	681	750	605	640	681	736	—	—	—	—	kg
Average weight													
Hand-made kg				1 170			1 320				995		
Machine-made kg				1 073			1 210				—		

Above information by the courtesy of Clay Products Technical Bureau of Great Britain. Allow approximately 2 cubic feet of mortar per square if bedded.

BATTENS REQUIRED PER SQUARE (9·29m²) OF TILING

Gauge	3	3¼	3½	3¾	4	4¼	inches
Battens per square	400	370	343	320	300	283	ft run
Gauge	83	89	95	102			metre run
Battens per square	113	103	98	92			
Nails	1·4	1·5	2·2	2	1·9	1·8	lbs

Add for waste: 10 per cent. timber, 5 per cent nails.

Description	Sizes in	Sizes mm	No. of Tiles	Lap (75mm) Battens ft run	Battens m run
Pantiles	13½× 9½	343×240	173	126	39
Roman Tiles	16½×13½	418×343	93	98	30
Somerset Interlocking Tiles	15½× 8¼	393×210	162	104	32

ALUMINIUM ROOFING SHEETS

CORRUGATED SHEETS

Pitch	Depth of Corrugation	Width in Corrugation
3 in	¾ in	
3 in	⅝ in	9½, 12 & 14
2⅔ in	17/32 in	9 & 11
2½ in	½ in	8, 10 & 12

LENGTHS

6—12ft, in 1ft increments. Sheets up to 35ft can be obtained to order.

TROUGHED SHEETS

	Pitch	Crown Width	Valley Width	Depth
A.	5 in	¾ in	2¼ in	1½ in
B.	5⅛ in	¾ in	¾ in	1¾ in
C.	5 in	1 in	2 in	1½ in

WIDTHS

A—2ft 2¾in: 2ft 7¾in: 3ft 0¾in.
B—2ft 4in: 2ft 9in: 3ft 2in.
C—2ft 6in: 2ft 11in.

LENGTHS

As for corrugated sheets, with the exception of C, which is obtainable in sheets 4—12ft, in 1ft increments.

MANSARD SHEETS

Have a 4in flat valley and a 2in half-round ridge. Widths are 2ft 6in and 3ft; Lengths are 4—12ft in 1ft increments.

RECOMMENDED LOADING FOR TROUGHED ALUMINIUM SHEETS

Span in ft Gauge	Loading, lb per sq. ft			
	18	19	20	22
5	—	—	74	57
$5\frac{1}{2}$	—	—	60	47
6	77	65	51	40
$6\frac{1}{2}$	66	55	43	34
7	57	48	37	29
$7\frac{1}{2}$	49	41	33	25
8	43	36	29	22
$8\frac{1}{2}$	39	32	25	20
9	34	29	23	18
$9\frac{1}{2}$	31	26	—	—
10	27	22	—	—
$10\frac{1}{2}$	23	19	—	—

Type C
5 in pitch
$1\frac{1}{2}$ in
2 in Valley

Type A & B
5 & $5\frac{1}{8}$ in

TILES
CONCRETE ROOFING TILES

	Plain or Cross Camber		Single Lap Interlocking				Bold Roll	Cladding	
Size of Tile	$10\frac{1}{2}" \times 6\frac{1}{2}"$		Standard 49 $15" \times 9"$	Double Roman $16\frac{1}{2}" \times 13"$	Grovebury Prantile $16\frac{1}{2}" \times 11"$	Inter: Slate $17" \times 15"$	Regent $15\frac{5}{8}" \times 10\frac{3}{4}"$	Orna-mental $10\frac{1}{2}" \times 6\frac{1}{2}"$	Brick-bond $9" \times 3"$
Lap in Inches	$3\frac{1}{2}$	$2\frac{1}{2}$	3	3	3	3	3	$1\frac{1}{2}$	—
Batten guage inches	$3\frac{1}{2}$	4	12	$13\frac{1}{2}$	$13\frac{1}{2}$	14	$12\frac{5}{8}$	$4\frac{1}{2}$	—
No. of tiles per square	590	550	150	90	110	77	127	492	534
Weight per square in lb	635	1480	820	896	930	1052	1052	1366	1533
	1700	1580							
Minimum Pitch (eaves)	30°		35°	30°	30°	$22\frac{1}{2}°$	30°	—	—
Usual rafter centres	16in		18in	18in	18in	18in	18in		
Batten size in inches	$1 \times \frac{3}{4}$		$1\frac{1}{2} \times \frac{3}{4}$	$1\frac{1}{2} \times \frac{3}{4}$	$1\frac{1}{2} \times \frac{3}{4}$	$1\frac{1}{2} \times \frac{3}{4}$	$1\frac{1}{2} \times \frac{3}{4}$	$1\frac{1}{2} \times \frac{3}{4}$	$1\frac{1}{2} \times \frac{3}{4}$
Nails	$1\frac{1}{2}$in $\times$ 12G		$1\frac{3}{4}"$ Alum Alloy	$1\frac{1}{2}"$ Alum Alloy	$1\frac{1}{2}"$ Alum Alloy	—	$1\frac{1}{2}"$ Alum Alloy	$1\frac{1}{2}"$ Alum Alloy	$1\frac{1}{4}" \times$ 12G

CONCRETE ROOFING TILES: METRIC SIZES

	Single lap interlocking						Bold Roll		Cladding
	Plain Tile	Delta Stonewold	Standard 49	Double Roman	Renown	Grovebury Pantile	Regent		Ornamental
Size of Tile	$10\frac{1}{2}'' \times 6\frac{1}{2}''$ 265×165mm	$17'' \times 15''$ 430×380mm	$15'' \times 9''$ 380×230mm	$16\frac{1}{2}'' \times 13''$ 420×330mm	$16\frac{1}{2}'' \times 13\frac{1}{8}''$ 420×332mm	$16\frac{1}{2}'' \times 13\frac{1}{8}''$ 420×332mm	$16\frac{1}{2}'' \times 13\frac{1}{16}''$ 420×332mm		$10\frac{1}{2}'' \times 6\frac{1}{2}''$ 265×165mm
							$22\frac{1}{2}°$ & over	under $22\frac{1}{2}°$	
Headlap	$2\frac{1}{2}''$/65mm	3"/75mm	3"/75mm	3"/75mm	3"/75mm	3"/75mm	3"/75mm	4"/100mm	$1\frac{1}{2}''$/38mm
Gauge (max)	4"/100mm	14"/355mm	12"/305mm	$13\frac{1}{2}''$/345mm	$13\frac{1}{2}''$/345mm	$13\frac{1}{2}''$/345mm	$13\frac{1}{2}''$/345mm	12"/320mm	$4\frac{1}{2}''$/114mm
No. of Tiles yd2/m2	50/60	6·9/8·2	137/16·5	2·8/9·5	8/9·5	8·2/9·7	8·2/9·7	8·8/10·4	45/53
Weight per 1,000 tiles tonnes/tons	1·290/1·3	Delta 7·060/6·95 Stonewold 6·270/6·2	2·800/2·76	4·500/4·4	4·430/4·3	5·100/5 Mk. I Mk. II	Mk. I 4·500/4·48 Mk. I	Mk. II 5·080/5 Mk. II	1·290/1·3
Min. Pitch	35°	$17\frac{1}{2}''$	30°	30°	30°	30° $22\frac{1}{2}°$	30°	$17\frac{1}{2}°$	
Nails Alum Aloy	$1\frac{1}{2}''$/38mm	—	$1\frac{3}{4}''$/45mm	$2\frac{1}{2}''$/65mm	$1\frac{3}{4}''$/45mm	$2\frac{3}{4}''$/ — 70mm	$2\frac{3}{4}''$/70mm	$1\frac{1}{2}''$/38mm	$1\frac{1}{2}''$/38mm

Fittings: All customary fittings are made to match the tiles; eaves, gables, bonnets, valleys, internal angle, mansards, half and third round ridge, Pascall roll and others for plain tiles, and half-tiles mansards, valleys and half third round ridge tiles for interlocking tiles.

BATTEN SIZES

		Single Lap	Plain Tiles
For rafters not exceeding	457mm/18" c/c	38×19mm/$1\frac{1}{2}'' \times \frac{1}{2}''$	25×19mm/$1'' \times \frac{3}{4}''$
,,	610mm/24" c/c	38×25mm/$1\frac{1}{2}'' \times 1''$	32×25mm/$1\frac{1}{4}'' \times 1''$
,,	914mm/36" c/c	38×38mm/$1\frac{1}{2}'' \times 1\frac{1}{2}''$	

CLIPS. Purpose made clips available for extra securement where necessary.
(Data by courtesy of Redland Tiles Ltd.)

MINIMUM PITCHES

Eaves Pitch, in Degrees	Type of Material for Roofing
45	Slates, 6 in wide (150 mm)
40	Plain tiles, slates 7 in wide (175 mm)
35	Single-lap tiles; slates 8 in wide (200 mm)
30	Fully interlocking tiles (on sides and ends); Slates 9 in and 10 in wide (225-350 mm)
25	Slates 12 in and 14 in wide; optimum pitch for corrugated roofing materials with one corrugation of side lap (300 and 350 mm).
17	Roman tiles; certain types of fully-interlocking tiles, imported from France.
15	Lowest limit for corrugated roofing materials, unsealed laps.
9	Corrugated roofing materials, end and side laps sealed.
Under 9	Flat roofs of concrete, asphalt or bitumen-felt, or lead, zinc or aluminium*.

The minimum slope for sheet metal roofs is 1in in to 6ft with a fall of 2in in 6ft.

RELATIONSHIP OF PITCH TO LAP, FOR SLATES AND PLAIN TILES

Eaves Pitch in Degrees	Lap
35 and less	$3\frac{1}{2}$ in (88 mm)
35—45	3 in (75 mm)
Over 45	$2\frac{1}{2}$ in (63 mm)

Note 1 Allowance should also be made for exposure of site: a very exposed site may justify a greater lap.

Note 2: Lap for plain tiles should not be less than $2\frac{1}{2}$in. In exposed sites 3in/75 mm is recommended.   Lap of plain tiles should never exceed one-third the length of the tile.

14. THE DRAINLAYER

Codes of Practice	296
British Standards	296
Recent Developments	301
Domestic Drainage Data	303
Drain and Sewer Falls	303
Lead and Yarn	305
Manhole Covers	305

CODES OF PRACTICE

CP 3: Chapter VII: 1950: Engineering and utility services.
Parts dealing with sanitation, sewage and rainwater disposal, etc.
CP 301: 1971: Building drainage.
Recommendations for design, layout and construction of foul-sewage, surface-water and ground-water drains.
CP 302: 1972: Small sewage treatment works.
Methods, equipment and installation for treatment of domestic sewage (small groups of houses etc.) for up to 300 persons.
CP 302: 200: 1949: Cesspools.
Guidance on the provision of cesspools.
CP 2005: 1968: Sewerage.
Design and construction of the whole system of sewers, manholes, storm overflows, siphons, pumping stations and mains, etc.
B.S. 5572: 1978: Code of practice for sanitary pipework (formerly CP 304).
Design, installation, testing and maintenance of above ground non-pressure pipework.

BRITISH STANDARDS

B.S. 65 and B.S. 540: Clay drain and sewer pipes including surface water pipes and fittings.
Part 1: 1971: pipes and fittings, dimensions and tolerances.

TABLE 1. DIMENSIONS OF BARRELS

1	2		3		4
Nominal bore	Limits of internal diameter				Permissible variation in thickness (See 5.1.2) ....
	British Standard		British Standard Surface Water		
	Min. dia.	Max. dia.	Min. dia.	Max. dia.	
mm	mm	mm	mm	mm	mm
75	73	80	71	81	2
100	98	105	96	107	2
*125	122	132	119	135	2
150	147	158	145	160	2
*175	173	183	170	185	2
*200	196	210	193	213	2
225	222	235	219	239	2
*250	246	262	242	266	2
300	296	313	292	317	2
375	371	391	366	396	3
450	447	467	442	472	3
525	520	547	514	553	3
600	596	623	590	629	3
675	666	705	657	715	3
750	736	788	736	801	3
825	812	864	801	876	3
900	889	940	876	953	3

* These are non-preferred sizes.

TABLE 2. DIMENSIONS OF SOCKETS (TYPE 2)

1	2	3
Nominal bore	Minimum internal depth of socket (B)	Minimum excess shoulder measurement (C - A)
mm	mm	mm
75	50	8
100	50	10
*125	55	10
150	55	10
*175	55	10
*200	65	13
225	65	13
*250	70	15
300	70	15
375	75	15
450	75	15
525	80	20
600	90	20
675	90	20
750	90	20
825	90	25
900	90	25

* These are non-preferred sizes.

Part 2: 1972: flexible mechanical joints, tests for deflection, straight draw, line displacement, shear loading and water tightness.
B.S. 437: Cast iron spigot and socket drain pipes and fittings.
Part 1: 1970: pipes, bends, branches and access fittings.

TABLE 1. DIMENSIONS OF DRAIN PIPES (M)

			Nominal bore (mm)				
			50	75	100	150	225
			mm	mm	mm	mm	mm
Pipe	Minimum bore	A	48	74	99	150	226
	External diameter (max.)	B	65	92	119	173	256
	....						
Socket	Minimum bore	F	81	108	136	190	276
	External diameter (nominal)	G	117	150	185	239	337
	....						
	Internal depth (min.)	J	76	76	76	89	114
	....						
	Caulking clearance (min.)		8	8	9	9	10
			kg	kg	kg	kg	kg
Nominal weight of pipe per 1·83 m length (single socket)			17·7	28·6	40·8	71·2	143·8
per 1·83 m length (double socket)			—	33·1	46·7	80·3	162·8
per 2·74 m length (single socket)			—	40·8	58·5	102·1	206·4
per 3·66 m length (single socket)			—	52·6	76·2	132·9	268·5

NOTE 1. For other sizes of pipe refer to BS 1211 (Centrifugally cast (spun) iron pressure pipes for water, gas and sewage), Class B.
NOTE 2. Length for 50 mm size   1·83 m effective
        Length for 75, 100, 150, 225 mm size   1·83, 2·74 and 3·66 m effective
        Length for 75, 100, 150, 225 mm size
        with double socket   1·83 m effective
NOTE 3. 3 in, 4 in, 6 in and 9 in centrifugally cast (spun) iron pressure pipes to BS 1211. Class B shall be deemed to comply with ... this Britsh Standard.

B.S. 497: Specification for manhole covers, road gulley gratings and frames for drainage purposes.
Part 1: 1976: cast iron and cast steel.
B.S. 539: 1971: Dimensions of fittings for use with clay drain and sewer pipes.
Design and dimensions of branch heads, channels, interceptors, gullies, traps and raising pieces, hoppers, yard or street gullies.
B.S. 556: Concrete cylindrical pipes and fittings including manholes, inspection chambers, and street gullies.
Part 1: 1966: imperial units.
Part 2: 1972: metric units.

TABLE 1. DIMENSIONS AND TOLERANCES FOR CONCRETE PIPES AND FITTINGS

1	2	3	4	5
Nominal bore	Permissible deviation (plus or minus) from mean bore	Permissible variation of wall thickness (plus or minus)	Minimum clearance between spigot and socket*	Minimum depth of socket* ....
A			B	C
mm	mm	mm	mm	mm
150	3	3	10	56
225	3	3	10	56
300	3	3	15	56
375	5	3	15	62
450	5	3	15	70
525	6	3	15	70
600	6	3	20	70
675	6	3	20	76
750	6	3	20	82
825, 900	6	3	20	88
975, 1050	6	5	20	95
1125	6	5	20	100
1200	10	5	20	100
1350	10	5	20	100
1500	10	5	20	100
1650	10	5	20	100
1800	10	5	20	100

* The minimum dimensions given in Columns 4 and 5 apply only to ordinary spigot and socket joints . . . .

B.S. 1143: 1974: Vitrified clay pipes and fittings with extra chemically resistant properties.

Pipes and fittings up to 900mm nominal bore for use under conditions of severe acidity.

B.S. 1194: 1969: Concrete porous pipes for under-drainage.

Cement and aggregate, composition and preparation. Form, dimensions and tolerances.

B.S. 1196: 1971: Clayware field drain pipes.

Quality, workmanship, dimensions 65mm to 400mm pipes.

B.S. 1247: 1975: Manhole step irons.

Tests and requirements, appendices.

B.S. 2494: 1976: Materials for elastomeric jointings for pipework and pipelines.
Suitable elastomers for jointing pipes.
B.S. 2760: 1973: Pitch-impregnated fibre pipes and fittings for below and above ground drainage.
Dimensions and requirements. Specifies materials and tests.

TABLE 1. NOMINAL BORES

Nominal bore, mm	50	75	100	125	150	200	225
Limits of internal diameter	mm	mm	mm	mm	mm	mm	mm
min.	51	76	102	127	152	203	229
max.	54	80	106	133	160	213	239

B.S. 3656: 1973: Asbestos-cement pipes, joints and fittings for sewerage and drainage.
Nominal bore range 100mm to 1050mm.

TABLE 1. CLASSIFICATION OF PIPES

Nominal bore	Minimum ultimate crushing load		
	Class L (AC)	Class M (AC)	Class H (AC)
mm	kg/m	kg/m	kg/m
100	—	—	3870
125	—	—	3870
150	—.	—	3870
175	—	—	3870
200	—	—	3870
225	—	—	3870
250	—	—	3870
300	—	3570	4760
375	—	3940	5360
450	3570	4460	5950
525	3870	*4840	6700
600	4300	5800	7900
675	4900	6550	8780
750	5200	7000	9370
825	5650	7600	10270
900	5950	8930	11000
975	6400	9670	11900
1050	6700	10120	12650

NOTE 1. In specifying the above crushing loads in kg/m it is assumed that the measurements will be made in conditions of standard gravity.
NOTE 2. The choice of class of pipe is determined by the pipeline design engineer who alone is qualified to judge the conditions of installation, laying and operation of the pipes. However, it is recommended that a class be selected such that, taking into account all the loads and the bedding adopted, the pipes in use give a factor of safety at crushing of at least 1.25.
NOTE 3. Occasional internal pressures are admissible provided that an adequate factor of safety be maintained in relation to the hydrostatic test pressure given in 2.4.2.1 and 4.4.2.

B.S. 3868: 1973: Prefabricated drainage stack units, galvanised steel.
For use above ground in plumbing systems for soil waste and rainwater.

TABLE 1. DIMENSIONS AND WEIGHTS

Class	Nominal bore (mm)							
	32	40	50	65	80	100	125	150
	Light	Light	Light	Light	Light	Light	Medium	Medium
Max. diameter (o.d.) mm	42.5	48.4	60.2	76.0	88.7	113.9	140.6	166.1
Thickness mm	2.65	2.90	2.90	3.25	3.25	3.65	4.85	4.85
Weight (ungalvanized) kg/m run	2.58	3.25	4.11	5.80	6.81	9.89	16.20	19.20

B.S. 4101: 1967: Concrete unreinforced tubes and fittings with ogee joints for surface water drainage.
Definitions, materials, testing, etc.
B.S. 4118: 1967: Glossary of sanitation terms.
Covers above and below ground drainage, including connection to a local authority's sewerage system.
B.S. 4514: 1969: Unplasticised PVC soil and ventilating pipe, fittings and accessories.
For use in above ground drainage systems.

TABLE 1. DIMENSIONS OF PIPES AND FITTINGS

Nominal size	Mean outside diameter*		Tolerance on mean outside diameter*		Minimum wall thickness			
					pipe		fittings	
	mm	(in)	mm	(in)	mm	(in)	mm	(in)
3	82.6	(3.250)	± 0.2	(± 0.008)	3.2	(0.125)	3.2	(0.125)
4	110.2	(4.339)	± 0.2	(± 0.008)	3.2	(0.125)	3.2	(0.125)
6	160.3	(6.312)	± 0.3	(± 0.012)	3.3	(0.130)	3.5	(0.138)

* The mean outside diameters and tolerances thereon apply to pipe, and to spigots of fittings for ring seal and solvent cement joints. ' Mean outside diameter ' of a pipe is defined as the arithmetic mean of any two perpendicularly opposed diameters.

B.S. 4660: 1973: Unplasticised PVC underground drain pipe and fittings. Nominal sizes 110mm and 160mm.

TABLE 1. DIMENSIONS OF PIPE AND FITTINGS

Nominal size	Mean outside diameter		Extreme individual outside diameter		Minimum wall thickness		
	min.	max.	min.	max.	Pipe	Fittings other than junctions	Junctions
mm	mm	mm	mm	mm	mm	mm	mm
110	110.0	110.4	108.0	112.4	3.4	3.4	3.8
160	160.0	160.6	157.1	163.5	4.1	4.1	4.1

NOTE 1. 110 mm and 160 mm pipes have previously been known as 4 inch and 6 inch nominal drain pipes respectively.
NOTE 2. The minimum wall thickness of the socket, where it is supported by the pipe, may be reduced by up to 10% of the specified value.
NOTE 3. Mean outside diameter applies to pipe and to spigots of fittings for ring seal and solvent cement joints. Mean outside diameter is defined as the arithmetic mean of any two diameters at right angles to each other.

B.S. 4962: 1973: Performance requirements for plastics pipe for use as light duty sub-soil drains.
Materials, construction and tests, etc.

B.S. 5178: 1975: Prestressed concrete pipes for drainage and sewage.
Pipes and fittings for use at atmospheric pressure. Nominal bores from 450 to 3000mm. Materials, design, tests, etc.

B.S. 5481: 1977: Specification for unplasticised PVC pipe and fittings for gravity sewers.
Nominal diameters 200 to 630mm. Materials, dimensions, construction, tests, etc.

RECENT DEVELOPMENTS

FLEXIBLE PIPELINES

Recent research carried out at the Building Research Station has established principles that will revolutionize the traditional methods of drainlaying. It has been found that ground movement, especially in shrinkable clay and some other soils, causes fractures in the usual type of drain—that is, pipes of ceramic or similar brittle material, laid with rigid joints of cement on a continuous bed of concrete. If, however, the rigid concrete bed is omitted, and in addition, flexible pipelines of pitch-fibre, steel or plastics are used instead of the usual stone-ware drainpipes, fractures are virtually eliminated. If brittle pipes are used, such as the ordinary stoneware drain pipes or cast-iron pipes, they can be jointed with flexible joints, using rubber jointing rings or flexible bitumen joints. They will then be as free from fracture as pitch-fibre or steel or plastics pipes. Rings for ceramic, concrete, asbestos-cement and spun iron pipes can now be bought.

This does not mean that the pipes may be laid direct on a trench bottom: this bottom should be of uniform hardness, and well consolidated. Nevertheless, these new materials and methods enable considerable savings to be made, notably by the omission of concrete beds, the speed of forming joints, the possibility of testing as soon as the line is laid and the advantage of being able to cover in as soon as testing is complete. Further details are given in Building Research Station Digests, Nos. 124, 125, 134 and (New Series) No. 6 (obtainable from H.M.S.O.).

PITCH-FIBRE PIPES

Pitch-fibre pipes, made from a cellulose material impregnated with bitumen, and long used in Germany, Canada and the U.S.A., are now manufactured in this country to B.S. 2760, and are accepted by most local authorities for drainage underground. They can be used to carry hot liquids, up to 120-130°F without fear of deterioration. The great advantage of using them is fast laying—up to 400ft run of 4in pipes per hour—due to long lengths, light weight and simple jointing. In addition, because of their flexibility, no concrete bedding or haunching is required. Long lengths can be made up, and lowered into the trenches as required. Pipes should be to B.S. 2760.

Pipes are obtainable in internal diameters of 2, 3, 4, 5, 6 and 8in, and in lengths of 5, 6, 8 and 10ft, with ends tooled to meet couplings. Couplings are available (straight and 5° angled) in all diameters. Junctions, male or female taper, in 90°, 95°, 112°, 120° and 130°, all sizes on all pipes, and long tailed septic tank junctions (90°) are available from stock, as are long-radius bends ($22\frac{1}{2}°$, 45° and 90°, radius from 1ft 4in to 3ft 6in are general, but a larger range is stocked) and tapers, adaptors and other fittings. Joints are coned, and jointing consists merely in hammering the pipes together, using a wood block to prevent deformation of the cone joint.

The pipe can be cut to length by a carpenter's saw, and tapers can be made on the cut ends by simple jointing tools that can be hired from the suppliers. Details of laying procedure are given in Building Research Digests 97, 124 and 125, and in publications of the Pitch Fibre Pipe Association. Intending users should verify, with their local authority, whether the use of these pipes is permitted, and under what conditions, before embarking on their use. The Association will gladly supply names and addresses of manufacturers. Pitch-fibre and plastics pipes are liable to flattening or other distortion if subjected to heavy loads.

Although these new methods and materials dispense, in the ordinary way, with concrete drain-beds, there are conditions where possible damage to the pipeline demands either a concrete bed or a concrete surround that may also be reinforced, or a concrete cover. Further design information is given in *The structural design of the cross-section of vitrified clay pipelines buried in trenches* by J. H. Walton, F.I.P.H.E., from the Clay Pipe Development Association. The MOHLG Working Party second report (see above) gives design methods for the rational structural design of pipelines and design tables, for use with these methods (by C. E. G. Bland, M.I.P.H.E.) are published, by the Clay Pipe Development Association, in a collection of papers on drainage.

DOMESTIC DRAINAGE MEMORANDA

Waste water allowances in gallons per person per day discharged in 12 hours are:

Domestic	35
Hospitals	50
Schools	10
Barracks	15

Example. For one house of 5 persons, the discharge is

$$\frac{35 \times 5}{12 \times 60} = \cdot 24 \text{ gal/min}$$

Stormwater allowances should be made for $\frac{1}{2}''$ to $1''$, say $\frac{3}{4}''$ of rainfall per hour for impermeable areas, and $\frac{1}{6}$, say $\frac{1}{8}$, from permeable areas.

Example. For one house at a density of 12 to the acre with about $\frac{1}{3}$ impermeable area (roofs, paths and roads), the requirement in gallons per minute would be as follows:—

$$\text{Impermeable } 1210 \text{ sq. ft} \times \frac{3}{4 \times 12} \times \frac{6 \cdot 25}{60} = 7 \cdot 88 \text{ gal/min}$$

$$\text{Permeable } 2420 \text{ sq. ft} \times \frac{1}{8 \times 12} \times \frac{6 \cdot 25}{60} = 2 \cdot 63 \text{ gal/min}$$

Stormwater for $\frac{1}{12}$ th acre or 3630 sq. ft = 10·51 gal/min
Add Domestic soil and waste (see above)   ·24 gal/min

Total per house...   ...   .   10·75 gal/min
Total per acre   129   gal/min

Note: 1 litre   equals   0·22 galls
      1 gallon   equals   4·54 litres
      1 acre   equals   0·40 hectares
      1 hectare   equals   2·47 acres

DRAIN AND SEWER FALLS

Drains, to be self-cleansing, should not be larger than necessary and should have a minimum velocity of 3 ft per second. Maguire's rule of a fall of 1 in 10d (d equals internal diameter of pipe) gives the following familiar falls: for a 4in pipe, 1 in 40ft; for a 6in pipe, 1 in 60ft; and for a 9in pipe, 1 in

90ft. These are today considered excessive and the following table (from Escritt's *Building Sanitation*, p. 91) should be used:

| Internal Diameter | | Fall: |
millimetres	inches	one in
101·6	4	90
152·4	6	150
177·8	7	190
228·6	9	265
304·8	12	385
381·0	15	520

DISCHARGE AND VELOCITY IN VITRIFIED CLAY PIPES

Diam of Drain	Area (full) sq. in	Fall			Velocity Ft per sec	Discharge Gal per min	
						Full	Half full
4"	12·57	1	:	60	3·02	99	49
		1	:	40	3·70	121	61
		1	:	30	4·28	140	70
6"	28·27	1	:	100	3·07	226	113
		1	:	80	3·44	253	126
		1	:	60	3·97	291	146
		1	:	40	4·86	357	179
9"	63·62	1	:	180	3·01	497	249
		1	:	120	3·68	609	305
		1	:	90	4·25	702	351
		1	:	60	5·21	860	430
12"	113·10	1	:	260	3·03	892	446
		1	:	120	4·46	1311	655

The velocity in the foregoing table is calculated from Manning's formula:

$B = C.R.^{67}S^{5}$, where $C = 128·83$ for stoneware drains, R is the Hydraulic Mean Depth in feet (= sectional area of flow ÷ wetted perimeter), and S is the sine of the slope.

The discharge (G) in gallons per minute is found from the formula $G = 2·6\,aV$, where aV = the sectional area of flow in inches.

LEAD AND YARN

APPROXIMATE AMOUNTS REQUIRED FOR JOINTS IN CAST IRON SOIL AND WATER PIPES

Size of Pipe	Caulking Space in Joint	Depth of Lead in Joint	Weight of Lead	Weight of Yarn	
in	in	in	lb	lb	oz
2	⎫	$1\frac{1}{2}$	$2\frac{3}{4}$	0	2
$2\frac{1}{2}$		$1\frac{1}{2}$	3	0	$2\frac{1}{2}$
3		$1\frac{3}{4}$	4	0	3
4		$1\frac{3}{4}$	$4\frac{3}{4}$	0	4
5		$1\frac{3}{4}$	$5\frac{3}{4}$	0	7
6		$1\frac{3}{4}$	$6\frac{3}{4}$	0	8
7		$1\frac{3}{4}$	8	0	9
8	$\frac{3}{8}$	2	10	0	11
9		2	$11\frac{1}{4}$	0	13
10		2	$12\frac{1}{2}$	0	14
12		2	$14\frac{1}{2}$	1	1
14		$2\frac{1}{4}$	$18\frac{1}{2}$	1	6
15		$2\frac{1}{4}$	20	1	7
16		$2\frac{1}{4}$	$21\frac{1}{2}$	1	9
18	⎬	$2\frac{1}{4}$	28	2	0
20		$2\frac{1}{4}$	31	2	3
22		$2\frac{1}{2}$	37	2	9
24		$2\frac{1}{2}$	40	2	13
26		$2\frac{1}{2}$	$43\frac{1}{2}$	3	1
27		$2\frac{1}{2}$	45	3	2
28		$2\frac{1}{2}$	47	3	4
30		$2\frac{1}{2}$	50	3	7
32		$2\frac{1}{2}$	53	3	11
33	$\frac{7}{16}$	$2\frac{1}{2}$	55	3	13
36		$2\frac{1}{2}$	$59\frac{1}{2}$	4	3
38		$2\frac{1}{2}$	68	4	5
39		$2\frac{3}{4}$	71	4	1
40		$2\frac{3}{4}$	73	4	3
42		$2\frac{3}{4}$	76	4	5
44		$2\frac{3}{4}$	79	4	7
45		$2\frac{3}{4}$	$81\frac{1}{2}$	4	9
46		$2\frac{3}{4}$	$82\frac{1}{2}$	4	11
48	⎭	$2\frac{3}{4}$	$86\frac{1}{2}$	4	14

MANHOLE COVERS

Manhole covers, to B.S. 497, are manufactured in three grades—Grade A, heavy duty (HD) taking wheel loads up to $11\frac{1}{4}$ tons (11·43 tonnes): Grade B, medium duty (MD) for use where heavy commercial vehicles would be exceptional and Grade C, light duty (LD) not suitable for wheeled vehicular traffic.

Grade	Description	TYPES AND SIZES British Standard reference	Clear opening inches	millimetres
HD	Double triangle	A1	20	508
HD	Double triangle	A1	22	558
HD	Single triangle	A2	19½	495
HD	Double triangle	A3	27	608
MD	Circular	B4	20 & 25	508 & 559
MD	Rectangular*	B5	24 × 18	610 × 475
LD	Single seal	C6		
LD	Double seal	C7	18 × 18	457 × 457
LD	Single seal recessed	C8	24 × 18	610 × 457
LD	Double seal recessed	C9	24 × 24	610 × 610

*Recessed top type also available. When ordering any of the above range, specify by the B.S. reference, to avoid confusion. A large range of manhole covers and access covers, not covered by a British Standard, are commonly used in the trade.

15. THE PLUMBER

Codes of Practice	308
British Standards	308
Water Data	313
Lead Pipes	314
Copper Tube	315
Water Pressure	317
External Diameters of Pipes	318
Composition Gas Pipe	319
Solder and Petrol Requirements	319
Rainwater Pipes and Gutters	319
The Single Stack System	320
Lead Sheets	320
Lead Soakers	321

CODES OF PRACTICE

CP 3: Chapter VII: 1950: Engineering utility services.
Information concerning services for most types of buildings. Parts dealing with cold and hot water supply—tables covering consumption, rates of flow and sanitary provision.

CP 308: 1974: Drainage of roofs and paved areas.
Covers surface water from roofs and paved areas. Design and installation of gutters and rainwater pipes.

CP 310: 1965: Water supply.
Deals with the supply of water to houses and most other buildings. Appendix for pipe sizing.

B.S. 5572: 1978: Code of practice for sanitary pipework (formerly CP 304).
Design, installation, testing and maintenance of above ground non-pressure pipework.

BRITISH STANDARDS

B.S. 417: Galvanised mild steel cisterns and covers, tanks and cylinders.
Part 1: 1964: imperial units.
Part 2: 1973: metric units.

TABLE 1. DIMENSIONS OF CISTERNS AND COVERS

1	2	3	4	5	6	7	8	9	10	11	12
	Dimensions of cisterns			Minimum thickness of steel, before galvanizing for cisterns and covers					Covers		
BS type Reference*	Length	Width	Depth	Capacity to	Distance of water	Grade A cisterns	Grade B cisterns	Covers for	No. of pieces Length	Size of pieces Width	
	L	W	D	water line	line from top						
	mm	mm	mm	litres	mm	mm	mm	mm		mm	mm
SCM 34	457	305	305	18	111	1.6	1.2	1.0	1	482	330
SCM 70	610	305	371	36	111	1.6	1.2	1.0	1	635	330
SCM 90	610	406	371	54	111	1.6	1.2	1.0	1	635	432
SCM 110	610	432	432	68	114	1.6	1.2	1.0	1	635	457
SCM 135	610	457	482	86	114	1.6	1.2	1.0	1	635	482
SCM 180	686	508	508	114	114	1.6	1.2	1.0	1	711	533
SCM 230	736	559	559	159	114	2.0	1.6	1.0	1	762	584
SCM 270	762	584	610	191	114	2.0	1.6	1.0	1	787	610
SCM 320	914	610	584	227	114	2.0	1.6	1.0	1	940	635
SCM 360	914	660	610	264	114	2.0	1.6	1.0	1	940	686
SCM 450/1	1219	610	610	327	114	2.0	1.6	1.0	1	1245	635
SCM 450/2	965	686	686	336	114	2.0	1.6	1.0	1	991	711
SCM 570	965	762	787	423	146	2.5	2.0	1.2	1	991	787
SCM 680	1092	864	736	491	146	2.5	2.0	1.2·	1	1118	889
SCM 910	1168	889	889	709	146	2.5	2.0	1.2	1	1194	914
SCM 1130	1524	914	813	841	146	2.5	2.0	1.2	1	1549	940
SCM 1600	1524	1143	914	1227	146	3.2	2.5	1.6	2	787	1175
SCM 2270	1828	1219	1016	1727	146	3.2	2.5	1.6	2	940	1251
SCM 2720	1829	1219	1219	2137	190	3.2	2.5	1.6	3	940	1251
SCM 4540	2438	1524	1219	3364	254	4.8	3.2	1.6	3	838	1556

* Cisterns should be ordered by the BS type reference to avoid confusion.
For welded cisterns, the lengths may be increased by not more than 25 mm (see Fig. 1).
These dimensions were determined for the depths necessary to accommodate ballvalve, inlet and overflow pipes in the sizes likely to be used, and in the position, to satisfy water byelaws. The dimensions have been used in calculating the capacities.
The dimension is for the lapping piece; the lapped piece is 5 mm less.

B.S. 460: 1964: Cast iron rainwater goods.
Pipes from 2in to 6in, fittings and accessories (imperial units).
B.S. 504: 1961: Drawn lead traps.
Imperial units, 'S' and 'P' lead traps.
B.S. 569: 1973: Asbestos-cement rainwater goods.
Quality, dimensions, coating and marking for pipes, gutters and fittings.

TABLE 1. DIMENSIONAL TOLERANCES: PIPES AND PIPE FITTINGS

Nominal bore of pipe and fitting	Internal diameter permissible deviation (plus or minus) from nominal bore	Thickness of pipe and fitting	Permissible deviation (plus or minus) from thickness of pipes and fittings	Outside diameter (see Note)	Permissible deviation (plus or minus) from outside diameter	Length of pipe only	Permissible deviation from length of pipes only
mm	mm	mm	mm	mm	mm	mm	mm
50	3.0	6.0	1.0	62.0	3.0	1800 2400 and 3000	0 12
65		6.0	1.0	77.0			
75		7.5	1.0	90.0			
100		7.5	1.0	115.0			
150		9.0	1.5	168.0			

NOTE. The sum of the permissible deviations on the nominal bore and on the thickness shall not exceed the limits of the outside diameter.

TABLE 2. DIMENSIONAL TOLERANCES: GUTTERS AND GUTTER FITTINGS

Type of gutter	Work size of gutter or fitting	Thickness of gutter or fitting	Permissible deviation (plus or minus) from thickness	Permissible deviation (plus or minus) from internal dimensions	Effective length of gutter	Permissible deviation from length
	mm	mm	mm	mm	mm	mm
Half round	75	7.5		2.0		
	100	9.0				
	115	9.0		3.0		
	125	9.0				
	150	9.0				
	200	9.0				
Valley No. 1 pattern	406 × 127 × 254	12.0		5.0		
	457 × 127 × 152	12.0				
	610 × 152 × 229	12.0				
North light valley	457 × 152 × 102	12.0	1.5	5.0	1800	+0 −12
Boundary wall No. 1 pattern	279 × 127 × 178	9.0		3.0		
	305 × 152 × 229	12.0		5.0		
	457 × 152 × 305	12.0		5.0		
Boundary wall No. 2 pattern	559 × 152 × 406	12.0		5.0		
Box	127 × 152	9.0		3.0		
	305 × 203	12.0		5.0		
	381 × 127	12.0		5.0		

B.S. 1091: 1963: Pressed steel gutters, rainwater pipes, fittings and accessories.

Light pressed, galvanised gutters, pipes etc., and heavy pressed gutters.

B.S. 1184: 1976: Copper and copper alloy traps.

'S' and 'P' traps for use with baths, basins, sinks, etc.

B.S. 1431: 1960: Wrought copper and wrought zinc rainwater goods.

Half-round, rectangular, ogee gutters, round and rectangular pipes and accessories.

B.S. 1968: 1953: Floats for ballvalves (copper).

Class A, soldered joints; Class B, solderless joints; Class C, brazed, welded or silver soldered joints.

B.S. 1972: 1967: Polythene pipe (Tyne 32) for cold water services.

Dimensions, physical and mechanical characteristics. (See table overleaf).

B.S. 2456: 1973: Floats (plastics) for hot and cold water.

Requirements for floats 102mm, 114mm, 127mm and 152 mm diameter. Performance tests specified.

TABLE 2. BOSSES

1	2	3	4	5
BS type reference of floats (see 1.3(2))	Minimum length of screw thread in boss	Diameter of BS Whitworth screw thread in boss	Size of BS 1212 ballvalve which boss thread will fit	
	mm	in	in	in
102 S	13	$5/16$	$1/2$ and $3/4$	
102 NS	13	$5/16$	$1/2$ and $3/4$	
114 S	13	$5/16$	$1/2$ and $3/4$	
114 NS	13	$5/16$	$1/2$ and $3/4$	
127 S	13	$5/16$	$1/2$ and $3/4$	
127 NS	13	$5/16$	$1/2$ and $3/4$	
152 S	19	$7/16$	1	—
152 NS	19	$7/16$	1	—
152 LS	—	—	—	—
152 LNS	—	—	—	—

B.S. 2997: 1958: Aluminium rainwater goods.

Range of rainwater goods in aluminium alloys. Quality of materials, dimensions, workmanship, jointing and fixing.

B.S. 3380: Wastes for sanitary appliances and overflows for baths.

Part 1: 1976: wastes (excluding skeleton sink wastes) and bath overflows; for baths, basins, sinks and bidets.

Part 2: 1962: skeleton sink wastes; three-piece brass wastes for fireclay sinks.

TABLE 1. PIPE DIMENSIONS

Nominal size	Outside diameter				Wall thickness							
	Min.		Max.		Class B (0.60 N/mm²*: 0.60 MPa*: 200 ft hd: 86.7 lbf/in²)				Class C (0.89 N/mm²*: 0.89 MPa*: 300 ft hd: 130 lbf/in²)			
in	mm	(in)	mm	(in)	Min.		Max.		Min.		Max.	
					mm	(in)	mm	(in)	mm	(in)	mm	(in)
⅜	17.0	(0.669)	17.3	(0.681)	—		—		2.2	(0.087)	2.5	(0.098)
½	21.2	(0.835)	21.5	(0.847)	—		—		2.7	(0.106)	3.0	(0.118)
¾	26.6	(1.047)	26.9	(1.059)	2.3	(0.091)	2.6	(0.102)	3.4	(0.134)	3.7	(0.146)
1	33.4	(1.315)	33.7	(1.327)	3.0	(0.118)	3.3	(0.130)	4.2	(0.165)	4.6	(0.181)
1¼	42.1	(1.657)	42.5	(1.673)	3.7	(0.146)	4.1	(0.161)	5.3	(0.209)	5.8	(0.228)
1½	48.1	(1.894)	48.5	(1.910)	4.3	(0.169)	4.7	(0.185)	6.1	(0.240)	6.7	(0.264)
2	60.1	(2.366)	60.6	(2.386)	5.3	(0.209)	5.8	(0.228)	7.6	(0.299)	8.4	(0.331)
3	88.6	(3.488)	89.3	(3.516)	7.8	(0.307)	8.6	(0.339)	11.2	(0.441)	12.3	(0.484)
4	113.9	(4.484)	114.7	(4.516)	10.0	(0.394)	11.0	(0.433)	—		—	

Nominal size	Class D (1.20 N/mm²**: 1.20 MPa*: 400 ft hd: 173 lbf/in²)			
	Min.		Max.	
in	mm	(in)	mm	(in)
⅜	2.8	(0.110)	3.1	(0.122)
½	3.4	(0.134)	3.7	(0.146)
¾	4.3	(0.169)	4.7	(0.185)
1	5.4	(0.213)	5.9	(0.232)
1¼	6.8	(0.268)	7.5	(0.295)
1½	7.8	(0.307)	8.6	(0.338)
2	—		—	
3	—		—	

* In the original standard these values are expressed in MN/m^2. Numerically they are identical.

B.S. 3943: 1965: Plastics waste traps.
Dimensions, design and construction.
B.S. 4118: 1967: Glossary of sanitation terms.
Covers water supply, sanitary appliances and drainage.
B.S. 4213: 1975: Cold water storage cisterns (polyolefin or olefin copolymer) and cistern covers.
Requirements, dimensions and tests.

TABLE 1 (METRIC UNITS)

BS type reference	Minimum capacity to water line	Distance of water line from top	Maximum height of cistern	Minimum thickness	Minimum mass
	l	mm	mm	mm	kg
PC 4	18	111	310	1.4	1.36
PC 8	36	111	380	1.4	1.81
PC 15	68	114	430	2.1	2.95
PC 20	91	114	510	2.1	3.18
PC 25*	114	114	560	2.1	3.40
PC 40	182	114	610	2.1	6.35
PC 50*	227	114	660	2.1	7.03
PC 60	273	114	660	2.1	7.26
PC 70	318	114	660	2.1	9.07
PC 100	455	114	760	2.1	12.70

B.S. 4576: Unplasticised PVC rainwater goods.

Part 1: 1970: half-round gutter and circular pipe; material, colour, dimensions, notes on fixing.

TABLE 1. PIPE DIMENSIONS

Nominal size	Mean outside diameter		Tolerance on outside diameter		Minimum wall thickness		Minimum bore	
	mm	(in)	mm	(in)	mm	(in)	mm	(in)
63 mm	63.1	(2.486)	± 0.15	(± 0.006)	1.8	(0.071)	58.4	(2.300)
2½ in	68.15	(2.696)	± 0.18	(± 0.007)	1.8	(0.071)	63.5	(2.500)
75 mm	75.2	(2.961)	± 0.20	(± 0.008)	1.8	(0.071)	70	(2.756)

NOTE. Three further sizes of pipe (3, 4 and 6 nominal) are not specified in this standard but are included in BS 4514, 'Unplasticized PVC soil and ventilating pipe, fittings and accessories', and are suitable for sealed systems for internal or external use.

TABLE 2. DIMENSIONS OF TRUE HALF-ROUND* GUTTERS

Nominal size	Minimum wall thickness		Minimum cross-sectional area		Minimum effective sealed cross-sectional area	
in	mm	(in)	mm2	(in2)	mm2	(in2)
4	2.2	(0.085)	4000	(6.2)	3200	(5.0)
4½	2.2	(0.085)	5100	(7.9)	4080	(6.3)
5	2.3	(0.090)	6300	(9.8)	5040	(7.8)
6	2.5	(0.100)	9100	(14.5)	7280	(11.3)

TABLE 3. DIMENSIONS OF NORMAL HALF-ROUND* GUTTERS

Nominal size	Minimum wall thickness		Minimum cross-sectional area		Minimum effective sealed cross-sectional area	
in	mm	(in)	mm2	(in2)	mm2	(in2)
4	2.2	(0.085)	3200	(5.0)	2560	(4.0)
4½	2.2	(0.085)	4000	(6.2)	3200	(5.0)
5	2.3	(0.090)	4900	(7.6)	3920	(6.1)
6	2.5	(0.100)	7200	(11.1)	5760	(8.9)

The minimum cross-sectional areas given in Tables 2 and 3 are calculated cross-sectional areas and no cutting allowance has been made.

B.S. 5254: 1976: Polypropylene waste pipe and fittings (external diameter 34.6mm, 41.0mm and 54.1mm).
Dimensions and performances.

B.S. 5255: 1976: Plastics waste pipe and fittings.
Dimensions and performance for pipes made from acrylonitrile butadiene styrene (ABS) modified unplasticised polyvinyl chloride (MUPVC), polypropylene (PP) and polyethylene (PE).

WATER DATA

1 gallon of water weighs 10 lb or 4·454 kg.
1 cubic foot of water equals 6·232 gallons or 28·293 kg.
1 hundredweight of water (50 kg) equals 1·8 cu. ft.
1 litre equals 1 kg of water or 1 cubic decimetre, or 0·22 galls.
1 cubic metre equals 1,000 cubic decimetres.
Water pressure is now in a new unit—the Bar. See page 320.

WATER CONSUMPTION FOR BUILDINGS OF VARIOUS TYPES

	per person per day	
	Galls	Litres
Schools	10	45·45
Office Buildings	15	60·24
Dwelling Houses	25*	112·50
Hotels	30	136·37

*This figure is made up as follows:

Drinking	¼	1·17
Food preparation	¼	1·17
Washing-up	1	4·54
House Cleaning	2	9·09
Laundry	3	13·60
Toilet Purposes	5	22·72
Baths and WC's	13½	61·35
Total	25	112·50

RELATIVE DISCHARGING POWER OF WATER PIPES
INTERNAL DIAMETER OF PIPES

Inches	4	3½	3	2½	2	1½	1¼	1	¾	½
mm	102	89	76	74	51	38	31	26	19	13
Lead Code numbers	27	26	25	24	22-23	19-21	17-18	13-16	7-12	4-6
	1=	1	2	3	5	11	18	32	65	181
		1=	1	2	4	8	13	22	47	129
			1=	1	2	5	8	15	31	88
				1=	1	3	5	9	20	55
					1=	2	3	5	11	32
						1=	1	2	5	19
							1=	1	3	9
								1=	2	5
									1=	2

LEAD PIPES

Lead pipe to B.S. 602: 1970 and B.S. 1085: 1972. These new sizes of lead pipe are now in code numbers.

APPENDIX A
STANDARD METRIC SIZES OF PIPES

Code No.	Nominal bore	Nominal wall thickness	Nearest arithmetical equivalent size in BS 602, 1085: 1956
	mm	mm	
1	10.0	5.00	⅜ in × 5 lb/yd
2	12.0	2.00	½ in × 2 lb/yd
3	12.0	3.60	½ in × 4 lb/yd
4	12.0	4.80	½ in × 6 lb/yd
5	12.0	5.60	½ in × 7 lb/yd
6	12.0	6.70	½ in × 9 lb/yd
7	16.0	2.30	⅝ in × 3 lb/yd
8	20.0	2.60	¾ in × 4 lb/yd
9	20.0	3.20	¾ in × 5 lb/yd

THE PLUMBER

Code No.	Nominal bore	Nominal wall thickness	Nearest arithmetical equivalent size in BS 602, 1085: 1956
10	20.0	5.30	¾ in × 9 lb/yd
11	20.0	6.40	¾ in × 11 lb/yd
12	20.0	8.50	¾ in × 15 lb/yd
13	25.0	3.00	1 in × 6 lb/yd
14	25.0	3.30	1 in × 7 lb/yd
15	25.0	8.00	1 in × 16 lb/yd
16	25.0	10.0	1 in × 21 lb/yd
17	32.0	2.80	1¼ in × 7 lb/yd
18	32.0	3.60	1¼ in × 9 lb/yd
19	40.0	2.50	1½ in × 7 lb/yd
20	40.0	3.10	1½ in × 9 lb/yd
21	40.0	4.00	1½ in × 12 lb/yd
22	50.0	3.10	2 in × 12 lb/yd
23	50.0	4.00	2 in × 16 lb/yd
24	65.0	2.50	2½ in × 12 lb/yd
25	75.0	2.70	3 in × 15 lb/yd
26	90.0	2.70	3½ in × 17 lb/yd
27	100.0	2.70	4 in × 19 lb/yd
28	115.0	2.70	4½ in × 22 lb/yd
29	125.0	3.20	5 in × 30 lb/yd

COPPER TUBE

TABLE Z. DIMENSIONS AND WORKING PRESSURES FOR HARD DRAWN THIN WALL COPPER TUBES

1	2	3	4	5
Size of tube	Outside diameter		Nominal thickness	Maximum working pressures*
	maximum	minimum		
mm	mm	mm	mm	bar†
6	6.045	5.965	0.5	113
8	8.045	7.965	0.5	98
10	10.045	9.965	0.5	78
12	12.045	11.965	0.5	64
15	15.045	14.965	0.5	50
18	18.045	17.965	0.6	50
22	22.055	21.975	0.6	41
28	28.055	27.975	0.6	32
35	35.07	34.99	0.7	30

1	2	3	4	5
Size of tube	Outside diameter		Nominal thickness	Maximum working pressures*
	maximum	minimum		
42	42.07	41.99	0.8	28
54	54.07	53.99	0.9	25
67	66.75	66.60	1.0	20
76.1	76.30	76.15	1.2	19
108	108.25	108.00	1.2	17
133	133.50	133.25	1.5	16
159	159.50	159.25	1.5	15

* Based on material in H condition .... at 65 °C.
† 1 bar = 0.1 N/mm² = 0.1 MPa.

Note: This tube must not be bent: use easy bends and pitcher tees.

TABLE X. DIMENSIONS AND WORKING PRESSURES FOR HALF HARD, LIGHT GAUGE COPPER TUBES

1	2	3	4	5
Size of tube	Outside diameter		Nominal thickness	Maximum working pressures*
	maximum	minimum		
mm	mm	mm	mm	bar†
6	6.045	5.965	0.6	133
8	8.045	7.965	0.6	97
10	10.045	9.965	0.6	77
12	12.045	11.965	0.6	63
15	15.045	14.965	0.7	58
18	18.045	17.965	0.8	56
22	22.055	21.975	0.9	51
28	28.055	27.975	0.9	40
35	35.07	34.99	1.2	42
42	42.07	41.99	1.2	35
54	54.07	53.99	1.2	27
67	66.75	66.60	1.2	20
76.1	76.30	76.15	1.5	24
108	108.25	108.00	1.5	17
133	133.50	133.25	1.5	14
159	159.50	159.25	2.0	15

* Based on material in ½H condition .... at 65 °C.
† 1 bar = 0.1 N/mm² = 0.1 MPa².

TABLE Y. DIMENSIONS AND WORKING PRESSURES FOR HALF HARD AND ANNEALED COPPER TUBES

1	2	3	4	5	6
Size of tube	Outside diameter		Nominal thickness	Maximum working pressures 1/2H condition*	Maximum working pressures O condition†
	maximum	minimum			
mm	mm	mm	mm	bar‡	bar‡
6	6.045	5.965	0.8	188	144
8	8.045	7.965	0.8	136	105
10	10.045	9.965	0.8	106	82
12	12.045	11.965	0.8	87	67
15	15.045	14.965	1.0	87	67
18	18.045	17.965	1.0	72	55
22	22.055	21.975	1.2	69	57
28	28.055	27.975	1.2	55	42
35	35.07	34.99	1.5	54	41
42	42.07	41.99	1.5	45	34
54	54.07	53.99	2.0	47	36
67	66.75	66.60	2.0	37	28
76.1	76.30	76.15	2.0	33	25
108	108.25	108.00	2.5	29	22

* Based on material in ½H condition .... at 65 °C.
† Based on material in O condition .... at 65 °C.
‡ 1 bar = 0.1 N/mm^2 = 0.1 MPa.

Note: Since the diameters in this table differ from the former (Imperial and metric) sizes, it is imperative that new formers for bending machines (except for ½in and 2in, which will serve) and new bending springs be obtained. Fittings for old sized pipe cannot be used on these tubes.

WATER PRESSURE

EQUIVALENTS OF THE BAR

1 atmosphere = 14.7 lb.f/sq. in = 1032 grams per cm^2.
10^5 Nm^2.
29·53in or 75·01cm of mercury (1in watergauge = 2·491 millibar).
1 million dynes per cm^2.

Note: This unit must not be confused with the Acoustic bar, which is the unit of acoustic pressure in a sound-wave, and 1 dyne/cm^2.

BAR AND POUNDS FORCE PER SQUARE INCH

bar		lb.f.sq.in.
0·0686	1	14·7
0·1373	2	29·4
0·2060	3	44·1
0·2756	4	58·8
0·3430	5	73·5
0·4120	6	88·2
0·4806	7	102·9
0·5512	8	117·6
0·6180	9	132·3
0·6860	10	147·0
1·030	15	220·5
1·373	20	294
2·060	30	441
2·756	40	588
3·450	50	735
6·890	100	1470

EXTERNAL DIAMETERS OF PIPES

Nominal internal diameters		Galvanised carbon steel, for central heating	Stainless steel, for central heating	Polythene, for cold water supply
in	mm	mm	mm	mm
$\frac{3}{8}$	9·5	11·96	11·96	17·3
$\frac{1}{2}$	12·7	15·14	15·14	21·5
$\frac{3}{4}$	19·0	21·49	21·49	26·9
1	25·4	28·24	28·24	33·7
$1\frac{1}{4}$	31·7	34·59	—	42·5
$1\frac{1}{2}$	38·1	40.94	—	48·5

Notes: Coated carbon steel pipes (B.S. 4182) are coming into use because they are considerably cheaper than copper. Soldered or compression joints can be used.

These tubes should not be used with copper cylinders, calorifiers or tanks, nor with copper or brass fittings.

Stainless steel tubes (B.S. 864) should be used with stainless steel compression fittings and with stainless steel cylinders (which are cheaper than copper) and should be used in those districts where the water supply attacks copper.

The range of polythene pipes extends to 2in, 4in and 6in pipes. These are three grades—B, C and D. Taper pipe threads can be cut on grades C and D.

Care should be taken not to cut too long a thread.

COMPOSITION GAS PIPE

Internal Diameter		Weight		Length of coil	
inches	mm	per yard lbs/ozs	per metre grams	yards	metres
$\frac{1}{4}$	6·35	15oz	466·5	73	66·7
$\frac{5}{16}$	7·94	18oz	559·8	67	61·2
$\frac{3}{8}$	9·53	22oz	684·2	53	48·4
$\frac{7}{16}$	11·11	23oz	785·3	43	39·3
$\frac{1}{2}$	12·70	1lb 11oz	795·2	32	29·3
$\frac{9}{16}$	14·29	2lb 1oz	937·0	29	26·4
$\frac{5}{8}$	15·88	2lb 4oz	1 030·4	29	26·4
$\frac{3}{4}$	19·05	2lb 10oz	1 217·0	25	22·8
$\frac{7}{8}$	22·22	2lb 4½oz	1 045·0	20	18·2
1	25·40	3lb 2oz	1 289·0	15	13·7

SOLDER AND PETROL REQUIREMENTS

LEAD JOINTS TO PIPES ETC

Lead pipe Code No.	Solder		Petrol	
	g	lb	lt	pt
4–6	227	$\frac{1}{2}$	0·056	0·1
7–12	340	$\frac{3}{4}$	0·11	0·2
13–16	450	1	0·13	0·2
17–18	568	$1\frac{1}{4}$	0·17	0·3
19–21	680	$1\frac{1}{2}$	0·23	0·4
24	1 133	2	0·28	0·4
25	1 359	$2\frac{1}{2}$	0·29	0·5
26	1 586	3	0·34	0·6
27	1 613	4	0·56	0·6
Cesspools (each)	1 613	4	0·56	0·2
Soldered dots (each)	450	1	0·11	0·2
Soldered seams & angles metre run	46	0·1	0·11	0·1
Stop ends to pipes:				
4–12	153	0·33	As for wiped joints	
13–21	227	0·50	As for wiped joints	
24	340	0·75	As for wiped joints	

Note: For Imperial Diameters and Imperial (old) sizes see page 000.

RAINWATER PIPES AND GUTTERS

To deal adequately with a storm rainfall, which is taken to be at the rate of 3in (75mm) per hour, 1in (6·45cm) square of the sectional area of the bore of a rainwater pipe is required for every square of roofing (measured on plan) in the area drained. The crucial item is, therefore, not the type of size of gutter but the size of the outlet or (if the gutter drains into a hopper-head) the size of the rainwater pipe. It is very rare, however, that a gutter drains directly into a hopper-head. In the vast majority of instances, the size of gutter to be employed, and the size of rainwater pipe will depend on the bore of the outlet. Calculate the number of squares of roofing to be drained, and select the nearest figure in the table below. On the line above will be read the size of gutter to employ and on the line beneath the size of rainwater pipe, swan-necks, bends, offsets, etc.

Gutter size half round or ogee	3 76	4 102	$4\frac{1}{2}$ 113	5 127	6 152	inches mm
Area of roof to be drained	$1\frac{7}{8}$ 17·42	$2\frac{5}{8}$ 24·38	$2\frac{5}{8}$ 24·38	$2\frac{7}{8}$ 26·71	$2\frac{7}{8}$ 26·71	squares m²
size of outlet	2 51	$2\frac{1}{2}$ 63	$2\frac{1}{2}$ 63	3 76	4 102	inches mm

Notes: The length of gutter should not exceed 20ft (6m) on each side of an outlet. These figures are in accord with CP 142 (1951) and therefore are deemed to satisfy the Building Regulations.

THE SINGLE STACK SYSTEM

This system has largely taken the place of the two-pipe system, and is now developed for up to 25-storey buildings. Soil and waste are accommodated in one stack, without venting the traps. To prevent syphonage all fittings are grouped together. 3in deep seal traps should be used. The slope of the waste pipes should be as flat as possible. Branches should not exceed 5ft for basin and bath, and up to 9ft for sink, in length. Connections to stack are made by plumbers' unions. "Specials", 3ft long, with inlet for W.C. waste and threaded bosses for unions, in various positions on the pipe, are now obtainable from all manufacturers of cast-iron drainpipes, and are included in B.S. 416.

LEAD SHEET
WEIGHTS AND EQUIVALENT THICKNESSES

Metric number	Wt in lb per sq. ft	Thickness (inches)		Metric equiva- lent in millimetres	Nearest S.W.G.
		Fractions	Decimals		
3	3	3/64 plus	0·050	1·270	18
4	4	1/16 plus	0·068	1·702	16
5	5	5/64 plus	0·084	2·130	14
6	6	7/64 minus	1·101	2·560	12
7	7	$\frac{1}{8}$ minus	0·118	2·990	11
8	8	9/64	0·135	3·420	10

THE PLUMBER

WEIGHTS OF LEAD SHEETS FOR VARIOUS POSITIONS

Position	Metric Code Numbers
Main Gutters (Box or Parapet)	5 or 6
Flats, Roofs, small Gutters	4 or 5
Hips, Valleys and Ridges	4 or 5
Flashings and Aprons	4 or 5
Soakers	3 or 4
Flat Roof, small, no traffic	4 or 5
Flat Roof, large, with traffic	5, 6 or 7
Swept and pitched Roofs	4, 5 or 6
Dormer Cheeks	4 or 5
Cloaks in Cavity Walls	3
Linings:	
Water Tanks, Up to 30 galls Capacity	4 or 5
Safes to Baths and Tanks	4
Water Tanks over 30 galls Capacity	5 or 6
Flower boxes	6

LEAD SOAKERS

LENGTHS OF LEAD SOAKERS TO SUIT DIFFERENT SIZES OF SLATES

Length of Slate	Length of Soaker					
	Head nailing			Centre nailing		
	2" lap	3" lap	4" lap	2" lap	3" lap	4" lap
24	$13\frac{1}{2}$	14	$14\frac{1}{2}$	14	$14\frac{1}{2}$	15
22	$12\frac{1}{2}$	13	$13\frac{1}{2}$	13	$13\frac{1}{2}$	14
20	$11\frac{1}{2}$	12	$12\frac{1}{2}$	12	$12\frac{1}{2}$	13
18	$10\frac{1}{2}$	11	$11\frac{1}{2}$	11	$11\frac{1}{2}$	12
16	$9\frac{1}{2}$	10	$10\frac{1}{2}$	10	$10\frac{1}{2}$	11
14	$8\frac{1}{2}$	9	$9\frac{1}{2}$	9	$9\frac{1}{2}$	10
13	8	$8\frac{1}{2}$	9	$8\frac{1}{2}$	9	$9\frac{1}{2}$
12	$7\frac{1}{2}$	8	$8\frac{1}{2}$	8	$8\frac{1}{2}$	9
11	7	$7\frac{1}{2}$	8	$7\frac{1}{2}$	8	$8\frac{1}{2}$
10	$6\frac{1}{2}$	7	$7\frac{1}{2}$	7	$7\frac{1}{2}$	8
9	6	$6\frac{1}{2}$	7	$6\frac{1}{2}$	7	$7\frac{1}{2}$
8	$5\frac{1}{2}$	6	$6\frac{1}{2}$	6	$6\frac{1}{2}$	7

Note: Soakers vary in width from 7in to 9in.

16. THE HEATING ENGINEER

Codes of Practice	323
British Standards	323
Conversion Tables	331
Relative Fuel Values	332
Central Heating	333
Pipes	337
Pipe Threads	337

CODES OF PRACTICE

CP 3: Chapter VIII: 1949: Heating and thermal insulation.

Conditions affecting temperature in dwellings. Recommends standards of warmth for rooms.

CP 324 and CP 202; 1951: Provision of domestic electric water-heating installations.

Design and installation of domestic water-heating systems.

CP 332: Selection and installation of town gas space heating.

Part 1: 1961: independent domestic applicances, selection and installation of radiant or convector fires.

Part 3: 1970: boilers of more than 150,000 Btu/h (44kw) and up to 2,000,000 Btu/h (586kw) output.

Part 4: 1966: ducted warm air systems.

CP 333: Selection and installation of town gas hot water supplies.

Part 1: 1964: domestic premises.

CP 341 and CP 300-307: 1966: Central heating by low pressure hot water.

Head-code and seven sub-codes covering boilers and calorifiers; storage vessels; pipework, fittings, valves, taps and cocks; appliances; unit heaters, circulating pumps, and thermal insulation.

CP 342: 1950: Centralised domestic hot water supply.

Installation of boilers, calorifiers, storage vessels, pipework and electrically driven circulators in central systems of domestic hot water supply.

CP 403: 1974: Installation of domestic heating and cooking appliances.

Installation of room heaters, independent boilers and fanned warm air heater units.

CP 403: 101: 1952: Small boiler systems using solid fuel.

Combined heating and hot water supply systems using solid fuel.

CP 1018: 1971: Electric floor-warming systems for use with off-peak and similar supplies of electricity.

Guidance on electric floor-warming systems in concrete and similar floors.

B.S. 5410: Code of practice for oil-firing.

Part 2: 1977: installations up to 44kW output capacity for space heating, hot water and steam supply.

B.S. 5449: Code of practice for central heating for domestic premises.

Part 1: 1957: forced circulation hot water systems.

BRITISH STANDARDS

B.S. 699: 1972: Copper cylinders for domestic purposes.

Cylinders for heating and storage of hot water. (See table opposite).

B.S. 779: 1976: Cast iron boilers for central heating and indirect hot water supply.

Performance requirements, materials, mountings and fittings, wiring, etc.

324 BUILDER'S REFERENCE BOOK

TABLE 1. DIMENSIONS AND DETAILS OF COPPER INDIRECT CYLINDERS

1	2	3	4	5	6	7	8	9	10	11	12	13	14	15	16	17	18	19
British Standard type reference*	External diameter	External height (over dome)	Storage capacity	Heating surface	Minimum thickness of copper sheet before forming							Height of screwed connections (see Figs. 2, 3, 4 and 5)					Size of screwed connections	Size of primary heater connections
					Grade 2 Test pressure: 0.220 N/mm²[a] (MPa)[a]. Maximum working head: 15 m†			Grade 3 Test pressure: 0.145 N/mm²[a] (MPa)[a]. Maximum working head: 10 m†										
					Body and top	Bottom	Primary heater	Body and top	Bottom	Primary heater	H‡	J_1	J_2	L	M	P	BSP.F internal threads§	BSP.F external threads
	A	B	min.	min.									(see 12)			(see 13)		
	mm	mm	litres	m²	mm	mm	mm	mm	mm	mm	mm	mm	mm	mm	mm	mm	in	in
1	350	900	72	0.40	0.9	1.4	0.9	0.7	1.2	0.7	700	140	270	100	430	150	1	1
2	400	900	96	0.52	0.9	1.6	0.9	0.7	1.2	0.7	700	140	270	100	560	150	1	1
3‖	400	1050	114	0.63	0.9	1.6	0.9	0.7	1.2	0.7	800	140	270	100	670	150	1	1
4	450	675	84	0.46	1.0	1.6	1.0	0.7	1.2	0.7	450	140	270	100	500	150	1	1
5	450	750	95	0.52	1.0	1.6	1.0	0.7	1.2	0.7	550	140	270	100	560	150	1	1
6	450	825	106	0.60	1.0	1.6	1.0	0.7	1.2	0.7	625	140	270	100	640	150	1	1
7‖	450	900	117	0.66	1.0	1.6	1.0	0.7	1.2	0.7	700	140	270	100	700	150	1	1
8‖	450	1050	140	0.78	1.0	1.6	1.0	0.7	1.2	0.7	800	150	270	100	840	150	1¼	1¼
9‖	450	1200	162	0.91	1.0	1.6	1.0	0.7	1.2	0.7	950	150	270	100	950	150	1¼	1¼
10	500	1200	190	1.13	1.2	1.8	1.2	0.9	1.6	0.9	950	180	300	150	960	200	1½	1½
11	500	1500	245	1.30	1.2	1.8	1.2	0.9	1.6	0.9	1200	180	300	150	1110	200	1½	1½
12	600	1500	280	1.60	1.4	2.5	1.4	1.2	2.0	1.2	950	190	330	150	880	200	2	2
13	600	1500	360	2.10	1.4	2.5	1.4	1.2	2.0	1.2	1200	190	330	150	1190	200	2	2
14	600	1800	440	2.50	1.4	2.5	1.4	1.2	2.0	1.2	1500	190	330	150	1490	200	2	2

* Cylinders should be ordered by the BS type references.
† The working head shall be the vertical distance from the bottom of the cylinder to the water line of the cistern supplying it (1 mH$_2$O ≈ 0.01 N/mm²[a]).
‡ Fitted to sizes 1–9 only when specified.
§ BSP.F external threads may be supplied when so ordered.
‖ Preferred sizes for new installations in dwellings.
[a] *In the original standard these values are expressed in kN/m².*

B.S. 843: 1976: Specification for thermal-storage electric water heaters (constructional and water requirements).
Requirements for stationary non-instantaneous electric water heaters.
B.S. 853: Calorifiers for central heating and hot water supply.
Part 1: 1960: mild steel and cast iron.

MINIMUM REQUIREMENTS FOR INSPECTION OPENINGS
(All dimensions are inside, i.e. ' clear hole ')

Internal diameter of Calorifier	SIZES OF OPENINGS	
	Storage Calorifiers	Non-storage Calorifiers
Below 18 in	Optional	Optional
18 in up to and including 24 in	Two oval openings each 4¼ in × 2¾ in or one circular opening 8 in diameter	Two 3 in diameter screwed bosses fitted with plugs or one circular opening 8 in diameter
Over 24 in up to and including 30 in	Two oval openings each 5 in × 3½ in or one circular opening 10 in diameter	Two circular openings each 5 in diameter or one circular opening 10 in
Over 30 in up to and including 36 in	Two oval openings each 8 in × 4 in or one circular opening 12 in diameter or one oval opening 14 in × 11 in	Two circular openings each 6 in diameter or one circular opening 12 in diameter
Over 36 in up to and including 42 in	Two circular openings each 9 in diameter or one circular opening 15 in diameter or one oval opening 15 in × 11 in	Two circular openings 9 in diameter or one circular opening 15 in diameter
Over 42 in	Two circular openings each 10 in diameter or one circular opening 18 in diameter or one oval opening 16 in × 12 in	Two circular openings each 10 in diameter or one circular opening 18 in diameter

Oval openings shall be fitted with internal doors. Circular openings other than screwed bosses shall be fitted with external covers secured by setscrew which shall comply with *the standard* as regards permissible stresses.

Part 2: 1960: Copper.
MINIMUM REQUIREMENTS FOR INSPECTION OPENINGS
(All dimensions are inside i.e. ' clear hole ')

Internal diameter of Calorifier	SIZES OF OPENINGS	
	Storage Calorifiers	Non-storage Calorifiers
Below 18 in	Optional	Optional
18 in up to and including 24 in	Two oval openings each 4¼ in × 2¾ in or one circular opening 8 in diameter	Two 3 in dia. screwed bosses fitted with plugs or one circular opening 8 in diameter
Over 24 in up to and including 30 in	Two oval openings each 5 in × 3½ in or one circular opening 10 in diameter	Two circular openings each 5 in diameter or one circular opening 10 in
Over 30 in up to and including 36 in	Two oval openings each 8 in × 4 in or one circular opening 12 in diameter or one oval opening 14 in × 11 in	Two circular openings each 6 in diameter or one circular opening 12 in diameter
Over 36 in up to and including 42 in	Two circular openings each 9 in diameter or one circular opening 15 in diameter or one oval opening 15 in × 11 in	Two circular openings 9 in diameter or one circular opening 15 in diameter
Over 42 in	Two circular openings each 10 in diameter or one circular opening 18 in diameter or one oval opening 16 in × 12 in	Two circular openings each 10 in diameter or one circular opening 18 in diameter

Oval openings shall be fitted with internal doors. Circular openings, other than screwed bosses, shall be fitted with external covers secured by setscrew which shall comply with *the standard* as regards permissible stresses.

B.S. 855: 1976: Specification for welded steel boilers for central heating and indirect hot water supply (rated output 44kW to 3mW).
Specifies minimum efficiencies for oil, gas and solid fuel firing.
B.S. 1250: Part 4: 1965: Domestic appliance burning town gas.
Space heating appliances, performance, and test methods.
B.S. 1565: Galvanised mild steel indirect cylinders, annular or saddleback type.
  Part 1: 1949: imperial units.
  Part 2: 1973: metric units.

TABLE 1. DIMENSIONS AND DETAILS OF GALVANIZED MILD STEEL INDIRECT CYLINDERS

BS size No.	Dimensions				Heights of screwed connections						Size of cylinder connections BSP.F internal threads	Size of primary heater connections BSP.F external threads
	Int. dia. A	Height over dome B	Approx. capacity	Min. heating surface	H	J_1	J_2*	L	M	P		
	mm	mm	litres	m²	mm	mm	mm	mm	mm	mm	in	in
BSG.1M	457	762	109	0.74	559	140	254	†140	610	165	1	‡1
BSG.2M	457	914	136	0.88	686	140	254	†140	749	165	1	‡1
BSG.3M	457	1067	159	1.11	813	152	254	†152	914	165	1¼	‡1¼
BSG.4M	508	1270	227	1.58	991	178	305	178	1041	178	1½	1½
BSG.5M	508	1473	273	1.86	1143	178	305	178	1194	178	1½	1½
BSG.6M	610	1372	364	2.51	1041	191	330	191	1143	191	2	2
BSG.7M	610	1753	455	3.16	1346	191	330	191	1524	191	2	2
BSG.8M	457	838	123	0.84	635	140	254	†140	686	165	1	‡1

BS size No.	Minimum thicknesses of material							
	Class B. Tested 0.276 N/mm²‡ (0.276 MPa)¶, Maximum permissible working head § = 18 m				Class C. Tested 0.138 N/mm²‡ (0.138 MPa)¶ Maximum permissible working head § = 9 m			
	Cylinder		Heater		Cylinder		Heater	
	Body and dome	Bottom	Inner wall‖	Outer wall‖	Body and dome	Bottom	Inner wall‖	Outer wall‖
	mm	mm	mm	mm	mm	mm	mm	mm
BSG.1M	3.2	3.2	3.2	3.2	2.0	2.5	2.0	2.0
BSG.2M	3.2	3.2	3.2	3.2	2.0	2.5	2.0	2.0
BSG.3M	3.2	3.2	3.2	3.2	2.0	2.5	2.0	2.0
BSG.4M	4.8	4.8	4.8	4.8	3.2	3.2	3.2	3.2
BSG.5M	4.8	4.8	4.8	4.8	3.2	3.2	3.2	3.2
BSG.6M	4.8	4.8	4.8	4.8	3.2	3.2	3.2	3.2
BSG.7M	4.8	4.8	4.8	4.8	3.2	3.2	3.2	3.2
BSG.8M	3.2	3.2	3.2	3.2	2.0	2.5	2.0	2.0

* This arrangement does not apply to cylinders with saddle-back heaters (see Clause 5).
† For saddle-back heaters each of these dimensions becomes 75 mm.
‡ For saddle-back heaters the heater connection nearest the externally domed end is internal.
§ The maximum permissible working head of water shall be measured from the bottom of the cylinder.
‖ If the immersion heater is mounted horizontally (see Clause 13) these thicknesses shall be increased as may be necessary to secure the required strength from the altered design of primary heater.
¶ In the original standard these values are expressed in kN/m².

B.S. 1566: Copper indirect cylinders for domestic purposes.
  Part 1: 1972: double feed indirect cylinders. (See table overleaf).
  Part 2: 1972: single feed indirect cylinders. (See table overleaf).

TABLE 1. DIMENSIONS AND DETAILS OF COPPER INDIRECT CYLINDERS

1	2	3	4	5	6	7	8	9	10	11	12	13	14	15	16	17	18	19
					\multicolumn{6}{c}{Minimum thickness of copper sheet before forming}		\multicolumn{5}{c}{Height of screwed connections (see Figs. 2, 3, 4 and 5)}			
		External height (over dome)	Storage capacity	Heating surface	\multicolumn{3}{c}{Grade 2 Test pressure: 0.220 N/mm² (MPa)¶ Maximum working head: 15 m†}	\multicolumn{3}{c}{Grade 3 Test pressure: 0.145 N/mm² (MPa)¶ Maximum working head: 10 m†}							Size of screwed connections	Size of primary heater connections
British Standard type reference*	External diameter				Body and top	Bottom	Pri-mary heater	Body and top	Bottom	Pri-mary heater	H‡	J₁	J₂	L	M	P	BSP.F internal threads§	BSP.F external threads
	A	B	min.	min.														
	mm	mm	litres	m²	mm	mm	mm	mm	mm	mm	mm	mm	mm	mm	mm	mm	in	in
1	350	900	72	0.40	0.9	1.4	0.9	0.7	1.2	0.7	700	140	270	100	430	150	1	1
2	400	900	96	0.52	0.9	1.6	0.9	0.7	1.2	0.7	700	140	270	100	560	150	1	1
3‖	400	1050	114	0.63	0.9	1.6	0.9	0.7	1.2	0.7	800	140	270	100	670	150	1	1
4	450	675	84	0.46	1.0	1.6	1.0	0.7	1.2	0.7	450	140	270	100	500	150	1	1
5	450	750	95	0.52	1.0	1.6	1.0	0.7	1.2	0.7	550	140	270	100	560	150	1	1
6	450	825	106	0.60	1.0	1.6	1.0	0.7	1.2	0.7	625	140	270	100	640	150	1	1
7‖	450	900	117	0.66	1.0	1.6	1.0	0.7	1.2	0.7	700	140	270	100	700	150	1	1
8‖	450	1050	140	0.78	1.0	1.6	1.0	0.7	1.2	0.7	800	150	270	100	840	150	1¼	1¼
9‖	450	1200	162	0.91	1.0	1.6	1.0	0.7	1.2	0.7	950	150	270	100	950	150	1¼	1¼
10	500	1200	190	1.13	1.2	1.8	1.2	0.9	1.6	0.9	950	180	300	150	960	200	1½	1½
11	500	1500	245	1.30	1.2	1.8	1.2	0.9	1.6	0.9	1200	180	300	150	1110	200	1½	1½
12	600	1200	280	1.60	1.4	2.5	1.4	1.2	2.0	1.2	950	190	330	150	880	200	2	2
13	600	1500	360	2.10	1.4	2.5	1.4	1.2	2.0	1.2	1200	190	330	150	1190	200	2	2
14	600	1800	440	2.50	1.4	2.5	1.4	1.2	2.0	1.2	1350	190	330	150	1490	200	2	2

* Cylinders should be ordered by the BS type references.
† The working head shall be the vertical distance from the bottom of the cylinder to the water line of the cistern supplying it (1 mH₂O ≃ 0.01 N/mm²⁶).
‡ Fitted to sizes 1–9 only when specified.
§ BSP.F external threads may be supplied when so ordered.
‖ Preferred sizes for new installations in dwellings.
¶ In the original standard these values are expressed in kN/m².

TABLE 1. DIMENSIONS AND DETAILS OF SINGLE FEED COPPER INDIRECT CYLINDERS

1	2	3	4	5	6	7	8	9	10	11	12	13	14	15
British Standard type reference[*]	External diameter	External height (over dome)	Storage capacity	Heating surface	Minimum thickness of copper sheet before forming						Height of screwed connections (see Figs. 2 and 3)		Nominal size of screwed connections	Nominal size of primary heater connections
					Grade 2 Test pressure: 0.220 N/mm² (MPa)‖			Grade 3 Test pressure: 0.145 N/mm² (MPa)‖						
					Maximum working head: 15 m†		Primary heater	Maximum working head: 10 m†		Primary heater	J_1	L	BSP.F internal threads‡	BSP.F external threads
					Body and top	Bottom		Body and top	Bottom					
	A	B	min.	min.							(See 12)			
	mm	mm	litres	m²	mm	mm	mm	mm	mm	mm	mm	mm	in	in
3§	400	1050	104	0.63	0.9	1.6	0.55	0.7	1.2	0.55	140	100	1	1
5	450	750	86	0.52	1.0	1.6	0.55	0.7	1.2	0.55	140	100	1	1
7§	450	900	108	0.66	1.0	1.6	0.55	0.7	1.2	0.55	140	100	1	1
8§	450	1050	130	0.78	1.0	1.6	0.55	0.7	1.2	0.55	150	100	1¼	1¼
9§	450	1200	152	0.91	1.0	1.6	0.55	0.7	1.2	0.55	150	100	1¼	1¼
10	500	1200	180	1.13	1.2	1.8	0.55	0.9	1.6	0.55	180	150	1½	1½

* Cylinders should be ordered by the BS type references.
† The working head shall be the vertical distance from the bottom of the cylinder to the water line of the cistern supplying it (1 mH₂O ≃ 0.01N/mm²).
‡ BSP.F external threads may be supplied when so ordered.
§ Preferred sizes for new installations in dwellings.
‖ In the original standard these values are expressed in kN/m².

B.S. 1846: Glossary of terms relating to solid fuel burning equipment.
   Part 1: 1968: domestic appliances.
B.S. 2767: 1972: Valves and unions for hot water radiators.
   Revised to cover non-rising stem angle valves, etc.
B.S. 2773: 1965: Domestic single-room space heating appliances for use with liquefied petroleum gases.
   For use with commercial butane or propane.
B.S. 3198: 1960: Combination hot water storage units (copper) for domestic purposes.
   Requirements and tests for direct and indirect storage units of 25 gallons capacity.
B.S. 3377: 1969: Back boilers for use with domestic solid fuel appliances.
   Materials and methods of construction.

TABLE 1. MATERIALS FOR DOMESTIC BOILERS
(Recommendations for use are given in Appendix A)

1	2	3	4	5	6	7
MATERIAL		MINIMUM THICKNESS			METHOD OF CONSTRUCTION	TEST PRESSURE
Common name		Tubes	Wall	At tappings		bar*
		mm	mm	mm		
Mild steel	....	4.0	6.0†	8.0	Welding	2.75
Cast iron			9.5‡	16.0	Casting	2.75‡
Copper			4.5	11.0	Welding	2.10
Aluminium bronze			2.5	8.0	Welding	2.10
Stainless steel			2.5	8.0	Welding	2.10

* 1 bar = 1.0 N/mm² = 14.50 lbf/in².
† For water backed surfaces in contact with the fire or with products of primary combustion; 5.0 mm elsewhere.
‡ The thickness of the walls of cast iron boilers may be less than 9.5 mm if every boiler is tested hydraulically or pneumatically to a pressure of not less than 5.2 bar and is unaffected after a period of not less than 5 min.

B.S. 3378: 1972: Room heaters burning solid fuel.
   Construction, performance and methods of test.
B.S. 3528: 1977: Specification for convection type space heaters operating with steam or hot water.
   Determining thermal output of heating applicances, including radiators.
B.S. 3955: Electrical controls for domestic appliances.
   Part 1: 1965: general requirements.
   Section 2A: 1967: manually operated appliance switches.

Section 2B: 1966: thermostats for electrically heated hot water supply.
Section 2F: 1967: room thermostats.
Part 3: 1972: general and specific requirements.

6.3 Function
Classification according to the function of the controls shall be as follows:
(1) Thermostat.
(2) Temperature limiter.
(3) Self-resetting thermal cut-out.
(4) Non-self-resetting thermal cut-out.
(5) Thermal link.
(6) Energy regulator.
(7) Motorized control.
(8) Manually operated switch.
(9) Current sensitive control.
(10) Pressure sensing control.
(11) Position sensing control.
(12) Light sensing control.
(13) Velocity sensing control.
(14) Liquid level sensing control.
(15) Humidity sensing control.
(16) Solenoid operated control.

B.S. 4256: Oil-burning air heaters.
Part 2: 1972: fixed, flued, fan-assisted heaters.
Part 3: 1972: fixed, flued, convector heaters.
B.S. 4433: Solid smokeless fuel boilers with rated outputs up to 45kW.
Part 1: 1973: boilers with undergrate ash removal.

TABLE 1. MINIMUM NOMINAL BORE FOR TAPPINGS

Rated output	Minimum nominal pipe bore	
	One flow and one return tapping	Two flow and two return tappings
kW	mm	mm
Up to 6	25	25
6–13	32	25
13–22	38	32
22–35	50	38
Over 35	63	50

If boilers are supplied with bushes or reducers in the flow tappings, these shall be of the eccentric pattern and fixed so that the reduced outlet is in the uppermost position.

The minimum depth of tappings shall be as given in Table 2.

* BS 21, ' Pipe threads for tubes and fittings where pressure-tight joints are made on the threads '.

Part 2: 1969: gravity feed boilers designed to burn small anthracite.

TABLE 1. MINIMUM NOMINAL BORE FOR TAPPINGS

Rated output		Minimum nominal pipe bore			
		One flow and one return tappings		Two flow and two return tappings	
kW	Btu/h	mm	in	mm	in
Up to 13	Up to 45 000	32	1¼	32	1¼
13–22	45 001– 75 000	38	1½	38	1½
22–35	75 001–120 000	50	2	38	1½
Over 35	Over 120 000	63	2½	50	2

TABLE 2. MINIMUM DEPTH OF TAPPINGS

Nominal pipe bore		Minimum depth of tappings	
mm	in	mm	in
Up to 38	Up to 1½	16	⅝
Over 38	Over 1½	19	¾

* BS 21, 'Pipe threads for tubes and fittings where pressure and tight joints are made on the threads '.

B.S. 5258: Safety of domestic gas appliances.
    Part 1: 1975: central heating boilers and circulators.
    Part 4: 1977: fanned-circulation ducted-air heaters.
    Part 7: 1977: storage water heaters.
B.S. 5384: 1977: Guide to the selection and use of central systems for heating, ventilating and air conditioning installations.

CONVERSION TABLES
BRITISH THERMAL UNITS (BTU) AND KILOCALORIES

Btu		k/cal
397	100	25
794	200	51
1 191	300	76
1 587	400	101
1 984	500	126
3 908	1 000	252
7 837	2 000	504
11 905	3 000	756
15 873	4 000	1 008
19 842	5 000	1 260
39 682	10 000	2 520
78 306	20 000	5 640

Btu		k/cal
119 050	30 000	7 560
158 733	40 000	10 080
198 416	50 000	12 900
396 532	100 000	25 199

Note: The values are rounded up to the nearest whole number.

BTU PER SQUARE FOOT PER HOUR AND WATTS PER SQUARE METRE

Btu/ft^2/h		Wm^2
0·31	1	3·15
0·63	2	6·30
0·95	3	9·46
1·26	4	12·61
1·58	5	15·71
3·1	10	31·5
6·3	20	63·0
9·5	30	94·6
12·6	40	126·1
15·8	50	157·1
31·7	100	315·4
63·4	200	630·9
95·0	300	946·3
126·7	400	1 261·8
158·5	500	1 557·3
316·9	1 000	3 154·5
633·9	2 000	6 309·1
950·9	3 000	9 563·6
1 269·9	4 000	12 618·2
1 584·9	5 000	15 778·0
3 168·0	10 000	35 145·0
6 339·0	20 000	63 091·0

Note: A rough conversion is to multiply BTU/ft^2/h by 10 to obtain Wm^2, and to divide Wm^2 by 10 to obtain BTU/ft^2/h.

RELATIVE FUEL VALUES

Unit	Fuel	BTU per unit	Efficiency %	Units per 100,000 BTU at 100%	Units per 100,000 BTU at % shown	Av cost per unit	Relative value
1 lb	Anthracite	14000	(10-50) 45	7·14	15·87	·64d.	10·2
,,	Coke	10000	(10-50) 45	10·00	22·20	·51d.	11·3
,,	Coal	13000	(10-30) 25	7·69	30·77	·48d.	14·8
,,	Oil	18800	(50-60) 55	5·32	9·68	·88d.	8·5
1 therm	Gas	100000	(55-65) 60	1·00	1·67	1·8d.	30·1
1 Unit	Electricity	3412	(75-95) 80	29·30	36·63	1¼d.	45·8

Note: 'Efficiency' is the product of combustion, boiler, installation and circulation efficiencies (excluding power station conversion efficiency) and varies over the range shown in brackets. Fuel costs vary with locality, quality, quantity and period. Installation, maintenance and attendance costs and relative flexibility also effect system comparisons.

THE HEATING ENGINEER

CENTRAL HEATING

Central heating is the maintenance of a fixed temperature inside a building, irrespective of the temperature outside it. There are several methods of so doing:

1. By hot air, circulated through ducts;
2. By hot water, circulated through pipes and radiators, and
3. Various electrically heated devices, such as strip heaters fixed to skirtings or heated floors or heated ceilings.

Of these, the most popular is heating by hot-water. If solid fuel, such as coke or anthracite be used, it is generally the cheapest, with oil, gas and electricity following in order of expensiveness. Hot-water heating has become extremely popular of late, since rival fuel interests have devised design and installation schemes the cost of which can be spread over ten years. All systems of hot water heating consist of:

1. The boiler that heats the water;
2. The pipe that conducts the hot water (called the "flow") from the boiler to;
3. Surfaces called radiators that give off large quantities of heat and so warm the rooms in which they are situated. The water has now done its job, and is returned by;
4. The return pipe to the boiler, to be reheated and used again.

A 'make-up tank', fitted with a ball-valve connected to a cold-water supply, provides the little additional water the system needs because of loss through evaporation or by boiling through over-heating.

This is the simplest form of hot-water heating, and is known as the gravity system. The freshly heated water ascends, and by doing so, allows the cooled water to descend and return to the boiler. There it is heated and when heated, ascends, and the cycle repeats itself over and over again. The 'small' bore system, now so popular, is a development of the simple system outlined above, and was developed by the Coal Utilization Research Station. The addition of a pump, that forces the water through the system, permits the use of very small bore piping ($\frac{1}{2}$in copper or ⅜in steel) for many of the runs. Using special wall drills, these small pipes are easily and unobtrusively installed, without spoiling decorations.

In addition to the pump, it is necessary to install a temperature-controlling device. This may be, in order of cheapness:

1. A hand-set mixing valve, which delivers water at a pre-determined temperature to the pump;
2. An automatic mixing valve, in which the temperature of the water delivered to the pump is regulated by the external temperature; or
3. A thermostatic radiator valve fitted to each radiator.

This last, although very expensive, is generally considered to be the best.

When designing a heating system, it is first necessary to determine how much heat in British Thermal Units (BTU) would be needed to maintain the required constant internal temperature in the worst set of external circumstances. In Britain, this greatest difference between outside and inside temperatures (generally called the temperature rise) is normally assumed to be 30 degrees F. Therefore, if the external temperature be 30 degrees F (two degrees below freezing point) the temperature inside will be a comfortable 60 degrees.

In every building there is, in winter, a constant movement of heat (called 'heat loss') through the walls of the building from the warm inside to the colder outside. In addition, the air inside the building will leak out through crevices around doors and windows. This leakage is called 'air change' and a certain amount of air change is vitally necessary to health. The combination of heat loss and air change will represent the total loss of heat from the building, expressed in BTU in one hour.

To arrive at the total heat loss, the heat loss of each room, passage and hall must be ascertained. This is done by taking each room, hall and passage separately and calculating separately the heatloss of each element of construction. (Stage 1) Take each wall separately, and all doors and windows separately, and differing types of walling separately, and the floor and ceiling separately. Take the "U" value (the thermal transmittance value) of each element of construction, as given on page 347, in the Insulation Section. These values are given in square feet and the area of each element of construction must be multiplied by its "U" value. If the ceiling is below a heated room, it may be ignored for the purposes of this calculation. (Stage 2) Total up the results of these calculations and (Stage 3) add them to the heat loss by air change. For this, take the number of air changes per hour from Table on page 339, multiply this by the cubic capacity in feet of the room and by the "U" value, usually assumed to be 0·02. Repeat this process for every room, hall and passage. The grand total of the heat loss of each room will give the total number of BTU's required each hour to keep the whole building at the predetermined degree of heat.

Radiator size can be calculated by dividing the output (in BTU) per square foot of radiating surface of the type of radiator used (see page 340) into the number of BTU's required to keep the room warm, as ascertained at the end of Stage 3 above. A separate calculation must be made for each room, to ascertain the exact size of radiator or radiators required.

The size of boiler required for the whole system is ascertained from the grand total of heat loss in the building, plus the allowance for friction in the pipes (this can be calculated, or an additional 25 per cent be made to the total heat loss) and it is usual to add a further factor of 25 per cent of total heat loss so as to be sure that the boiler is not working 'flat out'.

The size of pipes for various parts of an installation can be taken from the table, and the size of the pump is determined by the circuit that offers the greatest resistance to flow, if there be more than one circuit.

The foregoing is a brief outline of the design of small-bore heating systems, and will serve at least to check heating calculations. Before attempting to design heating systems, the readers should consult one or more of the following booklets. They explain the subject in greater detail and also contain a number of valuable tables that will take much of the toil out of the calculations. The combination of hot water heating and domestic hot water has not been treated, partly for the sake of simplicity and partly because it is better and cheaper to have a completely separate means of heating small quantities of water for domestic use.

1. *Principles of Forced Circulation in Hot-water Heating Systems:* Sigmund Pumps Ltd, Team Valley Trading Estate, Gateshead 11.
2. *Copper Tubes for Small-bore Heating Installations;* The Copper Development Association.
3. *The Design and Installation of Small-pipe Heating Systems:* Coal Utilization Council.

HOT-AIR CENTRAL HEATING

This type of installation is cheaper to install than the usual system of radiators and pipes. It is necessary to make the calculations detailed above, to arrive at the hourly requirements of BTU. An appliance (fuelled by solid fuel, gas or oil) can be chosen from a manufacturer's range of sizes. Hot air as usually ducted to rooms by trunking, and released through gratings. The siting of the ducts and of the gratings that serve the rooms plays a vital part in the efficiency of the system and most manufacturers, if sent a plan and elevation of the building, are able to suggest the best layout for efficient working.

MICROBORE INSTALLATIONS

Pioneered by the Wednesbury Tube Co. Ltd, this type of forced circulation system has become popular. It employs very small tubes, 6, 8,

10mm external diameter, with special fittings. Normal boilers and standard pumps with special heads are used. Since the tubes are supplied in long lengths (coils of 90 to 200 metres maximum) fewer connectors are needed. The tube can be bent by hand or, for small radii, with an external spring bender, without cutting the tube. The advantages of the system can be summarised as follows:

1. Reduced labour costs and reduced materials costs;
2. A neater installation, with less disturbance to house occupiers;
3. Reduced water content provides greater sensitivity to thermostatic control;
4. A faster heat recovery and a more flexible installation, allowing ease of extension if required.

Calculation and design procedure for microbore systems are more critical than for small-bore installations: it cannot be over-emphasised that detailed and correct calculations are essential. The Wednesbury Tube Co. Ltd. has produced the *Microbore Installation Manual*, describing design and installation in detail. This is an essential for all designers and installers of such systems and copies can be had from the Company's Publications Department, Oxford St, Bilston, Staffs WV14 7DS.

HEAT EMISSION FROM PIPES, PER 100 PER CENT TEMPERATURE DIFFERENCE

Pipe size in inches	Heat emission in BTU per hour	
	copper pipe	steel pipe
$\frac{1}{2}$	25	56
$\frac{3}{4}$	36	70
1	43	80
$1\frac{1}{4}$	50	98

AIR CHANGES AND RECOMMENDED TEMPERATURES FOR ROOMS

Room	Air changes per hour	Recommended temperature degrees F
Halls, staircases and passages	2–3	60
Living rooms		
(a) fireplace sealed	$1\frac{1}{2}$	65
(b) with convector fire	$2-2\frac{1}{2}$	65
(c) with open fire	3–5	65
Kitchens	2–3	60
Bedrooms	$1\frac{1}{2}$	50
Bed-sitting rooms	2	65
Bathrooms and lavatories	2	60

RADIATION FROM RADIATORS IN BTU PER HOUR PER SQUARE FOOT

Type of Radiator	Width in inches	Emission per sq. ft per hr in BTU
Column radiators	$2\frac{1}{2}-3\frac{3}{4}$	185
	$5-5\frac{5}{8}$	170
Window radiators	$13-13\frac{1}{2}$	158
Cast-iron wall radiators	—	171
Pressed steel wall radiators	—	192

PIPES
PIPE SIZES

Emission of circuit or radiator, in BTU per hour	Size of pipe required, in inches
Up to 15,000	$\frac{1}{2}$
15,000 to 40,000	$\frac{3}{4}$
40,000 to 72,000	1
72,000 to 120,000	$1\frac{1}{4}$

PIPE THREADS

There are two systems in use in England—Whitworth Threads and British Standard Pipe Threads. Whitworth threads are almost universally used. The difference between the two systems is very slight: a matter of a few ten-thousandths at the top and bottom of the thread. Whitworth-threaded pipes fit easily into BSP threaded fittings, but BSP threaded pipes will only fit with difficulty into Whitworth-threaded fittings. BSP threads extend up to 6in, whereas Whitworth threads stop at 4in. It is very rare to go beyond that with screwed connections. The International Standards Organization has agreed, in principle, on a system of pipe threads, but these may not be introduced for many years.

17. INSULATION

Metrication of Thermal Insulation	339
Conversion Table	339
Thermal Insulation	340
Sound Insulation	344
Vibration Insulation	347

METRICATION OF THERMAL INSULATION

The change to metric involves the use of new units. These are as follows. For the 'K' value (thermal conductivity) the new unit is watts per square metre of surface area, with a temperature difference of 1 degree C, and is written W/m deg C. The reciprocal of this ($1/k$) is the *thermal resistivity:* if the thickness of a material is known (in metres and decimal fractions) $1/k$ times the thickness equals the *thermal resistance*, which is written m^2 deg C/W. The 'U' value, or *thermal transmittance* is derived as follows.

'A property of any material is its thermal conductivity (k): the reciprocal of this is resistivity ($1/k$). For materials of known thickness, the resistance (R) can be calculated (resistivity × thickness) and from the resistances of the various layers (of material) comprising a construction and the resistances of cavities and of inner and outer surfaces the 'U' value can be calculated'.

(From the BRS Digest No. 108, Second Series—'Standardised 'U' Values'—which is recommended as a clear and detailed account of the new system, including tables of the new, metric 'U' values.)

The calculation is, therefore, the same as used for Imperial quantities.

U = internal surface resistance + external surface resistance + resistance of cavities

+ resistances of each material in the construction.

In this section, metric and Imperial values are given in all the thermal insulation tables, and a conversion table for K values is included. It should be noted that the U values given in some of the tables do not exactly correspond. This is because the new metric values (called 'standardised values') are calculated from slightly different and more accurate basic data.

CONVERSION TABLE

K VALUES

IMPERIAL BTU in sq. ft h deg F	METRIC Wm deg. C
0·1	0·0144
0·2	0·0288
0·3	0·0432
0·4	0·0576
0·5	0·0720
0·6	0·0864
0·7	0·1000

K VALUES

IMPERIAL BTU in sq. ft h deg F	METRIC Wm deg. C
0·8	0·1150
0·9	0·1300
1·0	0·1440
2·0	0·2880
3·0	0·4320
4·0	0·5760
5·0	0·7200
6·0	0·8640
7·0	1·0080
8·0	1·1520
9·0	1·2960
10·0	1·4400
20·0	2·8800
30·0	4·3200

Insulation is a form of protection against heat, cold, sound and vibration. These are, in the main, diseases of civilization: part of the price we pay for our modern amenities, and for living at a standard of comfort far beyond that ever known by any people as a whole. It has been established, however, that we need not pay anything so great a price in discomfort, or in cash. Since the war, the activities of scientific workers in building subjects has shown how our homes, offices and factories may be made warmer, quieter and more comfortable—sometimes at little or no increased cost.

THERMAL INSULATION

The importance of thermal insulation in modern building cannot be over-estimated. All the heat that we put into houses, factories, offices and other buildings leaks out through the various parts of these structures—through walls, windows, floors and roofs. We are constantly compelled to renew the heat thus lost. The coal consumed every year in heating houses and small industrial buildings amounts to approximately 65,000,000 tons. The cost of this enormous amount of fuel is something like equivalent to that of the 200,000 to 300,000 new houses that we build every year. Large industrial buildings cost very much more to heat.

It is plain that, if the amount of heat escaping from buildings can be reduced, savings can be made because less fuel would be needed. Alternatively, a greater degree of comfort can be obtained or a balance may be struck by having some saving of money and some degree of increased comfort. In the normal house, the effect is a relatively long-term one, because the extra cost of preventing the escape of a sufficient amount of heat—a relatively large sum—anything over £200. It would take a considerable time to regain this by the accumulation of the amount saved on the purchase of fuel. In some shed-type factory buildings, with walls and roof of very light construction, the fuel bill is so huge that the cost of installing thermal insulation is repaid in two or three years, after which there is a definite saving in fuel and therefore a lower heating cost. In new buildings of this type a smaller heating plant is required if the structure is insulated. The saving on the heating plant pays for the insulation. Here there is an immediate reduction in the cost of heating.

Thermal insulation may be defined as the adding to an existing structure, or incorporating in a new one, materials that have the property of slowing down the rate of leakage of heat from the structure. These materials are usually of low density, that is, light in weight and cellular (those having a large number of pockets of air in them). In this type of material, the air pockets act as poor conductors of heat. To a lesser extent, air-spaces, such as the cavity in a cavity wall, perform the same function. The other type of material acts by reflecting back the heat that strikes it, and is always some sort of bright metal surface. Polished aluminium foil is practically the only material of this type in use.

The unit of thermal transmittance ('U' for short) is used to denote the suitability or otherwise of a material for insulating purposes. The lower the thermal transmittance, the slower heat goes through it, the better the material is as a thermal insulator.

This 'U' value is ascertained by a somewhat roundabout process. It is based on the conductivity of a material per inch of thickness, of which the unit is 'k', and from which is derived the resistivity of the material per inch of thickness, which is the reciprocal of 'k'. For instance, the conductivity of stone is 12, and the resistivity is therefore $\frac{1}{12}$ or 0·08.

When in practice it is desired to find the 'U' value of any particular wall, floor or roof, proceed as follows:

1. Take the external resistance of surface material.
2. Add the resistance of each element of the construction, and
3. The internal surface resistance of the material that forms the inner face.

Do not forget to include, under 2, the termal resistance of any air-space, such as the cavity in an 11in cavity wall.

Example.—What is the 'U' value of a 9in wall, in common bricks, rendered externally with 1in of rendering and plastered internally with $\frac{5}{8}$in of plaster?

Commence by setting out the elements of the walling, with the exterior surface resistance in front of them and the internal surface resistance after them, thus:

1	2	3	4	5
External resistance	1in of rendering	9in of common brickwork	⅜in of plaster	Internal resistance

Then, from the tables fill in the resistance value of each element. Do not forget that the resistances given are per inch of thickness and that, therefore, the resistance value for brickwork must be multiplied by 9 to give 3., while the plaster 4., is obtained by taking ⅜in only of the resistance for plaster per inch of thickness as given by the table. Set out the resistances thus obtained under the elements above, thus:

1	2	3	4	5
0·30	0·25	1·12	0·15	0·70

These five items, added together, make 2·52, which is the total resistance of the wall. The reciprocal of this (*i.e.* 1 over this number) is $\frac{1}{\cdot 252}$ or $\frac{100}{252}$, or 0·39, which is the 'U' value of this type of walling.

Thermal insulation is of value in preventing some undesirable and expensive phenomena, such as 'pattern staining' on ceilings and excessive condensation in buildings. This latter, in factories containing machinery, especially knitting and similar delicate mechanisms, is a major problem that can only be avoided by adequate insulation. True, it would not occur if the heating were continuous and were thermostatically controlled, but the expense of heating a building in this manner would at least quadruple the annual fuel bill.

RESISTIVITY AND CONDUCTIVITY OF VARIOUS BUILDING MATERIALS

Material	Thermal Conductivity (k) Per in thickness	Thermal Resistance $\frac{1}{k}$ Per in thickness	Metric (k)
"Asbestolux" insulating board	0·8	0·125	0·216
Asbestos insulating board	1·8	1·250	0·576
Asbestos cement sheets, plain or corrugated	1·9	0·527	0·576
Asphalt	8·7	0·115	1·210
Bricks (common)	8·0	0·125	1·470
„ (engineering)	5·5	0·182	1·440
Concrete (ballast, 1-2-4)	7·0	0·140	0·430
„ (clinker or breeze)	2·3	0·440	0·051
Glass	7·3	0·137	0·401
Plaster and renderings	4·0	0·250	1·530

INSULATION

Material	Thermal Conductivity (k) Per in thickness	Thermal Resistance $\frac{1}{k}$ Per in thickness	Metric (k)
Stone	12·0	0·080	0·576
Slate	10·4	0·096	0·853
Tiles (roofing)	5·8	0·170	0·216
Hardboard	0·7	1·410	0·065
Fibreboard	0·3	2·860	0·158
Plasterboard	1·1	0·910	0·049
Cork	0·7	1·370	0·138
Timber	1·0	1·000	0·093
Wood-wool	0·58	1·720	0·084

SURFACE RESISTANCES

Element of construction	Internal surface resistance		External surface resistance	
	Imperial	Metric	Imperial	Metric
Walls	0·70	0·12	0·30	0·053
Roofs (flat or pitched)	—	—	0·25	0·440
Ceiling (heat movement upwards)	0·60	0·15	—	—
Floor (heat movement downwards)	0·84	0′11	—	—

THERMAL RESISTANCES OF AIR SPACES

Type of air-space	Thermal resistance	
	Imperial	Metric
Corrugated and plain surfaces in close contact	0·5	0·16
Over ¾in between materials	0·9	0·14
Over ¾in between materials with bright metallic foil	2·2	0·25
Hollow blocks, 4in thick	1·0	—
Bright foil, 1in air-space each side	4·0	0·12

'U' VALUES OF DIFFERING CONSTRUCTIONAL ELEMENTS

Construction	Thermal Transmittance (U)	
	Imperial	Metric
Pitched Roofs		
Corrugated iron	1·50	5·4
Corrugated asbestos	1·40	5·3
Corrugated asbestos lined with $\frac{1}{4}$in asbestos sheet with air-space	0·55	1·8
Corrugated asbestos lined with $\frac{1}{2}$in fibreboard with air-space	0·32	1·3
Tiles on battens, plaster ceiling	0·56	—
Tiles on battens, felted, plaster ceiling	0·46	—
Tiles on battens, felted. Aluminium foil or 1in glass silk quilt on joists, plaster ceiling	0·15	1·5
Flat Roofs		
Asphalt on 6in concrete, plastered soffit	0·52	3·4
Asphalt on 6in concrete, 2in light weight concrete screed, plastered soffit	0·34	1·4
Asphalt on concrete, hollow tile roof with $\frac{1}{2}$in fibreboard insulation and plaster ceiling	0·20	0·90
Walls		
$\frac{1}{4}$in asbestos cement sheet	0·89	5·3
9in brick	0·47	3·3
11in cavity brick wall (unventilated), plastered	0·30	1·5
11in cavity brick with $\frac{1}{2}$in fibreboard on battens	0·18	2·8
Corrugated asbestos lined with $\frac{1}{2}$in fibreboard with air-space	0·30	0·78
Floors		
1in tongued and grooved boarding on timber joists ventilated under covered linoleum	0·35	—
1in wood block on concrete on ground	0·15	—
Windows		
Single glazing	1·0	—
Double glazing	0·5	—

SOUND INSULATION

Some degree of sound insulation is desirable in almost every building that is being used to its fullest extent. This desirability becomes a necessity in many modern buildings which consist of lightweight claddings suspended on a structural frame through which sound may be conducted from one part of the building to the other. Even in traditional buildings with load-bearing walls excessive noises from cisterns and other internal services may constitute a major annoyance as may traffic and other noises from a busy street. In special cases, such as doctors' consulting rooms, the intensity of the sound may not be great, but its transmittance to, say, an adjacent waiting-room must be prevented.

There are two units of sound measurement: the *phon* and the *decibel*. For all practical purposes, they can be regarded as equivalents for a large range of everyday noises. The decibel is the unit of sound intensity, while the phon is the unit of sound loudness. Put another way, the decibel is a unit of energy, measured on an instrument, while a phon is a unit of loudness as perceived by the ear and measured by comparison with other known sounds.

RECOMMENDED STANDARDS OF QUIET

Activity	Loudness in phons
Study or sleeping	15
Reading or writing	20
Boardroom	30
Sedentary office, or quiet conversation	35
Average office, telephone work, restaurant	40
Noisy office	60

Average airborne noises should be reduced to these levels of loudness, or as near to them as possible, within, say, 5 decibels or phons. The following notes show how this may be done.

WINDOWS

The sound reduction of an ordinary single pane of 24oz glass reduces external noise by 28 decibels, and so cuts down noise made by a noisy sports car or a pneumatic drill to the level of average music or loud public speaking. If, however, the room is used for any of the activities given in this table, the noise penetrating the room through the window will still be too much, and the following table gives the insulation value of various types of double-glazing.

SOUND REDUCTIONS OF GLAZING

Type of glazing	Sound reduction in decibels
Double glazing at following spacing:	
1 in	42
2in	46
5in	49
$\frac{1}{4}$in plate, $5\frac{3}{4}$in apart	58
As above, but interspace lined 1in felt or glass wool, or cork, or other absorbent	63

It will therefore be seen that a noise equivalent to a pneumatic drill constantly at work can be cut down, by the last method shown, to little more than quiet conversation, and providing an atmosphere where sedentary office work can be carried on in comfort.

WALLS

Here, again, the following table gives the amount of airborne sound reduction for typical constructions.

SOUND REDUCTIONS OF TYPICAL CONSTRUCTIONS

Material or construction	Average sound reduction in decibels
$\frac{1}{2}$in fibreboard	20
$\frac{3}{8}$in plasterboard	25
$\frac{1}{4}$in plate glass	30
$\frac{3}{8}$in plasterboard plastered $\frac{3}{8}$in both sides	35
3in clinker concrete block plastered	44
4$\frac{1}{2}$in brickwork or concrete plastered	43
9in brick work plastered	48
3in woodwool slab $\frac{3}{4}$in plastered both sides	48
Timber stud partition with metal lath and plaster both sides	35
Cupboards used as partitions	25–35
Double partition of 3in hollow clay blocks. 2in cavity plastered externally. Strip metal ties	40–43
Double partition of 3in breeze block. 2in cavity plastered. No wall ties	50
Two leaves of 2$\frac{1}{2}$in breeze blocks plastered and separated by a 2in air-space with suspended ceiling	50–60
As above, with 4$\frac{1}{2}$in brick walls	60–70

SOUND REDUCTIONS OF TIMBER FLOORS

Construction	Average sound reduction (phons)
Boarding on timber joints with plasterboard ceiling and plaster skim coat	30–35
As 1, but with metal lath and plaster ceiling	35–40
As 1, but with solid pugging 20lb per sq. ft	40–55
As 2, but with floating floor of boarding on battens on resilient quilt	55

FLOORS

Impact noises on floors are difficult to reduce, except by:

1. Carpet on underlay of rubber or felt.
2. Pugging between joists.
3. Suspended ceiling, each of which gives approximately 10 phons reduction.

Noises transmitted through a frame can be prevented by insulating the floor of each room from the building itself. This always means a 'floating' floor, resting on felt or rubber pads.

Absorption of sound, by lightweight cellular materials such as those used in thermal insulation, can do much to prevent reflection of sound, which causes echoes, but such absorbents are of little value in reducing airborne sound and even less in reducing impact sound.

VIBRATION INSULATION

In industrial and commercial buildings, vibration can present definite and urgent problems, sometimes of considerable magnitude. For instance, a drying machine running in a cleaner's shop has been known to cause bottles on the shelves in the adjacent shop to vibrate and chatter together. Vibration of machinery can cause a whole building to tremble unpleasantly and can seriously interfere with precision work being done on other machines, such as jig-borers, and with other delicate operations. On a large scale, as when immense forging-hammers are in operation, impact vibrations can cause something like minor earthquakes, which bring down plaster ceilings for 400 yards around, and literally shake adjacent buildings to pieces.

The whole subject of vibration engineering is too complex to summarize here, but, for the relatively simple problems that come within normal building practice, the following notes will indicate possible solutions.

GROUND VIBRATIONS

The transmission of vibrations from the ground, caused by external causes not under the control of the owner of the site can only be prevented from being transmitted through the structural frame of a building by insulating the frame itself. Special foundations will be required, incorporating rubber pads or springs, or lead pads, or a combination of these between the frame itself and the foundation.

HEAVY MACHINERY

Ground vibrations are often due to heavy machinery and here the remedy is to construct foundations for such machines that are isolated from both the structure and the ground. Normally, this is done by forming a pit, lined with sheeting of rubber or cork, into which the foundation concrete for the machine is poured. It is most important that the floor and sides of the pit be lined. Vibrations caused by very heavy impacts may need further 'damping', by providing systems of springs, as well as rubber and other pads, between the machine foundation block and the pit in which it is constructed.

Much can be done to overcome this trouble of vibration by using machinery driven by electricity, in preference to internal combustion engines. Each machine sholuld be independently motorized, not powered from shafting and belts actuated by one central source of power. Shafting and belts are one of the principal causes of vibration in the old-fashioned workshops where the centralized power system is still used. Another source of vibration is old machinery, badly maintained, and yet another is failure to anchor adequately the machinery to the floor, while the siting of machinery on floors not specially constructed for the purpose gives rise to much vibration trouble. In these latter instances, the strengthening and stiffening of the floor often remedies the evil.

LIGHT MACHINERY

It is often sufficient to mount light machines on special rubber pads: There is a range of proprietary rubber machine pads that can, and should be, used when installing every machine, whether vibration be anticipated or not. Where vibration still persists, the machine should be installed on an isolated concrete foundation as described above.

OTHER VIBRATION SOURCES

Pipelines that transmit vibrations may be rendered silent by inserting a short length (up to 2ft or 3ft at most) of special rubber tube in the pipe-line. Overhead travelling cranes and hoists may have their tracks isolated from the structure by pads or springs.

Very delicate instruments, such as chemical balances and optical measuring devices, can be successfully insulated from the small amounts of vibration that impair their efficiency by mounting them on specially designed rubber pads. It will often be found more economical, when only one or two precision machines suffer from relatively small vibration interference, to provide insulated foundation blocks for these machines in preference to installing every other machine in the engineering or other workshop on insulated foundations.

It is not generally realized that noises are caused at times by vibration, as when a WC pan is mounted on an ordinary wooden floor in a small house. Considerable lessening in the transmission of noise, and its propagation by the floor, which acts as a sounding-board, can be obtained by mounting the WC pan on a rubber pad, mounting the supply tank (the WWP) in a similar manner, and insulating the supply and overflow pipes by rubber or felt sleevings at the points where they pass through walls.

The insulation of vibrating machinery should on no account be undertaken without the advice of an expert. It is possible to produce a resilient support that at certain frequencies of the machine will become resonant and may well double or even treble the amount of vibration imparted to the structure.

18. THE ELECTRICIAN

Codes of Practice	351
British Standards	351
Electrical Regulations	354
Cables	355
Ring and Radial Circuits	356
Cable Supports	357
Earthing—A Warning	358

Normally, this trade is sublet to electrical contractors, and in a sense is not a building, but an engineering trade. However it is as well that everyone in a responsible position in building should have read the Wiring Rules issued by the Institution of Electrical Engineers and be acquainted with their provisions. Local Electricity Boards accept only work executed in accordance with these Wiring Rules as being fit to receive supplies of current.

CODES OF PRACTICE
CP 3: Chapter 1: Lighting.
Part 2: 1973: artificial lighting, aims and processes of design.
Chapter VIII: 1949: heating and thermal insulation.
CP 324: 202: 1951: Provision of domestic electric water heating installations.
Design and installation.
CP 1013: 1965: Earthing.
Methods for earthing electrical systems.
CP 1014: 1963: The protection of electrical power equipment against climatic conditions.
Classifies climates into three classes; 'slight', 'medium' and 'extreme',
CP 1017: 1969: Distribution of electricity on construction and building sites.
Recommendations for supplies operating at voltages up to 650V. Includes 110V systems for portable and hand-held tools and lamps.
CP 1018: 1971: Electric floor-warming systems, for use with off-peak and similar supplies of electricity.
Design of systems and electrical requirements.
CP 1019: 1972: Installation and servicing of electrical fire alarms.
Covers electrical fire alarms in buildings.

BRITISH STANDARDS
B.S. 31: 1940: Steel conduit and fittings for electrical wiring.
Class A; plain conduit; Class B; screwed conduit, corresponding fittings, couplers, boxes, elbows, etc.
B.S. 37: Electricity meters.
Part 1: 1952: general clauses.
Part 2: 1969: watt hour meters.
Part 3: 1970: prepayment watt hour meters.
B.S. 546: 1950: Two-pole and earthing pin plugs, socket-outlet adaptors for circuits up to 250V.
As used in domestic premises, offices, etc.
B.S. 731 Flexible steel conduit and adaptors for the protection of electric cable
Part 1 1952: flexible steel conduit and adaptors.
Part 2: 1958: flexible steel tubing to enclose flexible drives for power driven tools for general purposes.
B.S. 816: 1952: Requirements for electrical accessories.
Electrical devices not covered by British Standards for single phase a.c. or d.c. circuits, etc.
B.S. 889: 1965: Flameproof electric light fittings.
Three temperature range classes.

B.S. 1361: 1971: Cartridge fuses for a.c. circuits in domestic and similar premises.
Requirements, ratings and tests for fuse links, fuse bases and carriers.
B.S. 1362: 1973: General purpose fuse links for domestic and similar purposes (primarily for use in plugs).
Performance, dimensions, markings and tests, etc.
B.S. 1363: 1967: 13A plugs, switched and unswitched socket-outlets and boxes.
    Part 1: requirements for mulitple flush and surface-mounted sockets.
    Part 2: boxes and enclosurers.
B.S. 1454: 1969: Consumers' electricity control units. Fuse and miniature circuit breakers types principally for use in domestic premises.
Rating 100A. Control of different types of domestic load.
B.S. 1778: 1951: 15A three pin plugs, socket outlets and connectors (theatre type).
Round pin type, dimensionally interchangeable with those in B.S. 546.
B.S. 3052: 1958: Electric shaver units.
Fixed supply units for flush or surface mounting.
B.S. 3456: Specification for safety of household electrical appliances.
Specification used by the British Electrotechnical Board for Household Equipment.
    Part 1: 1969: and **Part 101: 1976:** general requirements.
    Section A1: 1966: general requirements for all household heating and cooking appliances.
    Section A4: 1971: electrically heated blankets.
    Part C: 1967: electrical refrigerators and food freezers.
    Part 2: particular requirements.
    Section 2.1: 1972: cooking ranges, cooking tables and similar appliances.
    Section 2.2: 1971: electrical vacuum cleaners and water suction cleaning appliances.
    Section 2.3: 1970: appliances for heating liquids.
    Section 2.4: 1970: electric irons, ironers and pressing machines.
    Section 2.5: 1970: shavers, hair clippers and similar appliances.
    Section 2.6: 1970: frying pans, grills, plate warmers, and other dry cooking appliances.
    Section 2.7: 1970: stationary non-instantaneous water heaters.
    Section 2.8: 1970: portable immersion heaters.
    Section 2.9: 1970: clothes drying cabinets and towel rails.
    Section 2.10: 1972: room heating and similar appliances.
    Section 2.11: 1970: electric clothes-washing machines.
    Section 2.12: 1970: spin extractors.
    Section 2.13: 1970: tumbler dryers.

Section 2.14: 1971: electric soldering irons.
Section 2.15: 1970: electric heaters for baby feeding bottles.
Section 2.16: 1971: clocks.
Section 2.17: 1970: electric sewing machines.
Section 2.18: 1970: massage appliances.
Section 2.19: 1970: electric firelighters.
Section 2.20: 1971: electric floor polishers.
Section 2.21: 1972: electric immersion heaters.
Section 2.22: 1972: electricaire heaters.
Section 2.23: 1971: cooking ventilating hoods.
Section 2.26: 1973: thermal storage electrical room heaters.
Section 2.27: 1973: dish-washing machines.
Section 2.28: 1973: food preparation machines including mixers, coffee grinders and coffee mills.
Section 2.29: 1971: ventilating fans.
Section 2.30: 1971: food waste disposal units.
Section 2.31: 1973: appliances for skin or hair treatment.
Section 2.32: 1974: mains operated electric lawn-movers.
Section 2.33: 1976: specification for microwave ovens.
Section 2.34: 1976: room air-conditioners.
Section 2.35: 1975: electrical pumps.
Section 2.36: 1973: battery chargers.
Section 2.37: sauna baths (in course of preparation).
Section 2.38: 1974: still projectors and viewers.
Section 2.39: 1973: room humidifiers.
Section 2.40: 1977: room heating and similar appliances for use in children's nurseries and similar situations.
Section 2.42: 1977: battery-operated lawn mowers.

B.S. 3676: 1963: Switches for domestic and similar purposes (for fixed or portable mounting.

Domestic in wall or ceiling mounting, and in cords and lampholders, and certain appliances.

B.S. 3999: Methods of measuring the performance of household electrical appliances.

In thirteen parts; (1) electric kettles, (2) thermal-storage electric water heaters, (3) food preparation machines, (4) electric clothes drying cabinets and racks, (5) electric cookers, (6) electrically heated blankets, (7) electric irons, (8) electric coffee percolators and similar appliances, (9) electric toasters, (10) food freezers, (11) dishwashing machines, (12) vacuum cleaners, (13) frozen food storage compartments in refrigerators and frozen food storage cabinets.

B.S. 4363: 1968: Distribution units for electricity supplies for construction and building sites.

Requirements for six types of distribution units to give a system of control and distribution on site.

B.S. 4491: 1969: Appliance couplers for household and similar general purposes.

Requirements and dimensions for connectors and appliance inlets for portable electrical appliances.

B.S. 4568: Steel conduit and fittings with metric threads of ISO form for electrical installation.

    Part 1: 1970: steel conduit, bends and couplers.

    Part 2: 1970: fittings and components.

B.S. 4573: 1970: Two-pin reversible plugs and shaver socket-outlets.

Requirements for rewireable or integrally moulded plugs and shuttered socket-outlets of 200mA.

B.S. 4607: Non-metallic conduits and fittings of electrical installations.

    Part 1: 1970: metric units, rigid PVC conduits and conduit fittings.

    Part 2: 1970: imperial units, rigid PVC conduits and conduit fittings.

    Part 3: 1971: pliable corrugated, plain and reinforced conduits of self-extinguishing plastics material.

    Part 5: 1973: rigid PVC conduit, fittings and components.

B.S. 4662: 1970: Boxes for the enclosure of electrical accessories. Recessed or surface mounting.

ELECTRICAL REGULATIONS

The Electricity Supply Regulations.

Deal mainly with the duties of the suppliers of current, that is, the various Area Boards of the Central Electricity Board. Regulations 26 to 35 are concerned with the suppliers duty to customers. Suppliers must ensure:

    1. That the insulation of the wiring is sufficiently good as to prevent more than 1/10,000 of the maximum current leaking through it.

    2. That the switches and other equipment are adequate in type and in installation to prevent danger.

    3. That single-pole switches are connected to the live leads.

    4. That all metal parts, other than the conductors, are earthed.

    5. That the voltage supplied must be maintained to plus or minus 6 per cent, and frequency to plus or minus 1 per cent.

I.E.E. Wiring Rules (properly called the Wiring Regulations of the Institution of Electrical Engineers).

Are generally accepted by supply undertakings as an adequate standard of workmanship. They are very detailed, and cover all aspects of electrical installation.

Requirements of the Electrical Supply Regulations do not apply to factories and building sites. Instead these are governed by the Electricity Regulations of the Factory Act.

For further details, consult: *The Electricity Supply Regulations*, and *The Electricity (Factory Acts) Special Regulations*, from H.M.S.O.

CABLES

CP 321.101 lays down that cables should be protected from excessive heat, dampness and mechanical injury, that external wiring should be weatherproof, and that adequate fuses be provided. The older method of fusing both the 'live' or 'phase' wire and the 'neutral' or 'earthed' wire has now been replaced by the custom of fusing the live or phase wire (this is usually brown-covered) only. It is imperative that switches and plug sockets be so connected that the circuit is 'dead' when the plug is withdrawn or the switch is in the 'off' position. The switch must always be in the 'live' or brown lead, not in the blue or 'neutral' one. It is advisable to control plug-sockets by switches, and to see that the type of plug-socket used incorporates some form of safety-shutter.

METRICATION

House wiring cables and flexible cords are now in metric dimensions. Various modifications in manufacture now give single wires instead of the three or five smaller wires in the old type of cable, which is no longer manufactured. Sizes of accessories have not changed, but dimensions are now expressed in metric units. Conduit remains unaltered in size, so that existing pipe dies may be used.

CABLE SIZES: NEW (METRIC) AND OLD

Metric (mm^2)	Old	Rating in amps	Type of Circuit
—	1/044	5	Lighting
—	3/044	10	
1·0	—	11	Lighting: small heaters
1·5	—	13	
—	7/029	15	Ring circuits, water-heaters, 3 kw heaters
2·5	—	18	
4·0	—	24	Radial circuits
—	7/036	30	
6·0	—	31	Cooker circuits
10·0	—	42	Large cookers internal mains from meter to consumers' control unit.
—	7/064	50	

It is usual to fit a SPN 60 amp switch-fuse near to the incoming main and the company's meter is connected to this by 16mm^2 cable. From the switch-fuse, 16mm^2 cable is taken to the consumer's control unit, which is an assembly containing a main switch and as many fuses as is considered necessary. The sizes of fuses should be as follows: 5 amp for lighting circuits, 10 amp for a lighting circuit incorporating a small heater (as in a bathroom), 15 amp for a single plug-point, or 3 kw water heater (either in a storage tank or in a washing machine) and 30 amp for a cooker. One extra fuse should always be installed, to allow for the addition of an extra circuit. Each ring-main (or, more properly, ring circuit) requires a 30 amp fuse.

METRIC SIZES OF FLEXIBLES AND CAPACITIES IN AMPERES

mm^2	amp
0·50	3
0·75	6
1·00	10
1·5	15
2·5	20

COLOUR CODE FOR 3-CORE FLEXIBLES

Old	New	Wire
Red	Brown	Live
Black	Blue	Neutral
Green	Green & Yellow stripes	Earth

RING AND RADIAL CIRCUITS

Modern practice favours the complete separation of wall and ceiling fixed lighting from circuits containing socket outlets. 3 amp and 5 amp socket outlets, often forming part of lighting circuits, are now obsolete. Instead, all socket outlets are grouped in separate circuits and serve for movable lighting fittings and all other appliances. This is made possible by employing ring circuits or radial subcircuits.

RING CIRCUITS

A ring circuit consists of a loop of cable, both ends of the wires in which are connected to their same terminals at the fuse or miniature circuit breaker in the customer's control box. In all ring circuits, there are three loops; one for line, one for neutral and one for earth.

Connection of socket-outlets is by looping-in line wire to line terminal, neutral wire to neutral terminal and earth wire to earth terminal. Each socket is of the 3 pin type, taking the rectangular pins of the plug. Each plug is fitted with a cartridge fuse on the line wire; these fuses are of 3 amp rating for standard or table lamps or 13 amp rating for other movable appliances, such as heaters or fans. This ensures that only the lamp or other appliance is disconnected by its fuse 'blowing'—not the rest of the appliances in circuit. A socket-outlet controlled by a switch gives added safety at very little extra cost: it cuts off the current to the appliance before the plug is removed.

When the total floor area of a dwelling is less than $100m^2$, an unlimited number of socket-outlets can be looped into a ring circuit. In larger premises, a ring circuit for each $100m^2$ is necessary. Appliances such as fixed wall heaters, fan heaters when fixed and sink-disposal units should be connected to the ring circuit by a fused spur-connector with switch either by direct looping-in or through a junction-box. For clocks, shavers and electric gas-lighters, special spur-connectors with 1 amp fuses should be used. For heaters that switch on and off automatically, such as water-heaters, a separate ring circuit, serving the appliance only, is recommended.

A single 3-core cable, called an unfused spur, can be taken by junction-box or direct looping-in, from a ring-circuit to serve one fixed appliance or two 13 amp socket outlets. Only one spur is allowed on each ring circuit. It should be remembered that the total loading on a ring circuit in $2.5\ mm^2$ wire should not exceed 7.2 kw.

RADIAL SUB-CIRCUITS

In some situations, where the load is particularly high, it is often advantageous to use a radial final subcircuit, as listed in table A3 of the I.E.E. Wiring Rules.

These circuits consist of single 3-core cables and typical such circuits are:

1. $4\ mm^2$ cable with a 30 amp fuse at the consumer's control box, serving five socket outlets and one fixed appliance.
2. A $2.5\ mm^2$ cable, with a 20 amp fuse, supplying two socket-outlets.
3. In a room not exceeding $30m^2$ in floor area, six socket outlets supplied by a $2.5\ mm^2$ cable, with a 20 amp fuse at the consumer's control box.

CABLE SUPPORTS

Tough rubber or PVC sheathed cable, either twin or three core, is often used without conduit for house-wiring, although it is advisable to provide some form of protection to all 'drops'. Cable so installed is likely to sag and so cause accidents, if the fixing clips are not spaced at the correct intervals. The following table gives the correct vertical and horizontal spacing of clips.

Cable	Maximum spacing of fixings			
	Vertical		Horizontal	
	mm	in	mm	in
1/044, 3/044, 7/029 : 1·0 & 1·5mm²	224	9	381	15
7/036 : 2·5 & 4·0 mm²	305	12	381	15
7/064 : 10·0 & 16·0mm²	381	15	457	18

SOCKET OUTLETS FOR DOMESTIC PREMISES

Room	Number of outlets	Remarks
Separate living room	3	Distributed so as conveniently to serve appliances in all parts of the rooms
Separate dining room	2	
'Through' living room	3 or 4	
Kitchen	3	Excluding one socket outlet on cooker control unit
Double bedroom	2	Placed with due consideration of likely furniture arrangements
Single bedroom	1	
Hall or landing	1	General service outlet for cleaning stairs, etc.
Store or workshop	1	For tools such as drills, soldering irons etc.

(*From Building Research Station Digest No. 22, 2nd series*).

EARTHING — A WARNING

It was formerly standard practice to earth electrical circuits and appliances to cold-water pipes. Plastic pipes, which are insulators, are often used nowadays in new work, and existing metal cold-water installations may be extended or repaired in plastics pipe. In instances where plastics piping is used, the householder should be warned that this does not provide an earth unless a metal insert has been provided for earthing purposes. The electricity supply undertaking should be requested, if necessary, to provide an earthing terminal. If this cannot be provided, an earth-leakage circuit breaker should be fitted.

Preferably, if expense permits, miniature circuit-breakers (one for each fuse in the consumer's control box) should be installed.

Metal water and gas pipes should still be bonded, as required by regulations D 10 and D 14 of the *I.E.E. Wiring Rules*.

NUMBERS AND SIZES OF 'COILED COIL' TUNGSTEN FILAMENT LAMPS FOR VARIOUS SITUATIONS

Situation	Lamp
Living room, 150-175 ft/14-16m² super floor area	1/150 watt or 2/75 watt, plus reading lamps.
Bathroom	1/60 watt, illuminating mirror over washbasin.
Bedroom—up to 120ft/11m² super floor area	60 watt direct lighting, or 100 watt indirect lighting.
Bedroom—over 120ft/11m² super floor area	100 watt direct or 150 watt indirect lighting.
(For bedrooms, allow two plug points in addition).	
W.C.	1/40 watt.
Stairs	1/40 watt at head of stairs.
Halls and passages	1/40 watt lamp every 20ft of passage length
Kitchens (general lighting only):—	
Up to 80 ft. super floor area 7·4m²	Lamps up to 100 watts
Up to 120 ft super floor area 11·0m²	Lamps up to 115 watts
Over 120ft super floor area 11·8m²	Lamps not less than 140 watts

For further details see CP 324.101.

The Electrical Development Association issues a series of data sheets and also publishes a number of technical notes, including one on typical specifications for house wiring and another on the design of simple heaters for domestic hot water. The Association is always ready to assist with technical advice without charge.

SIZES OF ELECTRIC HEATERS REQUIRED FOR ROOM HEATING

Radiant Fires

Floor area of room 8 ft celiing		Loading of fire	
m²	sq. ft	A	B
9·2	100	1·5 kW	1·5 kW
13·3	140	2·0 kW	1·5kW
15·8	170	2·25 kW	1·75 kW
19·8	210	2·75 kW	2·0 kW
22·3	240	3·0 kW	2·25 kW

Column A: 11 in cavity brick wall plastered, suspended floor, and plaster ceiling.
Column B: More efficient construction giving average 'U' value of 0·20.

Convectors

m²	Capacity of room cu ft		Halls, landings, living rooms, loadings in kW			Bedrooms loadings in kW	
		15·6 C 60° F	18·3 C 65° F	12·8 C* 60 F°	10/ 12·8 C 50/55° F	18·3 C 60/65° F
		10/ 12·8 C 50/55° F					
17	600	0·80	0·90	—	0·80	1·10
20	700	0·90	1·00	—	0·80	1·20
23	800	1·00	1·10	—	1·00	1·30
25	900	1·10	1·20	0·75	1·10	1·50
28	1,000	1·20	1·30	0·90	1·20	1·60
31	1,100	1·30	1·40	1·00	1·30	1·70
34	1,200	1·45	1·60	1·00	1·45	1·90
38	1,300	1·60	1·70	1·00	1·60	2·00
41	1,400	1·70	1·80	1·20	1·70	2·20
43	1,500	1·80	1·90	1·30	1·80	2·30
46	1,600	1·90	2·00	1·40	1·90	2·40
49	1,700	2·00	2·20	1·40	2·00	2·60
52	1,800	2·20	2·30	1·50	2·20	2·80
55	1,900	2·30	2·50	1·60	2·30	3·00
58	2,000	2·40	2·60	1·70	2·40	3·20

For intermediate sizes, take average of the two nearest sizes above and below. For rooms with outside walls facing north or east, add 10 per cent. The cubic capacity of halls must include staircase and top landing. Column headed * is for living rooms where the heating is auxiliary to a coal fire in constant use. Where electric fires are used as well as heaters, the latter should be sized for 50/55°F.

This table is only applicable to buildings of sound construction with an average 'U' value of 0.4 or better.

19. THE PAVIOR AND TILER

Codes of Practice	362
British Standards	362

CODES OF PRACTICE

CP 201: 1951: Timber flooring.
    Part 1: 1967: Wood flooring (board, strip, block and mosaic).
    Part 2: 1972: metric units.
CP 202: 1972: Tile flooring and slab flooring.
CP 203: 1969: Sheet and tile flooring (cork, linoleum, plastics and rubber).
    Part 2: 1972: metric units.
CP 204: In-situ floor finishes.
    Part 1: 1965: imperial units.
    Part 2: 1970: metric units.
CP 209: Care and maintenance of floor surfaces.
    Part 1: 1963: wooden flooring.

These Codes of Practice cover very diverse materials and equally diverse methods of laying, involving a large number of specialist trades. The provisions for work on site are so numerous as to be impossible to summarize here. Some points, however, need emphasis:

1. Make sure that the material is suitable for the situation in which it will be laid: e.g. magnesium oxychloride flooring, with sawdust as filler, is unsuitable for kitchens, as are corks tiles, Grease-resistant tiles are necessary in kitchens.
2. The screed on which linoleum, vinyl, rubber and other sheet or thin tile materials are laid must be
    a. solid with the base,
    b. free from dust and contamination by plaster, oil, etc.
3. All flooring materials of this type (and some others, such as magnesium oxychloride) are liable to fail if laid on solid ground floor slabs unless the slabs have a damp-proof membrane incorporated. If a floor, where a failure has occurred, has no damp-proof membrane, the recommended remedy is to lay a $\frac{3}{4}$ in (19 mm) layer of asphalt, to which the flooring is glued.
4. If laid at the same time as the mass concrete of the slab screeds may be $\frac{3}{4}$ in (19 mm) thick; otherwise a minimum of $1\frac{1}{2}$ in (38 mm) is recommended.

BRITISH STANDARDS

B.S. 776: Materials for magnesium oxychloride (magnesite) flooring.
    Part 1: 1963: imperial units.
    Part 2: 1972: metric units.
B.S. 810: 1966: Sheet linoleum (calendered types), cork carpet and linoleum tiles.

B.S. 882: and **B.S. 1201:** Aggregates from natural sources for concrete (including granolithic).
  Part 1: 1965: imperial units.
  Part 2: 1973: metric units.
Note: B.S. 1201 covers aggregates for granolithic concrete floor finishes, also '10 per cent fines' value of coarse aggregate.
B.S. 988, 1076, 1097, and **1451: 1973:** Mastic asphalt for building.
Covers flooring, coloured flooring and grades are given to flooring according to usage.
B.S. 1162, 1418 and **1410: 1973:** Mastic asphalt for building (natural rock asphalt aggregate).
  Three types, flooring type may be used as an underlay to other floor coverings as a protection against rising damp.
B.S. 1187: 1959: Wood blocks for floors.
B.S. 1197: Concrete flooring tiles and fittings.
  Part 1: 1955: imperial units.
  Part 2: 1973: metric units.
B.S. 1286: 1974: Clay tiles for flooring.
B.S. 1297: 1970: Grading and sizing of softwood flooring.
B.S. 1450: 1963: Black pitch mastic flooring.
B.S. 1711: 1975: Solid rubber flooring.
B.S. 1863: 1952: Felt backed linoleum.
B.S. 2592: 1973: Thermoplastic flooring tiles.
B.S. 2604: Resin-bonded wood chipboard.
  Part 1: 1963: imperial units (withdrawn).
  Part 2: 1970: metric units.
B.S. 3187: 1978: Specification for electrically conducting rubber flooring.
B.S. 3260: 1969: PVC (vinyl) asbestos floor tiles.
B.S. 3261: Unbacked flexible PVC flooring.
  Part 1: 1973: Homogeneous flooring (continuous lengths or tile form).
B.S. 3672: 1963: Coloured pitch mastic flooring.
B.S. 4050: Specification for mosaic parquet panels.
  Part 1: 1977: general characteristics.
  Part 2: 1966: classification and quality requirements.
B.S. 4131: 1973: Terrazzo tiles.

EXTERIOR PAVING
B.S. 435: 1975: Dressed natural stone kerbs, channels, quadrants and setts.
  There are three standard 'dressings' or finishes:
  A: fine picked.
  B: fair picked, single axed or nidged.
  C: rough punched.

Single dressed edge kerb[1]				Straight dressed flat kerb[2]			
Width (inches)	Depth	Length ft	in	Width (inches)	Depth	Length ft	in
8	12	2	6	12	8	2	6
6	12	2	6	12	6	2	6
6	10	2	0	10	6	2	0
6	9	1	9	9	6	1	9
6	8	1	9	8	6	1	9
5	10	2	0				
4	10	2	0				

Straight dressed channel[1]				Straight rough-punched kerb[2]			
Width (inches)	Depth	Length ft	in	Width (inches)	Depth	Length ft	in
12	6	2	0	6	12	2	6
10	6	2	0	5	10	2	6
				4	9	1	3

Setts[3]	
Section (in)	Length (in)
4×4	4
3×5	
3×6	5–10
4×4	
4×5	
4×6	
5×4	6–10
5×6	

Note 1: Top and front face to depth of 5in to be A or B dressing, with 1in chisel-drafted margins to top surface and to top edge.

Note 2: Standard dressing, with ends hammer dressed or punched.

Note 3: Setts are to be square, hammer-dressed.

Stone kerbs, setts etc. and occasionally, flagstones may be re-dressed and re-used two or three times. In addition, many new products in non-standard sizes are produced and are in fairly general use.

B.S. 340: 1963: Precast concrete kerbs, channels edgings, and quadrants.
This Standard had not, at the time of going to press, been metricated. It provides dimensioned sections of all the following.
Straight kerbs: 3ft long-12in by 6in; 10in by 5in; 10in by 4in; and 5in by 6in with ⅜in or ¾in radius on one top edge.
Radius kerbs and channels; length, between 2ft and 3ft.
External radii; 3, 6, 8, 10, 15, 20, 25, 30, 35 and 40ft.
Internal radii; as above, except 15, 25 and 35ft.
Straight channels; 3ft long by 12in by 6in; 10in by 5in and 10in by 4in.
Edging: 3ft long by 2in by 6, 8, and 10in, with plain, radiused or half-round top edges.
Quadrants: 12in and 18in radii, by 6, 8, and 10in deep, to match kerbs.
B.S. 368: 1971: Precast concrete flags.

TABLE 1. FLAG DIMENSIONS

Flag type	Co-ordinating size	Work size	Maximum limit of manufacturing size	Minimum limit of manufacturing size
	mm	mm	mm	mm
A	600 × 450	598 × 448	600 × 450	596 × 446
B	600 × 600	598 × 598	600 × 600	596 × 596
C	600 × 750	598 × 748	600 × 750	596 × 746
D	600 × 900	598 × 898	600 × 900	596 × 896

NOTE. When ordering, it will be necessary only to specify the type followed by the thickness, e.g. ' A 50 '.

B.S. 802: 1967: Tarmacadam with crushed rock or slag aggregate.
See table opposite.

APPROXIMATE COVERING CAPACITY OF TARMACADAM

The table on page 370 shows the approximate ranges of cover of various compacted thicknesses of tarmacadam. The tolerances given take into account the fact that the covering capacity is influenced by such factors as specific gravity of aggregate, grading of aggregates, weather conditions, length of haul, whether the material is laid in one or more courses, method of laying, weight of roller, etc.

TABLE 8. TARMACADAM FOR VARIOUS TYPES OF CONSTRUCTION

The lower limits of thickness given below apply only where the accuracy of finish of the surface on which the tarmacadam is to be laid is within the limits specified in *the standard*.

Number of courses	Range of total compacted thickness	Average thickness of compacted course or courses should be between the limits specified below (see A.3)	Nominal size and type of tarmacadam
One	¾ in (19 mm) to 3 in (76 mm)	2¼ in (57 mm) to 3 in (76 mm) 1¼ in (32 mm) to 1½ in (38 mm) ¾ in (19 mm) to 1 in (25 mm) ¾ in (19 mm)	1½ in (38 mm) single course ¾ in (19 mm) open or medium texture ½ in (13 mm) open or medium texture ⅜ in (10 mm) open or medium texture
Two	3 in (76 mm) to 5 in (127 mm)	Basecourse 2¼ in (57 mm) to 3 in (76 mm) Wearing course 1¼ in (32 mm) to 1½ in (38 mm) ¾ in (19 mm) to 1 in (25 mm) ¾ in (19 mm)	1½ in (38 mm) basecourse or 1½ in (38 mm) single course ¾ in (19 mm) open or medium texture ½ in (13 mm) open or medium texture ⅜ in (10 mm) open or medium texture
Multi (more than two)	Greater than 5 in (127 mm)	Upper course or courses as above Lower courses not greater than 3½ in (89 mm) each	Upper course or courses as above 1½ in (38 mm) single course (hot-laid)

Thickness of course		Nominal size of material (see Tables 1a to 4)	Approximate covering capacity	
in	mm		yd²/ton	m²/tonne
¾	19	⅜ in (10 mm) open texture ½ in (13 mm) or ⅜ in (10 mm) medium texture	25–30	20–25
1	25	½ in (13 mm) open texture ½ in (13 mm) medium texture	20–24	17–20
1¼	32	¾ in (19 mm) open texture ¾ in (19 mm) medium texture	17–20	14–17
1½	38	¾ in (19 mm) open texture ¾ in (19 mm) medium texture	15–18	13–15
2	51	¾ in (19 mm) open texture	11–14	9–12
2¼	57	1½ in (38 mm) basecourse 1½ in (38 mm) single course	10–13	8–11
2½	64	1½ in (38 mm) basecourse 1½ in (38 mm) single course	9–12	8–10
3	76	1½ in (38 mm) basecourse 1½ in (38 mm) single course	8–10	7–8

These coverages apply to materials to be laid on a foundation, the profile of which complies with the requirements of this specification. For more irregular foundations, lower coverages must be expected. The table is given without implying obligation on any of the parties concerned.

B.S. 802 also provides standards of workmanship in laying, and composition of tarmacadam mixes.

CLAY FLOORING TILES
B.S. 1286: 1974 Clay tiles for flooring.

Quality, shape, range of ceramic floor tiles and clay floor quarries, sills and associated fittings. Full range of fittings illustrated for each type. Sampling methods, tests for warpage, curvature and water absorption are covered.

Preferred sizes for modular ceramic floor tiles 100 by 100 by 9.5mm, and 200 by 200 by 9.5 mm, and for modular clay floor quarries 200 by 100 by 19 mm, and 100 by 100 by 19 mm. Predominant range in non-modular clay floor quarries from 229 by 229 by 32 mm, to 150 by 150 by 15 mm.

CONCRETE FLOORING TILES
B.S. 1197: Concrete flooring tiles and fittings.
Part 2: 1973: metric units.

TABLE 1. SIZES OF SQUARE TILES

Length of each side	Thickness
Work size	Work size
mm	mm
150	15
200	20
225	20
300	30
400	35
500	40
Permissible deviation ±1 mm	Permissible deviation ±3mm

When laid, diagonal and rectangular half-tiles shall conform to the above sizes.

TABLE 2. SIZES OF SKIRTINGS AND FITTINGS

Dimension (see Figs. 1, 2 and 3)		Work size	Permissible deviation
		mm	mm
Length of skirting	(l)	150	±1
		200	±1
		225	±1
		300	±1
		400	±1
		450	±1
		500	±1
		600	±1
		675	±1
		750	±1
		800	±1
		900	±1
Base width	(w)	40	±1
Height	(h)	100	±1
		150	±1
Thickness	(t)	12	±3
Radius of cove		25	±3
Radius of bull-nose (if made)		8	±4

TABLE 3. DIMENSIONALLY CO-ORDINATED SIZES FOR TILES

These sizes may be supplied by agreement between the purchaser and the supplier.

Length of tile-side co-ordinating size	mm 300	mm 400	mm 500
Joint clearance	3	3	3
Work size	297	397	497
Permissible deviation on work size	±	±	±
Maximum limit of manufacturing size	298	398	498
Minimum limit of manufacturing size	296	396	496
Thickness, work size (not dimensionally co-ordinated)	30 ± 3	35 ± 3	40 ± 3

CERAMIC INTERNAL WALL TILES

B.S. 1281: 1974: Glazed ceramic tiles and tile fittings for internal walls.

Provides specification for quality, shape, range of sizes, physical and chemical requirements. Sampling methods, tests for warpage, curvature, wedging, water absorption, crazing, resistance to chemicals and impact.

Preferred sizes for modular tiles 100 by 100 by 5mm, and 200 by 100 by 6.5mm. Predominant range in non-modular tiles from 152 by 152 by 9.5mm to 108 by 108 by 4mm.

Round edge, attached angle and angle bead tile fittings, cushion edges, and spacer lugs.

B.S. 1281 contains illustrations for application of tiles and tile fittings.

Warning: These tiles are not suitable for external work.

NUMBER OF TILES OF VARIOUS SIZES PER SQUARE METER

Size of tiles in	No. of tiles per m^2 Exact	To nearest whole tile
12 × 12	10·764	11
12 × 10	12·644	13
12 × 9	13·445	14
10 × 10	14·054	15

Size of tiles	No. of tiles per m^2	
in	Exact	To nearest whole tile
9×9	17·963	18
8×8	24·219	25
6×6	42·856	43
4×4	96·876	97
3×3	161·424	162
6×3	85·712	86
mm		
*200×200	25·000	25
*250×250	16·000	16

*These are the two principal sizes of Continental concrete tiles. Small quantities have been imported for some time.

The above table makes no allowance for mortar joints, nor for waste. Only rectangular tiles are included: a few hexagonal or rhomboidical tiles are occasionally used. Allow an extra 33 per cent over the nearest rectangular tiles.

20. THE PLASTERER

Codes of Practice	372
British Standards	372
Plastering Data	373
Modern Plastering Materials	373
Gypsum Building Plasters	374
Fillers	378
Composition of Lime Plasters	379
External Renderings	380

CODES OF PRACTICE

B.S. 5492: 1977: Code of practice for internal plastering.
Covers plastering on all common forms of background. Supersedes CP 211.

BRITISH STANDARDS

B.S. 890: 1972: Building limes.
In four parts; general; hydrated lime (powder); quicklime; lime putty.
B.S. 1191: Gypsum building plaster.
  Part 1: 1973: excluding premixed lightweight plaster.
  Part 2: 1973: premixed lightweight plaster.
B.S. 1230: 1970: Gypsum plasterboard.
Quality and dimensions (with appendices).

TABLE 1. SIZES*

	Length	Permissible deviation	Width	Permissible deviation	Thickness	Permissible deviation
	m	mm	mm	mm	mm	mm
Gypsum wallboard	1800, 2350, 2400, 2700, 3000	+0 / −6	600, 900, 1200	+0 / −5	9.5, 12.7	±0.50, ±0.60
Gypsum plank	2350, 2400, 2700, 3000	+0 / −6	600	+0 / −5	19.0	±0.75
Gypsum lath	1200	+0 / −6	406	±3	9.5, 12.7	±0.50, ±0.60
Gypsum baseboard	1200	+0 / −6	914	±3	9.5	±0.50

TABLE 2. WEIGHT*

Thickness	Gypsum plasterboard	
	Min	Max
mm	kg/100 m^2	kg/100 m^2
9.5	650	1000
12.7	950	1450
19.0	1400	2050

B.S. 1369: 1947: Metal lathing (steel) for plastering.
Five types of mild steel lathing.

B.S. 4049: 1966: Glossary of terms applicable to internal plastering, external rendering and floor screeding.
Defines terms in current use.

PLASTERING DATA

1 ton equals 1 metric tonne (approx).
1 bushel equals 8 gallons, or 36 litres, or 1.28 cubic feet, or 0.036 cubic metres.
21 bushels equal 1 cubic yard: 30 bushels equal 1 cubic metre.
1 ton or tonne of lime equals 42 bushels or $1\frac{10}{11}$ cubic yards, or 51 cubic feet or 1.444 cubic metres.
1 cubic foot (0.282 cu. metres) of lime weighs 44lb (19.950 kg).
1 cubic foot of sand weighs 100 lb (approx) or 45.35 kg.
A convenient bushel measure is a box with internal measurements of $13\frac{1}{4}$in $\times$ $13\frac{1}{4}$in $\times$ $13\frac{1}{4}$ (346.5 mm $\times$ 346.5 mm $\times$ 346.5 mm).
1 cubic yard is very approximately three-quarters of 1 cubic metre.

MODERN PLASTERING MATERIALS

Gypsum plasters have almost wholly replaced the old plastering system based on lime. Hydrated lime in bags is universally used for 'compo' mixes and for lime putty. Hydraulic limes, such as blue lias, are nowadays very difficult to obtain, because backing coats of lime and sand alone are no longer used. Nor is lime putty alone used as a finishing coat. Lime plastering is confined today to repair and restoration work. Plasterboard or expanded metal lathing have replaced wooden laths, which today are difficult to obtain and, when obtainable, very expensive.

LIGHTWEIGHT PLASTERS AND RENDERINGS

In recent years, a number of firms have introduced ready-gauged plaster, consisting of lightweight aggregates such as vermiculite or perlite, combined with gypsum plaster. Others sell these lightweight aggregates, and give directions for compounding mixes for various uses and situations. For many years, a patented process of applying, by spray-gun, mixtures of asbestos, or mixtures of vermiculite or perlite with Portland cement (the 'Pyrok' system) has met with considerable success for exterior renderings, for condensation control in factories and for fire-resisting coatings on all manner of building materials. These lightweight plasters, when used in domestic work, give increased thermal insulation. Local authorities accept them as internal plastering on 9in solid brickwork (used internally) to bring it up to the thermal insulation requirements of the Building Regulations. They are, however, more expensive than normal plastering.

PLASTERING MACIHNES
These machines require special mixes, some of which are based on Portland cement. The machine manufacturers supply recommended mixes, as does the Gypsum Products Development Association.

BONDING LIQUIDS
These materials are:
1. Bitumen solutions that, while tacky, are dashed with concreting sand to provide a key, or.
2. A type of adhesive.

They eliminate all hacking of the background and allow any surface, such as glazed tiles, to be plastered upon. When applied to old, cracked surfaces, they eliminate hacking-off. They also allow surfaces such as concrete to be plastered without using special grades of plaster. While it is now general to use a Portland cement backing coat, with a gypsum plaster skim coat in these instances, a lightweight plaster should be used if possible, because it puts less weight, and therefore less strain, on the adhesive.

'DRY LINING SYSTEMS'
These consist of plasterboard, stuck to the brickwork or other backing with 'dabs' or lines of plaster. The joints are filled with a special filler, with sometimes a scrim tape incorporated. Alternatively, after filling, a tapered paper strip is used to mask the joints. These methods have not become popular, except with prefabricated systems of construction, despite the elimination of wet plastering and its attendant mess. Cost and a slightly substandard finish have been the deciding factors.

SUSPENDED CEILING SYSTEMS
These consist of plasterboard panels set in a light metal grillage of 'T' section aluminium or steel. They are largely used in factory work, for thermal insulation. The work is not plastered.

GYPSUM BUILDING PLASTERS

Internal plastering employing gypsum plasters is described in B.S. 5492 and additional practical information is given in the Department of the Environment Advisory Leaflet No. 2 'Gypsum Plasters Used in Building', obtainable from H.M.S.O. These documents describe the range of plasters, including plaster of paris, retarded hemihydrate gypsum plaster, anhydrous gypsum plaster, Keenes and premixed lightweight plaster types, with a full list of trade names and the purposes for which they may be employed and the mixes to use.

The following is an outline of the uses of the types of gypsum plasters now available.

GYPSUM BUILDING PLASTERS MANUFACTURED TO B.S. 1191: PART 1. EXCLUDING PREMIXED LIGHTWEIGHT PLASTERS.

Type	Variety	Uses
Class A Plaster of paris	One type only: coarse pink, grey or white generally used in building work	For gauging lime finishing coats, use up to $\frac{1}{4}$ plaster to 1 of lime. For repair work and for fibrous plaster, use neat without lime.
Class B Retarded hemihydrate	(i) Undercoat type for all materials, including boards	Use 1 part of plaster to from 1 to 3 parts of sand. Use lower sand proportions on low-suction backgrounds, such as plasterboard and concrete.
	(ii) Final coat type may be 'finish' or 'board finish'	For finishing coats in 2 or 3 coat work, use neat or with lime. For single coat work on plasterboard, fibreboard or other boards, use neat. Do not retemper Class B plasters.
Class C Anhydrous	Final coat type only	May show a double set and first set can be 'killed' if desired. Use neat or with lime up to $\frac{1}{4}$ of plaster by volume. Use from $\frac{1}{4}$ to 1 part of plaster to 1 of lime with fine sand if desired for gauged lime finishing coats. Not recommended for direct application to boards. Do not retemper after $\frac{1}{2}$ hour from mixing.
Class D Keenes	Final coat type only	Use neat for finishing coats and external angles. Not suitable for use outside. Not recommended for direct application to boards. Do not retemper after 1 hour from mixing.

GYPSUM BUILDING PLASTERS MANUFACTURED TO B.S. 1191: PART 2. PREMIXED LIGHTWEIGHT PLASTERS.

Type	Variety	Uses
Premixed lightweight gypsum plaster	Undercoat type	Requires the addition of clean water only
Browning plaster	(i) containing expanded perlite aggregate	Use on normal suction backgrounds such as brick, clinker and most proprietary partition blocks. Do not use on plasterboard or in situ concrete
Bonding plaster	(ii) containing exfoliated vermiculite aggregate	Use on plasterboard and in situ concrete and other low suction backgrounds.
Metal lathing plaster	(iii) containing expanded perlite and exfoliated vermiculite aggregates	Use on expanded metal lathing and other backgrounds where there may be a risk of rusting. Also on expanded polystyrene and wood wool slabs.
	Final coat type	Requires the addition of clean water only.
Finish plaster	Containing exfoliated vermiculite aggregate	Use as a finishing coat for 2 or 3 coat work to any lightweight undercoat plaster.

GYPSUM BUILDING PLASTERS WITH SPECIAL APPLICATIONS

Type	Variety	Use
Acoustic plaster	Final coat type	Requires the addition of clean water only. Use in two coats on suitably prepared gypsum plaster undercoats. To obtain maximum sound absorption, follow manufacturer's instructions.
Premixed Gypsum X-Ray Plaster	Undercoat type containing barytes aggregate	Requires the addition of clean water only. Use in two or three coats depending upon thickness specified and type of background. Thickness can only be specified by the National Radiological Protection Board.
Finish plaster	Final coat type containing barytes aggregate.	Requires the addition of clean water only. Use as a finishing coat on barytite premixed gypsum X-ray plaster undercoats.
Thin coat finish plaster	Final coat type	Requires the addition of clean water only. Use on suitably smooth and straight backgrounds, regardless of suction in one coat to a minimum thickness of $\frac{1}{16}$ in.

MIX PROPORTIONS AND FINISHED THICKNESSES OF GYPSUM PLASTER UNDERCOATS

Background	Grade of plaster	Proportions by volume plaster / sand		Total thickness of finished plasterwork in / mm	
Clay brick, tiles, etc. Partitions	Browning	1	3	$\frac{1}{2}$	13
Concrete bricks Clinker blocks Lightweight concrete blocks	Browning	1	2	$\frac{1}{2}$	13
Wood wool slabs	Metal lathing	1	2	$\frac{1}{2}$	13
Metal lathing (render)	Metal lathing	1	$1\frac{1}{2}$	$\frac{1}{2}$ (from face of lath)	13
Metal lathing (float)	Metal lathing	1	2		
Plasterboard and Plaster lath	Haired	1	$1\frac{1}{2}$	$\frac{3}{8}$	9

Note: Cement-sand screeds are very often used instead of browning and metal lathing plasters. The sand should be to B.S. 1191, type 1.

(By courtesy of British Gypsum Ltd.)

AVERAGE COVERING CAPACITY OF GYPSUM PLASTERS. PER TONNE AND TON

Background	Coating	sq. yds	
1. Class B (retarded hemi-hydrate plasters)			
Clay brick, tiles, etc.	Floating coat (1–3)	280	234
Concrete blocks, clinker blocks, lightweight concrete blocks and wood-wool slabs	Floating coat (1–2)	217	181
Plasterboards	Floating coat (1–1$\frac{1}{2}$)	246	206
Metal lath	Rendering and floating (1–1$\frac{1}{2}$ & 1–2)	138	115
Neat finish (walls or ceilings)		467	390
Neat finish (plasterboards —one coat)		197	165
2. Class D (Keenes) setting coat only.		276	231

Background	Coating	sq. yds	
3. Special plasters supplied in bags, pre-mixed with aggregate.			
(a) X-ray (barium-gypsum)			
Floating coat, ($\frac{7}{8}$"–22mm)		26	21
Floating coat on metal lath ($\frac{7}{8}$"–22mm)		21	17
Setting coat ($\frac{1}{8}$"–3mm)		180	151
(b) Lightweight plasters			
Bricks, blocks, etc.	Floating coat ($\frac{7}{16}$"–11mm)	157–177	131–148
Metal lath	Floating coat	79	661
Wood wool slabs	Floating coat ($\frac{7}{16}$"–11mm)	143	120
Concrete and plasterboards	Floating coat ($\frac{5}{16}$"–8mm)	177–200	148–167
Finish		492–591	411–494

Note: Portland cement screeds are very often used instead of plaster for rendering and floating coats in section 1 above.

Neat finish on walls etc. can contain up to 25 per cent lime putty. For one-coat on plasterboards, no lime should be used.

FILLERS

A material under such trade names as 'Polyfilla' or 'Fleetfilla', originally produced for amateur use, has been adopted by the trade. It is gypsum based and contains a water-soluble adhesive. Its cost precludes its use as a skim coat, for which it is suitable despite some difficulty in application. For patches, especially damaged external angles, it should be applied so as to stand a little 'proud' of the surface and, when set and dry, be glasspapered level. Cracks should be cut out and left undercut, as for plaster and again the filler should be left 'proud' and be glasspapered level.

The material adheres well, even to dry surfaces, but being water-soluble, it should not be applied to permanently damp surfaces. It can be used as a hardener and binder in lime putty and also as a retarder for class A gypsum plaster ('plaster of paris') although fairly large quantities are required for this purpose. The material is tenacious and resists vibration: it will take nails and screws but only for light loads. An external quality is also available, based on Portland cement.

MORTAR MIXES IN TONS OR TONNES

	Cement mortar		Lime mortar		
Ratios	Cement tons	Sand tonnes	Lime putty cu. yd	m^3	Sand tonnes
1–1	0·80	0·875	$\frac{4}{5}$	0·607	0·875
1–2	0·55	1·200	$\frac{1}{2}$	0·382	1·200

Ratios	Cement mortar		Lime mortar			
	Cement tons	Sand tonnes	Lime putty cu. yd	m^3	Sand tonnes	
1–3	0·40	1·350	$\frac{2}{5}$	0·303	1·350	
1–4	0·30	1·350	$\frac{3}{10}$	0·218	1·350	
1–5	0·28	1·350	$\frac{1}{4}$	0·191	1·350	

PORTLAND CEMENT SCREEDS COVERING CAPACITY PER BUSHEL

Cement	Sand	in $\frac{1}{2}$	mm 13	in $\frac{3}{4}$	mm 19	in $\frac{7}{8}$	mm 23	in 1	mm 25
neat	—	2·8	2·4	2·1	1·5	1·7	1·2	1·4	1·0
1	1	4·6	3·5	3·6	2·7	2·8	2·1	2·3	1·7
1	2	6·7	5·1	5·0	3·8	4·2	3·2	3·4	2·5
1	3	9·0	6·8	6·7	5·1	5·6	4·2	4·5	3·4
		yd²	m²	yd²	m²	yd²	m²	yd²	m²

LIME MORTAR COVERING CAPACITY PER BUSHEL

Surface and No. of Coats		Total Thickness		Coverage					
				1st Coat		2nd Coat		3rd Coat	
		in	mm	cu. yd	m²	cu. yd	m²	cu. yd	m²
Brick or Similar	2	$\frac{5}{8}$	15·8	3·42	2·85	6·85	5·72	—	—
	3	$\frac{7}{8}$	22·2	3·42	2·85	4·57	3·82	13·71	11·3
Rubble	2	2	15·8	4·76	3·97	6·85	5·72	—	—
	3	$\frac{7}{8}$	22·2	4·76	3·97	4·47	3·82	13·71	11·3

For covering capacities per cu. yd, multiply by 21 ; for m³ multiply by 30.

COMPOSITION OF LIME PLASTERS
USING NON-HYDRAULIC AND SEMI-HYDRAULIC LIMES

Coarse stuff: One part of lime putty to three parts of sand.
Undercoats of lime plaster gauged with cement: One part of cement, one part of hydrated lime, six parts of sand.

Undercoats of lime plaster gauged with gypsum plaster: One part of lime putty to 2-4 parts sand. Immediately before use, add retarded hemi-hydrate plaster equal to $\frac{1}{3}$ by volume of the lime putty, with the additional water required to mix. This should not be retempered once it has stiffened.
Floating Coat: This is used for the intermediate coat in 'three coat work', and has the same composition as the undercoats described above.

Setting Coat: One volume of lime putty, up to one volume of fine sand, and $\frac{1}{4}$ volume of retarded hemi-hydrate plaster added just before use.

Or, one volume of lime putty gauged with $\frac{1}{4}$ to 1 volume of retarded hemi-hydrate plaster.

USING HYDRAULIC LIME

Coarse stuff: One part of lime to 2—3 parts sand.

Haired coarse stuff: 9lb hair to every cubic yard of coarse stuff.

EXTERNAL RENDERINGS

The following data have been extracted from: **CP 221: 1960**. The code covers workmanship, definitions and much other matter: it should be read by every builder and craftsman.

Table 1. Recommended mixes for external renderings in relation to background material, exposure conditions and finish desired. (See page 384.)

Mix: Type 1. 1 part Portland cement: 0 to $\frac{1}{4}$ part lime: 3 parts sand by volume.

Type 2. 1 part Portland cement: $\frac{1}{2}$ part lime: 4 to $4\frac{1}{2}$ parts sand by volume.

Type 3. 1 part Portland cement: 1 part lime: 5 to 6 parts sand by volume.

Type 4. 1 part Portland cement: 2 parts lime: 8 to 9 parts sand by volume.

(For mixes consisting of masonry cement and sand, or hydraulic lime and sand, see Clauses 313d and e. For mixes for machine-applied finishes, see Clause 313g.) Type of mix recommended for the given exposure conditions (see Clause 313a).

THE PLASTERER

Background material (see Clause 303)	Type of Finish (see Clause 302)	First and subsequent undercoats			Final Coat		
		severe	moderate	sheltered	severe	moderate	sheltered
(A) Dense, strong and smooth and (B) Moderately strong and porous	Wood-float Scraped or textured Roughcast Dry-dash	1, 2 or 3 3 1, 2 or 3 1 or 2	1, 3 or 4 3 or 4 1, 2 or 3 1 or 2	1, 3 or 4 3 or 4 1, 2 or 3 1 or 2	1 or 3 3 As undercoats 2	1, 3 or 4 3 or 4 As undercoats 2	1, 3 or 4 3 or 4 2
(C) Moderately weak and porous	Wood-float Scraped or textured Roughcast Dry-dash	3 3 2 or 3	3 or 4 3 2 or 3	3 or 4 3 2 or 3	As undercoats		
(D) No-fines concretes (see Clause 303b(iv))	Wood-float Scraped or textured Roughcast Dry-dash	1, 2 or 3 1, 2 or 3 1, 2 or 3 1 or 2	1, 2, 3 or 4 2, 3 or 4 1, 2 or 3 1 or 2	1, 2, 3 or 4 2, 3 or 4 1, 2 or 3 1 or 2	1, 2 or 3 3 1, 2 or 3 2	1, 3 or 4 3 or 4 1, 2 or 3 2	1, 3 or 4 3 or 4 1, 2 or 3 2
(E) Metal lathing or expanded metal	Wood-float Scraped or textured Roughcast Dry-dash	1, 2 or 3 1 or 2	1, 2 or 3 1 or 2	1, 2 or 3 1 or 2	1 or 3 3 2	1 or 3 3 2	1 or 3 3 2

FOR BRICKWORK HIGHLY CONTAMINATED BY SOLUBLE SALTS (GENERALLY SULPHATES)
In place of type 1 mixes (in above table) use 1 high alumina cement: 1 sand (by volume.)
In place of type 2 mixes, use 1 high alumina cement: $\frac{1}{2}$—$\frac{3}{4}$ ground chalk: 4—$4\frac{1}{2}$ sand.
In place of type 3, use 1 high alumina cement: 1—$1\frac{1}{2}$ ground chalk: 5—6 sand.
For spatterdash: 1 high alumina cement: $1\frac{1}{2}$—2 parts sand.

21. THE PAINTER

Codes of Practice	383
British Standards	383
Conversion Tables	385
Choice of Paint	385
Mixing Paints	386
Painting Data	387
Coverage of Various Materials	387
Priming Paints for Metal	390
Priming Paints for Softwood	390
Priming Paints for Plaster and Cement	391
The Causes of Paint Failures	391
Recent Developments	392
Coverage of Wallpapers	392

CODES OF PRACTICE

CP 231: 1966: Painting of buildings.
Covers design, organisation, supervision, paint systems, substrates, materials, surfaces, preparation and maintenance for a wide range of materials and surfaces.

BRITISH STANDARDS

B.S. 217: 1961: Red lead for paints and jointing compounds.
Red lead, non-setting for paints and ordinary for paint mixed on the site.
B.S. 239, 254, 296, 338 and 637: 1967: White pigments for paints.
White lead (239); zinc oxide (254); lithopones (296); antimony oxide (338); and basic lead sulphate (637).
B.S. 242, 243, 259 and 632: 1969: Linseed oil.
Raw, refined and boiled linseed oil.
B.S. 244 and 290: 1962: Turpentine for paints.
Gum spirits, wood and sulphate.
B.S. 245: 1976: Specification for mineral solvents (white spirit and related hydrocarbon solvents) for paint and other purposes.
Types A and B solvents based on two ranges of aromatic content.
B.S. 283: 1965: Prussian blues for paints.
Sampling, composition, colour, staining power etc.
B.S. 284-6: 1952: Black (carbon) pigments for paints.
Carbon black (284); bone black (285); and lamp (or vegetable) black (286).
B.S. 303: 1978: Specification for lead chrome green pigments for paints.
Definition, required characteristics, etc.
B.S. 311: 1936: Gold size for paints.
Quality, drying times, etc.
B.S. 314: 1968: Ultramarine pigments.
Requirements and methods of test for three types, etc.
B.S. 318: 1968: Chromic oxide pigments.
Requirements for two types, etc.
B.S. 332: 1956: Liquid dryers for oil paints.
Colours, mixing properties, etc.
B.S. 388: 1972: Aluminium pigments.
Powder, paste, leafing and non-leafiing, etc.
B.S. 391: 1962: Tung oil.
Two types of oil derived from *Aleurites Fordii* and *A. Montana*.
B.S. 1070: 1973: Black paint (tar-based).
Requirements, etc.
B.S. 1215: 1945: Oil stains.
Mainly for builders' joinery.
B.S. 1336: 1971: Knotting.
Solid content, appearance, drying time, resistance to white spirit, etc.

B.S. 1795: 1976: Specification for extenders for paints.
Grades and descriptions of various types, etc.
B.S. 1851: 1978: Titanium pigments for paints.
Description, classification, requirements, etc.
B.S. 2015: 1965: Glossary of paint terms.
Definitions of over 500 terms used in the paint industry.
B.S. 2521 and 2523: 1966: Lead-based priming paints.
Composition, samples, consistency, drying times, etc.
B.S. 2524: 1966: Red oxide-linseed oil priming paint.
Scope, composition, drying times, properties, etc.
B.S. 3357: 1961: Glue size for decorators' use.
Animal glue, in granular or powder form, for use in sizing walls, ceilings and similar surfaces.
B.S. 3416: 1975: Black bitumen coating solutions for cold application.
Bitumen coating solutions.
B.S. 3634: 1963: Black bitumen oil varnish.
Coating for the protection of iron and steel.
B.S. 3698: 1964: Calcium plumbate priming paints.
Requirements for two types.
B.S. 3699: 1964: Calcium plumbate for paints.
Requirements for use in paints.
B.S. 3722: 1964: Machine-made shellac.
Specifies four main types.
B.S. 3761: 1970: Non-flammable solvent-based paint remover.
Description and requirements, etc.
B.S. 3981: 1976: Iron oxide pigments for paints.
Requirements for five colour groups.
B.S. 3982: 1966: Zinc dust pigment.
Requirements, methods of test, etc.
B.S. 4232: 1967: Surface finish of blast-cleaned steel for painting.
First, second and third qualities of surface finish (cleanliness and roughness).
B.S. 4310: 1968: Permissible limit of lead in low-lead paints and similar materials.
Gives an upper limit and method for detection.
B.S. 4313: 1968: Strontium chromate for paints.
Requirements, etc.
B.S. 4652: 1971: Metallic zinc-rich priming paint (organic media).
Three types and requirements, etc.
B.S. 4725: 1971: Linseed stand oils.
Requirements and methods of test, etc.
B.S. 4756: 1971: Ready mixed aluminium priming paints for woodwork.
Two types, properties and finish, etc.

B.S. 4764: 1971: Powder cement paints.
Requirements, qualities, recommendations for use, etc.
B.S. 4800: 1972: Paint colours for building purposes.
Illustrates 86 colours on 8 cards and refers to black and white. Supersedes B.S. 2660.
B.S. 5082: 1974: Water-thinned priming paints for wood.
Specifies materials suitable for site application and for application to factory-primed joinery.
B.S. 5358: 1976: Specification for low-lead solvent-thinned priming paint for woodwork.
Methods for outdoor exposure and recoatability after outdoor exposure, etc.

CONVERSION TABLES

SQUARE YARDS AND SQUARE METRES			GALLONS AND LITRES		
sq. yd		m²	gallons		Litres
36	30	25	0·22	1	4·5
42	35	30	0·44	2	9·0
48	40	34	0·65	3	13·6
54	45	38	0·88	4	18·2
60	50	42	1·09	5	22·7
66	55	46	1·31	6	27·2
72	60	50	1·53	7	31·8
84	70	59	1·75	8	36·3
96	80	67	1·97	9	41·0
108	90	76	2·19	10	45·0
120	100	84			
239	200	168			
479	400	335			

(Rounded up to nearest whole numbers)

CHOICE OF PAINT

Formulae for mixing paint to-day have nothing more than academic interest. Paint is now bought ready-mixed and, since the war, has been made from materials other than the natural oils and resins and thinners and the traditional lead and zinc bases. There is no indication, except by chance, of the contents of the can of paint, distemper or varnish: all that a painter can do, to achieve a good job, is to follow slavishly the directions printed on the can and in the advertising literature concerning the material. It is not always enough to adhere to these except in very straightforward jobs. For a manufacturer to accept full responsibility, in new work, the whole painting system, from primer to finish coat, must be of his material, and of a good grade of that material. On old work, where repainting is undertaken every care should be given to adequate cleansing and filling; otherwise any defect in the finish will be put down to poor preparation.

The first essential is to choose a good paint manufacturer, and to use the better class paints and other materials that he manufactures. In anything but run-of-the mill jobs, consult his technical department, and request them to suggest a suitable painting system. Never forget that there are now primers other than red-lead priming, and special primers for non-ferrous metals, new cement and plaster, that sealers and primer-sealers are frequently used on stonework or concrete and that phosphating treatments have revolutionized the treatment of ironwork. Emulsion paints have brought with them their own problems, and the newer ranges of bituminous paints are very different from the crude materials of some few years ago.

Doubtless we all deplore the passing of the skilled painter and the coming of those who merely apply something of which they know nothing to surfaces of which they know less. It is, however, a regrettable fact that must be taken into account when undertaking painting work.

It is possible, for the large user of paint, such as a railway undertaking or a very large firm that employs paint in great quantities, to have paint compounded to a specification. The writing of such specifications is, however, no easy matter, since a knowledge of paint chemistry is a necessity. There are, however, signs that some knowledge of paint chemistry is reaching the skilled decorator, and firms are being forced to give, if not the composition, at least the type of paint they offer for sale. This classification appears to be by the type of synthetic resin or binder used in these preparations, although the presence of such expensive pigments (or bases) as titanium dioxide is often mentioned.

MIXING PAINTS

Although this procedure is not recommended by paint manufacturers, most painting materials of the same type can be mixed together without ill effects ensuing. Emulsion paints, of the three most usual types, can be mixed together, even if made by different manufacturers. This can be done either to alter the colour or to use up small batches of material. Emulsion paints of any type, or mixtures thereof, can be tinted with colour in oil or colour in water, taking care to work the colour to a consistency of cream with some of the emulsion paint before adding it to the bulk of the paint to be tinted. Colour in lumps or in powder form should never be added: these are most difficult to incorporate thoroughly and often cause streaks of strong unmixed colour to appear on the finished work.

The majority of high-gloss or hard-gloss paints are based on a synthetic gum known as alkyd resin. Such paints, of differing colours and by different manufacturers, can readily be mixed together without mishap; indeed, the product is often a better paint. They may be tinted with colours in oil: for small quantities, artists' colours in small tubes (students' quality will serve) are excellent. Generally it is not necessary to add driers, such as terebene, but thinning with white spirit is always permissible. For indoor use, high gloss can be mixed with undercoat to produce any degree of matt or semi-matt finish desired. High-gloss paints should not be mixed with chlorinated rubber paint, nor with two-to-four hour 'enamels' or one-hour 'finishes', because livering almost always results.

PAINTING DATA

One gallon of ready mixed paint weighs approximately 27 lbs or 12.5 kg, or 4.5 litres.

One book gold leaf contains 25 leaves size $3\frac{1}{2}$ in $\times$ $3\frac{3}{4}$ and covers approximately one superficial foot, or 0.09 sq. m.

The percentage of increase in area of corrugated sheets when painted is $12\frac{1}{2}$ per cent (Divide Ft Sup. by 8 for Yd Sup.).

COVERAGE OF VARIOUS MATERIALS

WARNING: All tables giving coverage of paints, varnishes and distempers should be used with caution. Exhaustive tests under scientific control, carried out in the U.S.A. have shown that the coverage of paints varies from painter to painter and that every craftsman gives his own personal coverage, which is different from every other painter's. The variations are very large and range from 516 to 896 ft super per gallon for primings (approx. 50 per cent variation) from 732 to 1,286 ft super per gallon for undercoating (a variation of 70 per cent) and from 705 to 1,335 ft super per gallon for finishing coats (a variation of almost 90 per cent).

COVERING CAPACITIES OF VARIOUS MATERIALS, FOR FIRST COAT PER 100 YDS SUPER AND PER 84 m²

Material	Surface	gal	lt
Creosote	Wood (rough)	5	23
Creosote	Wood (wrot)	2	9
French Polish	"	$\frac{3}{4}$	3·5
Oil	"	4	19
Stain	"	$\frac{1}{2}$	2·25
Tar	Wood (rough)	10	45
Tar	Wood (wrot)	5	23
Knotting	"	$\frac{1}{8}$	0·6
		pounds	kg
Size	Plaster	3	1·36

24 sheets of glasspaper or 3lb (1·36kg) of pumice are needed to rub down 100 yds super (84m).

SCHEDULE OF AVERAGE COVERAGE (IN YARDS SUPER) OF PAINTS AND DISTEMPERS
(THIS SCHEDULE CANCELS THAT ISSUED IN NOVEMBER 1951)

SURFACE COATINGS	Surfaces											
	Hard Wall Plaster	Lime Plaster	Brickwork (Fair Faced)	Hard Board	Soft Fibre Insulating Board	Asbestos Sheets	Steelwork	Metal Sheeting	Joinery	Primed Surfaced	Undercoated Surfaces	
Wood primer	—	—	—	—	—	—	—	—	65	—	—	per gallon
Metal primer	—	—	—	—	—	—	55	70	—	—	—	"
Plaster primer	60	50	35	—	—	50	—	—	—	—	—	"
Building board primer	—	—	—	60	35	50	—	—	—	—	—	"
Undercoat	—	—	40	—	40	—	—	—	—	65	70	"
Gloss finish	—	—	55	—	55	—	—	—	—	—	70	"
Eggshell finish	—	—	55	75	40	65	—	—	—	65	70	"
Emulsion paint	75	70	50	500	300	400	—	—	—	75	—	"
First grade water paint	500	450	350	500	300	400	—	—	—	600	—	per cwt
Oil bound water paint (B.S. 1053—Type A)	400	350	250	350	200	300	—	—	—	400	—	"
Distemper non-washable oil free (B.S. 1053-Type C)	250	225	175	250	150	225	—	—	—	250	—	"

METRIC

SURFACE COATINGS	Hard Wall Plaster	Lime Plaster	Brickwork (Fair Faced)	Hardboard	Fibreboard	Asbestos Cmt Board	Steelwork	Metal Sheeting	Joinery	Primed Surfaced	Unclassified Surfaces	
Wood primer	—	—	—	—	—	—	—	—	59	—	—	Square metre per 5 litres
Metal Primer	—	—	—	—	—	—	50	65	—	—	—	"
Plaster primer	35	46	33	—	—	46	—	—	—	—	—	"
Board primer	—	—	—	55	33	—	—	—	—	—	—	"
Undercoat	—	—	37	—	37	—	—	—	—	59	65	"
Gloss finish	—	—	50	—	50	—	—	—	—	—	65	"
Eggshell finish	—	—	50	—	50	—	—	—	—	59	65	"
Emulsion paint	69	65	46	69	37	59	—	—	—	69	—	"
First grade distemper	419	359	275	419	251	333	—	—	—	503	—	"
Oil bound distemper	333	275	210	275	168	211	—	—	—	335	—	square metres per 50kg
Distemper	510	184	148	210	126	184	—	—	—	275	—	"

Note 1: Differing conditions, such as absorbency, temperature of storage and at time of application, unevenness of surface, choice of colour, etc., can vary the figures on pages 391 and 392 by approximately 15 per cent either way.

Note 2: Thixotropic Paints. Whilst it is agreed that if well "brushed out", their spreading capacity is comparable with that of ordinary paints, their characteristics enable a thicker coating to be applied, when their coverage would be from 10-15 per cent lower.

(Issued by the Paint and Painting Industries' Liaison Committee representing the leading associations in the paint manufacturing and in the decorating industries, by whose courtesy this table is here reproduced.)

PRIMING PAINTS FOR METAL

IRON AND STEEL
1 Red oxide and zinc chromatic;
2. Red lead;
3. Calcium plumbate;
4. Zinc ('galvanizing') paint;
5. Metallic lead primer.

ZINC (including sheradized-zinc-sprayed and hot-dipped steel):
Any primer recommended for iron and steel. For hot-dipped galvanizing, and zinc sheet, use an etching primer or a mordant solution first.

ALUMINIUM
Castings may be primed direct with a zinc chromatic or other leadless primer. Extended sections should be de-greased with turps-substitute or solvent naphtha before priming. Use an etching primer on polished aluminium.

LEAD AND COPPER
Any type of primer may be used, subject to the general rule that for nonferrous metals, primers containing lead or graphite should not be used.

PRIMING PAINTS FOR SOFTWOOD

Tests by the Timber Research and Development Association have shown that an aluminium primer affords the best protection, and moreover, obviates one process, i.e. knotting. This does not, however, protect nails, screws and ironwork from rusting. If an aluminium primer is used externally or in bathrooms and kitchens, galvanized or sheradized nails and screws (or non-ferrous metal nails and screws) should be used. The next best is a properly compounded red lead primer, containing not less than 5 per cent of red lead. This type of primer should always be used if plain ferrous metal nails and screws are employed.

PRIMING PAINTS FOR PLASTER AND CEMENT

Gypsum plaster, if it is certain that it contains no free-lime, may be painted, preferably with a 'sharp' undercoat as primer.

Plaster containing lime, or mixtures of lime and Portland cement, or Portland cement concrete, must be treated with an alkali-resisting primer before painting. In an emergency, a coat of emulsion paint will serve. It should be remembered, however, that plaster and concrete must be allowed to dry out before priming; otherwise soluble salts may force off the primer.

THE CAUSES OF PAINT FAILURES

Cause	Resulting Defects
1. Insufficient or unskilled preparation of surface	Defective adhesion, delayed or imperfect drying; leading to peeling, blistering, discoloration.
2. Unskilled treatment of knots in wood.	Blistering, exudations, discoloration.
3. Unsuitable composition or unskilled application of priming coat.	Failure to exclude moisture, and consequent deterioration of surface painted. Defective adhesion of subsequent coats.
4. Unsuitable composition or unskilled application of undercoats and finishing coat.	Defective adhesion between coats; delayed drying; insufficient opacity; blistering, brush marks, cracking, loss of gloss in finishing coat. "Sheariness" or "flashing" (i.e. glossy patches on paint intended to dry flat).
5. Insufficient drying between coats.	Delayed or imperfect drying; cracking or peeling; loss of gloss.
6. Unsuitable treatment of absorbent surfaces.	
(a) Insufficient treatment.	Disintegration of paint or distemper; paint absorbed unequally, causing non-uniform finish.
(b) Excessive treatment.	Defective adhesion, resulting in peeling.
7. Redistemper over old distemper without either complete removal or application of effective sealing coat,	Flaking.
8. Unsuitable or insufficient preparation of ironwork.	Corrosion, leading to cracking and lifting of the paint film.
9. Improper use of alkaline materials for: (a) cleaning paint, (b) removal of paint.	Softening and loss of gloss. Defective drying and adhesion of the fresh paint.
10. Painting over creosote, bitumen or colours soluble in oil, without applying sealing coat.	Discoloration and defective drying due to 'bleeding' (i.e. substances on the under surfaces dissolve in and penetrate through the paint).

11. Exposure of newly-painted surfaces to frost, fog or moisture.	Delayed or defective drying: "blooming" (i.e. permanent or temporary dulling of the gloss of finishing coat).
12. Unsuitable or insufficient sterilization of mildewed surfaces.	Continued growth of moulds, causing disruption of paint film.
13. Exposure of new paint to excessive heat.	Blistering and cracking.
14. Premature painting on new cement and plasters.	Blistering or peeling of the paint due to pressure of escaping moisture.
15. Neglect to use alkali-resisting primer for oil paints on new cement and plasters.	Destruction of the paint by "saponification" (i.e. alkali in the surface converts oil of the paint medium into soap).
16. Allowing Keene's cement and similar plasters to dry too quickly in the early stages.	Powdering of the plaster surface, causing the paint to lose adhesion.
17. Painting old cement or plasters while damp.	Peeling due to defective adhesion, saponification or efflorescence due to alkaline salts brought to surface by escaping moisture.
18. Internal treatment of dampness arising from external causes.	Increases trouble by driving moisture elsewhere.
19. Painting on unseasoned new wood, or on old wood whilst damp.	Cracking, due to shrinkage of wood, blistering and peeling, due to defective adhesion. Possibly moulds, growths and decay of woodwork.

(*From 'Post-War Building Study No. 5'*)

RECENT DEVELOPMENTS

These include 'wet-on-wet' paints, of which the second coat can be applied while the first is still wet, and paints that can be applied with a 6 in to 8 in brush, without laying off. White and light-coloured paints, with very high obliterating power, that will literally cover black in one coat, are now available. The success of polyurethane varnish has led to the introduction of polyurethane gloss paints that are as heat resistant as the varnish. Emulsion paints incorporating a vinyl resin are now on the market: they wash as well as oil paint and have a slight sheen. For some time, emulsion paint has been used as an undercoat on cheap commercial work. The introduction of a water-thinned gloss paint now completes the tendency for distempers to form a complete painting system, containing no oil and requiring only water as thinners.

COVERAGE OF WALLPAPERS

Although the Standard Method of Measurement now stipulates that wallpapering be measured by the square metre, wallpaper is sold by the piece.

Estimators who calculate from the total of square metres only are liable to make costly mistakes. The following table supplies data on most types of wallpaper obtainable in this country. When ordering foreign wallpapers, check the contents of the roll with the suppliers.

Type of Paper	Size of piece		Feet super per piece	Average Coverage in sq. metres per piece inc. waste
	Width (inches)	Length (yards)		
English..	21*	11	58	5·0
English (lining)	22½	12	67	6·2
French	18	9	40½	3·7
American	18	8	36	3·3
Continental (i)	28	11⅔	81	7·5
Continental (ii)	22	11⅔	62½	5·75

*Ready trimmed, 20¾ inches

Some authorities reckon waste as 15 per cent of area to be covered, but this and the approximations given in the above table must vary with the size of the pattern, the type of match and the situation: staircases give rise to more waste than rooms.

The first table opposite is based on the number of lengths that can be cut from one piece or roll of paper. An addition of half-a-metre (1 ft 8 in) has been made to every length, to allow for waste in matching. Deductions can be made from the perimeter for doors and windows that extend from floor to ceiling. The room heights are divided into groups, according to the number of lengths that can be cut from a piece. In the shorter room-heights of each group, there will be greater waste than in the taller room-heights, and in these shorter heights further deduction from the perimeter can be made (up to 20 per cent) for doors and windows, since the waste will be enough to infill over doors and over and below windows.

It is necessary to hang ceiling papers across the room, working from the window, backwards, so as to avoid shadows that make the seams very prominent. The seams should be rolled down, and be butted. Take the width of the wall containing the window, and then find the other dimension under the length of the room. Where the two lines of figures coincide is the number of pieces required. For example, if a room is 14 ft long and 16 ft wide, 5 pieces will be required for the ceiling. See the second table opposite.

PAPERHANGING: ESTIMATING PAPER (PIECES) FOR WALLS

NUMBER OF PIECES REQUIRED

Height from skirting to cornice		Lengths in one piece																
feet	metres																	
up to 6½	2	4	5	5	6	6	7	8	8	9	10	11	12	13	13	14	14	15	16
over 6½ to 10	2·5–3	3	6	7	8	8	9	10	11	12	13	14	16	17	17	18	19	20	20
10–16	3·5–5	2	9	10	11	12	14	15	16	17	18	20	21	22	23	25	26	27	28	29	30
16–20	6	1	17	20	22	24	27	29	32	34	36	39	42	44	46	49	51	54	56	58	61

Distance around walls, including doors & windows —PERIMETER																		
feet		28	32	36	40	44	48	52	56	60	64	68	72	76	80	84	88	92	96	100
metres		8·5	9·75	11	12	13·5	14·5	15·75	17	18	19·5	20·75	22	23	24·3	25·5	27	28	29	30·5

ESTIMATING PAPER FOR CEILINGS

Width across wall containing window		Number of lengths per piece	Length of room		Number of pieces required						
Feet	Metres		Feet	Metres							
up to 6½	2	4			2	2	2	3	3	3	3
over 6½ to 10	2·5–3	3			2	3	3	3	4	4	4
10 to 16	3·5–5	2			3	4	5	5	5	6	6
16 to 20	6	1			6	7	9	10	11	12	
			10	3	12	14	16	18	20		
				3·6	7·25	4·75	5·5	6			

22. THE GLAZIER

Codes of Practice	396
British Standards	396
Glass Dimensions and Weights	397
Measuring	398
Minimum Thicknesses of Glass	398
Multiple Glazing	400
Plastics in Building	402
Thermoplastic Plastic	402
Thermosetting Plastics	403

CODES OF PRACTICE

CP 122: Part 1: 1966: Hollow glass blocks.
Hollow glass blocks forming walls and partitions, including fixing, etc.
CP 152: 1972: Glazing and fixing of glass for buildings.
Numerical values in SI units. Number of glass factor tables reduced to two.
B.S. 5516: 1977: Code of practice for patent glazing (formerly CP 145, Part 1).
Single and double patent glazing. Systems using four-edge support.

BRITISH STANDARDS

B.S. 952: 1964: Classification of glass for glazing and terminology for work on glass.
Types of glass available for building in four classes; transparent, translucent, opal and special.
Part 1: 1978: classification for building purposes in three groups; annealed flat glass, including float or polished plate, sheet, cast and wired glass; processed flat glass including toughened and laminated glass, and insulating units; miscellaneous glasses.
B.S. 1207: 1961: Hollow glass blocks.
Dimensions, finish, etc.
B.S. 3447: 1962: Glossary of terms used in the glass industry.
Classifies and defines a comprehensive list of terms.

THICKNESS AND WEIGHTS OF SHEET, HEAVY DRAWN AND FLOAT GLASS

Inches		Ounces	Millimetres
Fractions	Decimals		
$\frac{1}{28}$	0·0357–0·0385	7½–8	0·9 –1·0
$\frac{1}{24}$–$\frac{1}{26}$	0·0385–0·0417	8 –8½	0·95–1·05
$\frac{1}{22}$–$\frac{1}{24}$	0·0417–0·0455	8½–9½	1·05–1·15
$\frac{1}{20}$–$\frac{1}{22}$	0·0455–0·05	9½–10½	1·15–1·25
$\frac{1}{18}$–$\frac{1}{20}$	0·05–0·0556	10½–11½	1·25–1·4
$\frac{1}{15}$–$\frac{1}{17}$	0·0588–0·0667	12–14	1·5–1·7
$\frac{1}{13}$–$\frac{1}{15}$	0·0667–0·0769	14–16	1·7–1·9
$\frac{1}{12}$	about 0·08	17–18	about 2·0
$\frac{1}{8}$	,, 0·125	24–26	,, 3·0
$\frac{5}{32}$	,, 0·16	32–34	,, 4·0
$\frac{3}{16}$	,, 0·19	38–42	4·5–5·0
$\frac{7}{32}$ (¼ in bare)	,, 0·22	44–50	5·0–6·0
¼	,, 0·25	50–58	6·0–7·0
$\frac{5}{16}$	,, 0·32	60–70	about 8·0
$\frac{3}{8}$	,, 0·375	75–85	,, 10·0
½	,, 0·5	100–110	,, 12·5

GLASS DIMENSIONS AND WEIGHTS

Type of glass	Thickness	Wt. per ft sup	Maximum sizes
	ins	oz	ins
Sheet glass	$\frac{1}{12}$	18±1	62×42
	$\frac{1}{10}$	24±1	66×42
	$\frac{1}{8}$	26±2	66×42
	$\frac{5}{32}$	32±2	66×42
Rolled and cast	$\frac{1}{8}$	26	144×44
	$\frac{3}{16}$	40	144×44
	$\frac{1}{4}$	54	144×44
Wired, rolled and cast	$\frac{1}{4}$		120×40
Georgian wired cast	$\frac{1}{4}$		120×40
Figured and cathedral	$\frac{1}{8}, \frac{3}{16}, \frac{1}{4}$		120×42
Wired arctic	$\frac{1}{4}$		120×24 or 100×42

Weight per Square Foot for each $\frac{1}{16}$ in of thickness is about 13 oz, or about 8.2 oz for each millimetre of thickness.

Sheet Glass Qualities: O.Q. = Ordinary Glazing Quality.
                                S.Q. = Selected Glazing Quality.
                                S.S.Q. = Special Selected Quality.

Note: 18 oz sheet is not suitable for windows. 24 oz is generally satisfactory for sizes up to 4ft × 2ft.

PLAIN GLASS

Nominal thickness (mm) sheet	Nominal maximum size	Approx weight kg per m²
3	2 030×1 220mm	7·25
4	2 030×1 220mm	10·00
5	4·65m²×9·03m²	12·25
6	4·65m²×9·03m²	15·00
Float		
3	1 270×1 270 and 1 620×1 010mm	7·50
5	2 540×2 280mm	12·50
6	4 560×3 170mm	15·00
10	7 100×3 300mm	25·00
12	7 100×3 300mm	30·00
15, 19 and 25	4 320×2 800mm	37·5, 47·5 and 63·5
32	2 740×2 280mm	79·25
38	2 740×2 280mm	95·25

OTHER GLASS

Type	Nominal thickness	Max. size in mm
Patterned; all patterns	3mm	2 140×1 280
	5mm	2 140×1 320
Roughcast	5, 6 and 10mm	3 710×1 270
	12mm	4 420×2 490
'Insulite'	6 +12 +6mm (12mm airspace)	2 400×2 400 and 1 930×3 500

PUTTY FOR GLAZING SHEET GLASS

Type	Per ft super of pane sizes shown					Per ft run
	Not exc. 1ft	1–2ft	2–4ft	4–6ft	Exc 6ft	
	lb	lb	lb	lb	lb	lb
Wood sashes	·6	·4	·3	·25	·2	·125
Metal sashes	·8	·6	·4	·33	·3	·156

MEASURING

Allow $\frac{3}{32}$ in/2.4 mm all round (i.e., 'tight size' less $\frac{3}{16}$ in/5 mm). For coloured glass, or painted surfaces (*not* stained glass, but glass as used in curtain walling) allow $\frac{1}{8}$in/3 mm all round if the longer dimension does not exceed 30 in/770 mm. For pieces of glass with one or both dimensions above 30 in/770 mm, allow $\frac{3}{16}$ in/5 mm all round. The edge cover for coloured glass should not exceed $\frac{3}{8}$ in/ 9.5mm.

GLAZING WITHOUT BEADS MAXIMUM SIZES

Exposure grading	Maximum size (length and breadth)
A & B	84 in—2·13m
C	72 in—1·82m
D	60 in—1·52m

Note: For E, F, G and all other more severe conditions of exposure, beads should always be used.

MINIMUM THICKNESS OF GLASS
(FOR ALL APERTURES IN VARIOUS SITUATIONS)

The following must be taken into account when calculating the least thickness of glass for any opening: type of glass, size of opening to be

glazed, height above ground, and exposure conditions. The following method, condensed from CP 1152, makes allowance for all these factors.
1. Read off from table A the exposure gradient (the Capital Letter) given (a) for the height above ground and (b) the exposure conditions: i.e. the type of site.
2. Ascertain the 'Glass Factor' by dividing the area of the opening by twice the length plus the breadth. Take the answer to two places of decimals.
3. Read down on Table B, under the exposure gradient found in 1. above, until reaching the first 'Glass Factor' larger than that obtained in 2. above.
4. On the same line, under 'Minimum Thickness of Glass' will be found the least thickness suitable for the opening.

Example: A pane of glass, 5 ft 0in (1.52 m) long by 5 ft 6 in (1.70 m) broad is required for a window on the first floor of a house in an exposed position in town. What thickness of (a) float, or (b) sheet of glass would be required?
1. For an exposed urban site, less than 40 ft (12 m) from the ground, table A shows the exposure gradient to be C.
2. (a) The area of glass is 1.52 m × 1.70 m = 5.16 m^2.
   (b) Twice the length plus the breadth = 4.74.
   (c) Dividing (a) by (b) the result is 1.10 (approx.).
3. The nearest glass factor greater than this in column C of Table B is, for Float glass, 1.12. Therefore 4.5 mm ($\frac{3}{16}$ in) glass is required.
   For sheet glass, the nearest glass factor in column C is 1.49, and therefore 4.5 mm ($\frac{3}{16}$ in) glass is required. The same result will be obtained if the calculations are in Imperial units.

EXPOSURE GRADINGS

Type of Site	Mean wind velocity in m.p.h.	GRADIENTS: by Height of Glass above ground		
		0–40 0–12	40–80 12–24	80–120 ft 24–36 m
Sheltered urban	45	A	B	C
Normal urban and sheltered elsewhere	54	B	C	D
Exposed urban and open country	63	C	E	E
All others except extremes*	72	D	F	F

*Extreme conditions seldom arise in the United Kingdom. If suspected consult the Meteorological Office.

GLASS FACTORS: FLOAT

Exposure gradients						Minimum thickness of glass	
A	B	C	D	E	F	in	mm
1·56	1·30	1·12	0·98	0·87	0·78	$\frac{3}{16}$	4·5–5
2·18	1·82	1·56	1·37	1·21	1·09	$\frac{1}{4}$	6–7
3·12	2·60	2·23	2·00	1·73	1·59	$\frac{5}{16}$–$\frac{3}{8}$	8–9·5
3·59	2·99	2·56	2·24	1·99	1·80	$\frac{3}{8}$	9·5

GLASS FACTORS: SHEET

Exposure gradients						Minimum thickness of glass	
0·76	0·71	0·66	0·62	0·58	0·54	24oz	3mm
0·95	0·88	0·81	0·75	0·69	0·64	26 oz	3mm
1·27	1·16	1·06	0·97	0·89	0·83	32 oz	4mm
2·08	1·73	1·49	1·30	1·16	1·04	$\frac{3}{16}$ in	4·5mm

MULTIPLE GLAZING

In recent years, double glazing has steadily come into fashion in Britain. Until recently it has laboured under the disadvantage of being uneconomical. To burn a little more fuel has been less costly than incurring the capital cost of double glazing. However, with the recent staggering increases in fuel costs, double glazing is now an economical proposition.

HOW DOUBLE GLAZING WORKS

It is almost true to say that the temperature on both sides of a pane of sheet glass is the same. All that the glass does is to protect one from the cooling effects of the wind; but a single sheet of glass is a cold spot in a room, and sometimes causes an unwanted circulation of air in the room, which one feels as a draught. This is why, in winter, it is uncomfortable to sit within 3 ft (0.9 m) of a single-glazed window, and why radiators are situated under windows. Double glazing for thermal insulation consists essentially of two panes of glass with an air-space between them. The size of the air-space is critical, because the thermal insulation is provided not by the glass panes but by the air imprisoned between them. This air must be still, without air-currents being set up in it and a spacing of $\frac{3}{4}$ in (20 mm) between the glasses has been found to be the optimum distance without air currents being created.

THERMAL TRANSMITTANCE DATA

Single glass in vertical frame	5.6 W/m^2 deg C
Double glass, with 20 mm air-space	2.94 ,, ,,
'Insulite' units, with 12 mm air-space	3.01 ,, ,,
'Insulite' units, with 6 mm air-space	3.43 ,, ,,
9 in brickwork (215 mm) unplastered	2.76 ,, ,,

The greater the thermal transmittance, the greater the heat loss.

DOUBLE GLAZING SYSTEMS

In practice, double-glazing can be effected in three ways:
1. By double sashes or casements, opening independently;
2. By sub-frames or plastics or metal, screwed to the existing lights; and
3. Factory-produced, sealed units consisting of two panes spaced 6 or 12 mm apart. Some variations have an inert gas or a vacuum instead of air between the panes.

Method 1 is the oldest system, but can suffer from the liability of condensation between the panes, thus obstructing the view. The inner lights, if made to open inwards, are unpopular because they interfere with window curtains. Condensation can be avoided by controlled ventilation and moisture absorbents, such as silica gel. Double casements can be made to open outwards, as one unit and the inner leaves need only be opened for cleaning.

Method 2, the most modern and least expensive relies on perfect installation and is best installed by a specialist subcontractor. The possibility of condensation in the air-space cannot be ruled out.

Method 3 is the only certain way of obtaining a sealed, condensation-free air-space. The units are purpose-made to required sizes, in the factory: they require extra-deep rebates in the lights. It is difficult to adapt existing lights to take these units.

The table above shows the gain in heat-retention in a room by various types of double glazing. It is interesting to see that the best type of double-glazing gives almost as much insulation as a 9 in (215 mm) brick wall.

In double-glazing, it is vital to see that the lights are as good a fit as possible. New windows should be weatherstripped and old, warped ones will need new or supplementary stops and beads, as well as weatherstripping. A poorly-fitting light can destroy much of the advantage of double-glazing.

TRIPLE GLAZING

In the Scandinavian countries and northern Russia, double-glazing does not provide enough thermal insulation, and triple glazing is general. Thermal insulation is thus greatly increased, to something like twice that provided by a 9 in (215 mm) wall. Triple glazing is an expensive business and presents some interesting problems in window design. It is not a practical proposition for much of England, but in exposed situations of northern England and Scotland could be worth consideration. For a large picture window in such situations, double-glazing might not be sufficient to prevent discomfort in winter, which triple glazing would remedy.

ADVANTAGES

The principal advantage of multiple glazing is that there are no unusable parts of the room. Heating is more even over the whole of the living space and there are no eddies to cause draughts. There will be some fuel saving: radiators can be sited where most advantageous and less heating surface will be required.

PLASTICS IN BUILDING

Plastics may be briefly described as organic substances, mostly derived from by-products of coal-gas manufacture and of the refining of mineral oil, that can be moulded into stable shapes. The substances are manipulated or combined with others, by modern chemistry, to form 'long chain' molecules, on which the plasticity of the material and final rigidity of the objects made from it depend. Indeed, the chemistry of these substances is so far advanced that it is virtually true to say that even their basic molecular structure-patterns are man-made, in order to produce substances with predetermined properties. Space does not allow further consideration of the chemical constituents in the method of manufacture: for further information on this subject, see *Plastics in the Service of Man* by Couzens and Yarsley (Penguin Books).

Plastics range from soft, pliable rubber-like substances to intensely hard and brittle ones. They are divisible into two main groups—thermoplastic and thermosetting plastics. Articles made from thermoplastic plastics can be bent under applied heat, and 'welded' under heat, usually in an atmosphere of inert gas, to prevent oxidization. An example is Polyvinyl Chloride (PVC). Thermosetting plastics cannot be altered in shape nor joined by the application of heat after they have once become moulded into shape. In general, plastics cannot be successfully glued. Thermoplastic plastics tend to be pliable, while thermosetting plastics are brittle.

THERMOPLASTIC PLASTIC

1. POLYETHELINE ('Alkathene', 'Cryntholene', 'Polythene', 'Visqueen') Very tough, obtainable in transparent or delicately tinted forms. Used in tubular form for water supply (usually confined to cold water) and jointed

by compression joints or welded joints. It can be moulded into baths, basins, etc. Also used in sheet form, as vapour barriers, underlay for concrete, and (plain or reinforced with nylon or cotton) as covers for buildings in course of erection and as damp-courses.

2. POLYVINYL CHLORIDE

Used as flooring sheets and tiles ('Marley-film', 'Vinyltiles') as a coating on washable wallpapers; as hessian or felt-backed floor-coverings ('Cresylene', 'Hardura'). Also used as coatings on steel and other metals, as cold-water tank linings, and in high-density forms such as ball-valve floats and rainwater goods.

3. POLYVINYL ACETATE

Its principal use in building is in an aqueous suspension, as 'emulsion paint'. Some rainwater goods are made from this plastic.

4. POLYSTYRENE

Used, in emulsion form, in emulsion paint; and in foamed form as an insulating material ('Jablite').

5. METHYL METHACRYLATE ('Perspex', 'Diakon')

Used for transparent corrugated sheeting on a flat sheet.

6. POLYAMIDES

Represented by 'Nylon', now used for washers and for parts of ball-valves and for ropes.

7. SILICONE-BASED PLASTICS

Used in various sealing compounds.

THERMOSETTING PLASTICS

1. PHENOL-FORMALDEHYDE PLASTICS ('Bakelite').

Are those employed in the production of electrical switch-gear and fittings, door furniture, rigid sheets, hollow partitions, ventilators, tubes and many other fittings: these are among the oldest plastic materials employed in building. They are dark in colour.

2. UREA FORMALDEHYDE PLASTICS

Have similar uses, but in their natural state are transparent. They are used for the production of fittings in light colours.

3. MELAMINE

Is similar to urea-formaldehyde, and is largely used in the production of paper-based laminates ('Formica', 'Wearite').

4. POLYESTER PLASTICS

Are used in conjunction with glass-fibre reinforcement, to produce translucent corrugated and plain sheets ('Filon', 'Cascalite').

Combinations of plastics, such as phenol-formaldehyde, and common resin, polyesters and natural oils — such as chinawood oil (called alkyd resins) and epoxy resins — form a large range of synthetic gums that have largely replaced the natural gums, such as copal, in paints, varnishes and enamels.

Similarly, in the field of adhesives, synthetic glues such as urea-formaldehyde (mildly water-resistant), melamine (water-resistant) and resorcinal (water-resistant) plastics have been employed, with acids to harden them, as adhesives that set more quickly in the cold state, or with the aid of heat. They have virtually replaced the older animal glue in factory and workshop production, and have given to the building industry glues that will withstand permanent exposure to the severest weather conditions.

Employers' and Trade Unions Organisations

Amalgamated Union of Asphalt Workers
Jenkin House, 173A Queens Road, Peckham, London SE15 2NF
(01-639 1669)

Amalgamated Union of Engineering Workers, Constructional Section
Construction House, 190 Cedars Road, Clapham, London SW4 0PP
(01-622 4451)

Amalgamated Union of Engineering Workers, Technical and Supervisory Section
Onslow Hall, Little Green, Richmond, Surey TW9 1QN
(01-948 2271)

Architects' Benevolent Society
66 Portland Place, London W1N 4AD
(01-580 2823)

Association of Painting and Decorating Employers Ltd
1 Sheffield Road, Wylde Green, Sutton Coldfield B75 5HA
(021-350 5209)

Builders' Merchants Federation
15 Soho Square, London W1V 5FB
(01-439 1753)

Building Employers Confederation
82 New Cavendish Street, London W1M 8AD
(01-580 5588)

Committee of Associations of Specialist Engineering Contractors
34 Palace Court, Bayswater, London W2
(01-229 2488)

Confederation of British Industry
Centre Point, 103 New Oxford Street, London WC1
(01-379 7400)

Electrical Contractors Association
ESCA House, 34 Palace Court, Bayswater, London W2
(01-229 1266)

Electrical, Electronic Telecommunication and Plumbing Union
Hayes Court, West Common Road, Bromley, Kent BR2 7AU
(01-462 7755)

Federation of Associations of Specialists and Subcontractors
Maxwelton House, Boltro Road, Haywards Heath, West Sussex RH16 1BJ
(0444 451836)

Federation of Building and Civil Engineering Contractors (NI) Ltd
143 Malone Road, Belfast BT9 6SU
(0232 661711)

Federation of Building Sub-Contractors
82 New Cavendish Street, London W1M 8AD
(01-580 5588)

Federation of Civil Engineering Contractors
Cowdray House, Portugal Street, London WC2A 2HH
(01-404 4020)

Federation of Master Builders
33 John Street, Holborn, London WC1N 2BB
(01-242 7583)

Furniture, Timber and Allied Trades Union
Fairfields, Roe Green, Kingsbury, London NW9 0PT
(01-204 0273)

General Municipal Boiler Makers and Allied Trade Unions
Thorne House, Ruxley Ridge, Claygate, Esher, Surrey KT10 0TL
(78 62081)

Heating and Ventilation Contractors Association
ESCA House, 34 Palace Court, Bayswater, London W2 4JG
(01-229 2488)

House Builders Federation
82 New Cavendish Street, London W1M 8AD
(01-580 4041)

Iron and Steel Trades Confederation
Swinton House, 324 Grays Inn Road, London WC1X 8DD
(01-837 6691)

Joint Industry Board for the Electrical Contracting Industry
Kingswood House, 47/51 Sidcup Hill, Sidcup, Kent DA14 6HJ
(01-302 0031)

Metal Roofing Contractors Association
The Secretary, Hill Crest, Hockering Road, Woking, Surrey GU22 7HP
(04862 70324)

National Association of Formwork Contractors
82 New Cavenfish Street, London W1M 8AD
(01-580 5588)

National Association of Plumbing, Heating and Mechanical Services Contractors
6 Gate Street, London WC2A 3HX
(01-405 2678)

National Association of Scaffolding Contractors
82 New Cavendish Street, London W1M 8AD
(01-580 5588)

National Federation of Building Trades Employers (Now BEC)
82 New Cavendish Street, London W1M 8AD
(01-580 5588)

National Federation of Demolition Contractors
6 Portugal Street, London WC2A 2HH
(01-404 4020)

National Federation of Glass Reinforced Plastics Cladding Contractors
82 New Cavendish Street, London W1M 8AD
(01-580 5588)

National Federation of Painting and Decorating Contractors
82 New Cavendish Street, London W1M 8AD
(01-580 5588)

National Federation of Plastering Contractors
82 New Cavendish Street, London W1M 8AD
(01-580 5588)

National Federation of Roofing Contractors
15 Soho Square, London W1V 5FB
(01-439 1753)

National Graphical Association
63–67 Bromham Road, Bedford, Bedfordshire
(0234 51521)

National Joint Council for the Building Industry
18 Mansfield Street, London W1M 9FG
(01-580 1740)

National Joint Council for Felt Roofing Contracting Industry
Maxwelton House, Boltro Road, Haywards Heath, West Sussex RH16 1BJ
(0444 451835)

National and Local Government Officers Association (NALGO)
1 Mabledon Place, London WC1
(01-388 2366)

National Union of Public Employees
Civic House, 20 Grand Depot Road, Woolwich, London SE18 6SF
(01-854 2244)

Scottish Building Contractors Association
13 Woodside Crescent, Glasgow G3 7UP
(041-332 7144)
Scottish Building Employers Federation
13 Woodside Crescent, Glasgow G3 7UP
(041-332 7144)
Scottish Decorators Federation
c/o 249 West George Street, Glasgow G2 4RG
(041-221 7090)
Scottish and Northern Ireland Plumbing Employers Federation
2 Walker Street, Edinburgh EH3 7LB
(031-225 2255)
STAMP (Supervisory, Technical, Administrative, Managerial and Professional Section of UCATT)
UCATT House, 177 Abbeville Road, London SW4 9RL
(01-662 2442)
Transport and General Workers Union, Building Crafts Section
Transport House, Smith Square, London SW1P 3JA
(01-828 7788)
Union of Construction, Allied Trades and Technicians
UCATT House, 177 Abbeville Road, London SW4 9RL
(01-662 2442)

GOVERNMENT AND PUBLIC ORGANISATIONS

Association of County Councils
66a Eaton Square, London SW1W 9BH
(01-235 1200)
Association of District Councils
9 Buckingham Gate, London SW1 6LE
(01-828 7931)
Association of Metropolitan Authorities
36 Old Queen Street, London SW1H 9HZ
(01-222 8100)
British Gas Corporation
59 Bryanston Street, London W1A 2AZ
(01-723 7030)
British Overseas Trade Board
1 Victoria Street, London SW1H 0ET
(01-215 7877)

British Steel Corporation
9 Albert Embankment, London SE1 RSM
(01-735 7654)
Building Economic Development Committee
National Economic Development Office, Millbank Tower, Millbank, London SW1P 4QX
(01-211 3555)
Central Electricity Generating Board
Sudbury House, 15 Newgate Street, London EC1A 7AU
(01-248 1202)
Central Office of Information
Hercules Road, London SE1 7DU
(01-928 2345)
Civic Trust
17 Carlton House Terrace, London SW1Y 5AW
(01-930 0914)
Civil Engineering Economic Development Committee
National Economic Development Office, Millbank Tower, Millbank, London SW1P 4QX
(01-211 3555)
Commission for the New Towns
Glen House, Stag Place, London SW1E 5AS
(01-828 7722)
Construction Industry Training Board
Radnor House, 1272 London Road, Norbury, London SW16 4EL
(01-764 5060)
Council for Environmental Conservation
Zoological Gardens, Regent's Park, London NW1 4RY
(01-722 7111)
Council for Small Industries in Rural Areas
141 Castle Street, Salisbury, Wiltshire SP1 3TP
(0722 6255)
Countryside Commission
John Dower House, Crescent Place, Cheltenham, Gloucestershire GL50 3RA
(0242 21381)
Department of Energy
Thames House South, Millbank, London SW1P 4QJ
(01-211 3000)
Department of the Environment
2 Marsham Street, London SW1P 3EB
(01-212 3434)

Department of Housing, Local Government (NI)
Stormont, Belfast BT4 3SS
(0232 63210)

Department of Planning (NI)
Commonwealth House, Castle Street, City Centre, Belfast
(0232 21212)

Department of Industry
Ashdown House, 213 Victoria Street, London SW1E 6RB
(01-212 3395)

Department of Trade
1-19 Victoria Street, London SW1H 0ET
(01-215 7877)

Development Commission
11 Cowley Street, London SW1P 3NA
(01-222) 9134

Electricity Council
30 Millbank, London SW1P 4RD
(01-834 2333

Historic Buildings Council for England
25 Savile Row, London W1X 2BT
(01-734 6010)

Housing Centre Trust
33 Alfred Place, London WC1E 7JU
(01-637 4202)

Housing Corporation
149 Tottenham Court Road, London W1P 0BN
(01-387 9466)

Housing Project Control, Costs and Contracts
c/o Department of the Environment, 2 Marsham Street, London SW1
(01-212 3434)

Land Registry
32 Lincoln's Inn Fields, London WC2A 3PH
(01-405 3488)

Local Authorities Management Services and Computer Committee (LAMSAC)
Vincent House, Vincent Square, London SW1P 2NB
(01-828 2333)

London Boroughs Association
Room 215, Camden Town Hall, Euston Road, London NW1 2RU
(01-278 4444)

National Association of Local Councils
108 Great Russell Street, London WC1B 3LD
(01-637 1865)

National Coal Board
Hobart House, Grosvenor Place, London SW1X 7AE
(01-235 2020)

National Housing and Town Planning Council
Norvin House, 45/55 Commercial Street, London E1 6BA
(01-247 5732/3)

National Investments and Loan Office
Royex House, Aldermanbury Square, London EC2V 7LT
(01-606 7321)

New Towns Association
Metro House, 57/58 St James Street, London SW1A 1LD
(01-409 2635)

Ordnance Survey
Romsey Road, Maybush, Southampton SO9 4DH
(0703 775555)

Property Services Agency
Lambeth Bridge House, Albert Embankment, London SE1 7SB
(01-212 3434)

Scottish Civic Trust
24 George Square, Glasgow G2 1EF
(041-221 1466)

Scottish Council (Development and Industry)
23 Chester Street, Edinburgh EH3 7ET
(031-225 7911)

Scottish Development Agency
120 Bothwell Street, Glasgow G2 7JP
(041-248 2700)

Scottish Office
New St Andrews House, St James's Centre, Edinburgh EH1 3SX
(031-556 8400)

Wales Investment Location
8th Floor, Pearl Assurance House, Greyfriars Road, Cardiff CF1 3AG
(0222 371641)

Water Authorities Association (WAA)
1 Queen Anne's Gate, London SW1H 9BT
(01-222 8111)

Welsh Development Agency
Treforest Industrial Estate, Pontypridd, Glamorgan CS37 5UT
(044-385 2666)

Welsh Office
Cathays Park, Cardiff CF1 3NQ
(0222 825111)

INFORMATION, RESEARCH, TESTING, STANDARDS AND APPROVALS ORGANISATIONS AND ACADEMIC INSTITUTIONS

Air Conditioning Advisory Bureau
30 Millbank, London SW1
(01-834 8827)

Asbestos Information Centre Ltd
Sackville House, 40 Piccadilly, London W1V 9PA
(01-439 9231)

Aslib (Association of Special Libraries and Information Bureaux)
3 Belgrave Squre, London SW1X 8PL
(01-235 5050)

Birmingham Engineering and Building Centre Ltd
16 Highfield Road, Edgbaston, Birmingham B15 3DU
(021-454 9137)

BNF Metals Technology Centre
Grove Laboratories, Denchworth Road, Wantage, Oxfordshire OX12 9BJ
(023 57 2992)

Brick Development Association
Woodside House, Winkfield, Windsor, Berkshire SL4 2DP
(0344 885651)

British Board of Agrément
PO Box 195, Bucknall's Lane, Garston, Watford, Herts WD2 7NG
(09273 70844)

British Ceramic Research Association
Queens Road, Penkhull, Stoke-on-Trent ST4 7LQ
(0782 45431)

British Coal Utilisation Research Association
Randalls Raod, Leatherhead, Surrey
(Leatherhead 379222)

British Electrotechnical Approvals Board
Mark House, The Green, 9-11 Queens Road, Hersham, Walton-on-Thames, Surrey KT12 5NA
(Walton-on-Thames 44401)

British Glass Industry Research Association
Northumberland Road, Sheffield S10 2UA
(0742 686201)

British Standards Institution
2 Park Street, London W1A 2BS
(01-629 9000)

BSI Centre
Maylands Avenue, Hemel Hempstead, Hertfordshire HP2 4SQ
(0442 3111)

British Standards Society
2 Park Street, London W1A 2BS
(01-629 9000)

Build Electric Bureau
The Building Centre, 26 Store Street, London WC1E 7BT
(01-580 4986)

Building Advisory Management Services
18 Mansfield Street, London W1M 9EG
(01-636 2862)

Building Centre
26 Store Street, London WC1E 7BT
(01-637 1022/8361 Information: 0344 884999)

Building Centre, Bristol
Stonebridge House, Colston Avenue, The Centre, Bristol BS1 4TW
(0272 277022)

Building Centre, Cambridge
15–16 Trumpington Street, Cambridge CB2 1QD
(0223 359625)

Building Centre, Manchester
113–115 Portland Street, Manchester M1 6FB
(061-236 9802/6933)

Building Centre, Scotland
3 Claremont Terrace, Glasgow G3 7PF
(041-332 7399)

Building Centre, Southampton
Grosvenor House, 18–20 Cumberland Place, Southampton SO1 2BD
(0703 27350)

Building Centre Trust
26 Store Street, London WC1E 7BT
(01-637 1022)

Building Conservation Trust
Apartment 39, Hampton Court Palace, East Moseley, Surrey KT8 9BS
(01-943 2277)

Building Cost Information Service
85–87 Clarence Street, Kingston-upon-Thames, Surrey KT1 1RB
(01-546 7554)

Building Regulations Division
c/o Department of the Environment, 2 Marsham Street, London SW1
(01-212 3434)

Building Research Advisory Service
Building Research Establishment, Bucknall's Lane, Garston, Watford WD2 7JR
(Garston 76612)

Building Research Establishment
(Department of the Environment), Bucknall's Lane, Garston, Watford WD2 7JR
(Garston 74040)

Building Research Establishment Scottish Laboratory
Kelvin Road, East Kilbride, Glasgow G75 0RZ
(035 52 33941)

Building Services Research and Information Association
Old Bracknell Lane, Bracknell, Berkshire RG12 4AH
(0344 26511)

Cement and Concrete Association
Terminal House, 52 Grosvenor Gardens, London SW1W 0AQ
(01-235 6661)

Central Statistical Office
Cabinet Office, Great George Street, London SW1P 3AQ
(01-233 3000)

Centre for Advanced Land Use Studies
College of Estate Management, Whiteknights Park, Reading RG6 2AW
(0734 861101)

Centre for Urban Studies
55 Gordon Square, London WC1H 0NT
(01-636 2537)

Concrete Society
Terminal House, 52 Grosvenor Gardens, London SW1W 0AJ
(01-730 8252)

CONSTRADO (Constructional Steel Research and Development Organisation)
NLA Tower, 12 Addiscombe Road, Croydon CR9 3JH
(01-688 2688/01-686 0366)

Construction Industry Information Group
26 Store Street, London WC1E 7BT

Construction Industry Research and Information Association
6 Storey's Gate, London SW1P 3AU
(01-222 8891)

Copper Development Association
Orchard House, Mutton Lane, Potters Bar, Hertfordshire EN6 3AP
(77 50711)

Coventry Building Information Centre
Department of Architecture and Planning, Council House, Earl Street, Coventry CV1 5RT
(0203 25555)

Department of Architecture, University of Sheffield
Arts Tower, Sheffield S10 2TN
(0742 78555)

Department of Architecture and Building Science, University of Strathclyde
131 Rottenrow, Glasgow G4 0NG
(041-552 4400)

Department of Building, Polytechnic of Central London
35 Marylebone Road, London NW1
(01-486 5811)

Department of Building, University of Manchester
Institute of Science and Technology, Sackville Street, Manchester M60 1QD
(061-236 3311)

Department of Building Science, University of Newcastle-upon-Tyne
Newcastle-upon-Tyne NE1 7RU
(0632 328511)

Department of Building Science, University of Sheffield
Western Bank, Sheffield S10 2TN
(0742 78555 Ext 4708)

Design Council
Design Centre, 28 Haymarket, London SW1Y 4SU
(01-839 8000)

Edinburgh Computer Aided Architectural Design Studies, University of Edinburgh
Department of Architecture, 20 Chambers Street, Edinburgh EH1 1JZ
(031-667 1011 Ext 4598)

Electrical Research Association Ltd
Cleeve Road, Leatherhead, Surrey KT22 7SA
(0372) 374151

Felt Roofing Contractors Advisory Board
Maxwelton House, 41–43 Boltro Road, Haywards Heath, Sussex RH16 1BJ
(0444 451835)

Fire Prevention Information and Publication Service
Aldermary House, 10–15 Queen Street, London EC4N 1TJ
(01-248 4477)

Fire Protection Association
Aldermary House, 10–15 Queen Street, London EC4N 1TJ
(01-248 4477)

Fire Research Station
Building Research Establishment, Melrose Avenue, Borehamwood, Herts WD6 2BL
(01-953 6177)

Glass Advisory Council
6 Mount Row, London W1Y 6DY
(01-629 8334)

Gypsum Products Development Association
Westfield, 360 Singlewell Road, Gravesend, Kent DA11 7RZ
(0474 534251)

Housing Research Foundation
58 Portland Place, London W1N 4BU
(01-637 1248)

Housing Research Foundation
58 Portland Place, London W1N 4BU
(01-637 1248)

Hydraulics Research Station Ltd
Department of the Environment, Wallingford, Oxon OX10 8BA
(0491 35381)

Institute of Advanced Architectural Studies, University of York
King's Manor, York YO1 2EP
(0904 59861)

Institute of Asphalt Technology
Staines Road, Chertsey, Surrey KT16 8PP
(09328 60172)

Institute of Clay Technology
c/o Butterley Building Materials Ltd, Wellington Street, Ripley, Derby DE5 3DZ
(0773 43661)

Institute of Wood Science
Premier House, 150 Southampton Row, London WC1B 5AL
(01-837 8219)

Institution of Corrosion Science and Technology
Exeter House, 48 Holloway Head, Birmingham B1 1NQ
(021-622 1912)

Lead Development Association
34 Berkeley Square, London W1X 6AJ
(01-499 8422)

National Association of Building Centres
26 Store Street, London WC1E 7BT
(01-637 8361)

National Building Specification
Mansion House Chambers, The Close, Newcastle-upon-Tyne NE1 3OE
(0632) 329594)

National Computing Centre Ltd
Oxford Road, Manchester M1 7ED
(061-228 6333)

National Engineering Laboratory
East Kilbride, Glasgow G75 0QU
(035 52 202222)

National GRP Cladding Association
82 New Cavendish Street, London W1M 8AD
(01-580 5588)

National House Building Council
58 Portland Place, London W1N 4BU
(01-637 1248)

National Housing and Town Planning Council
Norvin House, 45/55 Commercial Street, London E1 6BA
(01-247 5732/3)

National Inspection Council for Electrical Installation Contracting
237 Kennington Lane, London SE11 5QJ
(01-582 7746)

National Materials Handling Centre, Cranfield Institute of Technology
Cranfield, Bedfordshire MK43 0AL
(0234 750323)

National Physical Laboratory
Queens Road, Teddington, Middlesex TW11 0LW
(01-977 3222)

National Research Development Corporation
101 Newington Causeway, London SE1 6BU
(01-403 6666)

Natural Environment Research Council
Polaris House, Northstar Avenue, Swindon, Wiltshire SN2 1EU
(0793 40101)

Paint Research Association
Waldegrave Road, Teddington, Middlesex TW11 8LD
(01-977 4427)

Plastics and Rubber Institute
11 Hobart Place, London SW1W 0HL
(01-245 9555)

Princes Risborough Laboratory
Building Research Establishment, Princes Risborough, Aylesbury, Bucks HP17 9PX
(08444 3101)

RIBA Services Ltd
66 Portland Place, London W1N 4AD
(01-580 5533)

Rubber and Plastics Research Association of GB
Shawbury, Shrewsbury, Shropshire SY4 4NA
(0939 250383)

Scottish Association for Building Education and Training
c/o Clydebank Technical College, Kilbowie Road, Clydebank, Dunbartonshire G81 2A
(041-952 7771)

Scottish Design Centre
72 St Vincent Street, Glasgow G2 5TN
(041-221 6121)

Solar Trade Association Ltd
26 Store Street, London WC1E 7BT
(01-636 4717)

Solid Fuel Advisory Service
Hobart House, Grosvenor Place, London SW1X 7AE
(01-235 2020)

Timber Research and Development Association
Stocking Lane, Hughenden Valley, High Wycombe HP14 4ND
(02424 3091)

Town and Country Planning Association
17 Carlton House Terrace, London SW1Y 5AS
(01-930 8903/5)

Transport and Road Research Laboratory
Departments of the Environment and Transport, Old Wokingham Road, Crowthorne, Berkshire RG1 6AU
(0344 773131)

Welsh School of Architecture, University of Wales
Institute of Science and Technology, Colum Drive, Cardiff CF1 3EU
(0222 42588)

Zinc Development Association
34 Berkeley Square, London W1X 6AJ
(01-499 6636)

PROFESSIONAL ORGANISATIONS AND SEMI-PROFESSIONAL ORGANISATIONS

Architects Registration Council of the United Kingdom
73 Hallam Street, London W1N 6EE
(01-580 5861)

Architectural Association (Inc)
34–36 Bedford Square, London WC1B 3ES
(01-636 0974)

Associated Master Plumbers and Domestic Engineers
Barnard Close, 107–113 Powis Street, Woolwich, London SE18 6JB
(01-855 1844)

Associattion of Consultant Architects
The Secretary, c/o 7 King Street, Bristol BS1 4EQ
(0272 293372)

Association of Consulting Engineers
Alliance House, 12 Caxton Street, London SW1H 1QL
(01-222 6557)

Association of Local Authorities Chief Architects
30 Kings Road, Brighton, BN1 1PF
(0273 29801 Ext 249)

Association of Local Authority Valuers and Estate Surveyors
City Estates Surveyor, Leicester City Council, Welford Place, New Walk Centre, Leicester LE1 6ZG
(0533 549922)

Association of London Borough Architects
Technical Services Group, 22/24 Uxbridge Road, Ealing W5
(01-579 2424 Ext 2520)

Association of Official Architects
17 Claverton Road, Saltford, Bristol BS18 3DW

Association of Public Service Professional Engineers
7 Elm Grove, Wimbledon, London SW19
(01-946 3777)

British Computer Society
13 Mansfield Street, London W1N 0BP
(01-637 0471)

British Decorators Association
6 Haywra Street, Harrogate, N. Yorkshire HG1 5BL
(0423 67292)

British Institute of Interior Design
Lenton Lodge, Wollaton Hall Drive, Nottingham NG8 1AF
(0602 701205)

British Property Federation
35 Catherine Place, London SW1E 6DY
(01-828 0111)
Building Societies Institute
Fanhams Hall, Ware, Herts SG12 7PZ
(0920 5051)
Chartered Institute of Arbitrators
75 Cannon Street, London EC4N 5BH
(01-236 8761)
Chartered Institute of Building
Englemere, King's Ride, Ascot, Berkshire SL5 8BJ
(0990 23355)
Chartered Institute of Building Services
Delta House, 222 Balham High Road, London SW1 9BS
(01-675 5211)
City and Guilds of London Institute
76 Portland Place, London W1N 4AA
(01-580 3050)
Construction Surveyors Institute
203 Lordship Lane, East Dulwich, London SE22 8HA
(01-693 0219/0210)
Contractors Mechanical Plant Engineers
The Secretary, 20 Knavewood Road, Kemsing, Sevenoaks, Kent TN15 6RH
(Otford 2628)
County Planning Officers Society
County Hall, Glenfield, Leicester LE3 8RH
(0533 871313)
County Surveyors Society
Northampton House, Northampton NN1 2HZ
(0604 34833)
Design and Industries Association
17 Lawn Crescent, Kew Gardens, Surrey TW9 3NR
(01-940 4925)
District Surveyors' Association
1 Myron Place, London SE13 5AT
(01-852 3253)
Engineering Council
Canberra House, 10–16 Maltravers Street, London WC2R 3ER
(01-240 7891)
Faculty of Architects and Surveyors
15 St Mary Street, Chippenham, Wiltshire
(0249 655398)

Faculty of Building
10 Manor Way, Borehamwood, Hertfordshire WD6 1QQ
(01-953 7053)

Faculty of Royal Designers for Industry
6–8 John Adam Street, London WC2N 6EZ
(01-839 2366)

Guild of Architectural Ironmongers
8 Stepney Green, London E1
(01-790 3431/6)

Guild of Bricklayers
1 Tamhouse, Orten Malborne, Peterborough, Cambridgeshire PE2 0NA
(0733 231594)

Guild of Surveyors
The Lodge, Eastbury Farm, 33 Batchworth Lane, Northwood, Middlesex HA6 3EQ
(65 27755)

Incorporated Association of Architects and Surveyors
Jubilee House, Billing Brook Road, Weston Favell, Northampton
(0604 404121)

Incorporated Society of Valuers and Auctioneers
3 Cadogan Gate, London SW1X 0AS
(01-235 2282)

Institute of Builders' Merchants
Parnall House, 5 Parnall Road, Staple Tye, Harlow, Essex CM18 7NG
(0279 419650)

Institute of Carpenters
45 Sheen Lane, London SW14 8AB
(01-876 4415)

Institute of Clerks of Works
41 The Mall, London W5
(01-579 2917)

Institute of Housing
12 Upper Belgrave Street, London SW1X 8BL
(01-245 9933)

Institute of Plumbing
Scottish Mutual House, North Street, Hornchurch, Essex RM11 1RU
(04024 51236)
(moving to 64 Station Lane, Hornchurch, Essex end 1983)

Institute of Professional Designers
1–5 Rosslyn Mews, Rosslyn Hill, London NW3
(01-794 3233)

Institute of Mechanical Engineers
1 Birdcage Walk, London SW1H 9JJ
(01-222 7899)

Institute of Quarrying
7 Regent Street, Nottingham NG1 5BY
(0602 411315)

Institute of Solid Wastes' Management
3 Albion Place, Northampton NN1 1UD
(0604 20426/7/8)

Institute of Water Pollution Control
Ledson House, 53 London Road, Maidstone, Kent ME16 8JH
(0622 62034)

Institution of Building Control Officers
The White House, 41 Carshalton Road, Sutton, Surrey SM1 4TA
(01-661 1360)

Institution of Civil Engineers
1–7 Great George Street, London SW1P 3AA
(01-222 7722)

Institution of Electrical Engineers
2 Savoy Place, London WC2R 0BL
(01-240 1871)

Institution of Engineering Designers
Courtleigh, Westbury Leigh, Westbury, Wiltshire BA13 3TA
(0373 822801)

Institution of Fire Engineers
148 New Walk, Leicester LE1 7QB
(0533 553654)

Institution of Gas Engineers
17 Grosvenor Crescent, London SW1X 7ES
(01-245 9811)

Institution of Mining Engineers
Hobart House, Grosvenor Place, London SW1
(01-235 3691)

Institution of Municipal Engineers
13 Grosvenor Place, London SW1X 7EN
(01-245 9778)

Institution of Occupational Safety and Health
222 Uppingham Road, Leicester LE5 0RG
(0533 768424)

Institution of Public Health Engineers
32 Eccleston Square, London SW1V 1PB
(01-834 3017)

Institution of Structural Engineers
11 Upper Belgrave Street, London SW1X 8BH
(01-235 4535/6841)

Institution of Works and Highways Technicians Engineers
Room 21, 4th Floor, 125 High Holborn, London WC1V 6QR
(01-242 5733)

The Landscape Institute
Nash House, 12 Carlton House Terrace, London SW1Y 5AH
(01-839 4044)

Property Consultants Society Ltd
5 Maple Parade, Walberton, Sussex
(0243 551 368)

Rating and Valuation Association
115 Ebury Street, London SW1W 9QT
(01-730 7258/9)

Royal Incorporation of Architects in Scotland
15 Rutland Square, Edinburgh EH1 2BE
(031-229 7205/6)

Royal Institute of British Architects
66 Portland Place, London W1N 4AD
(01-580 5533)

Royal Institute of Chartered Surveyors
12 Great George Street, London SW1P 3AD
(01-222 7000)

Royal Society
6 Carlton House Terrace, London SW1Y 5AG
(01-839 5561)

Royal Society of Arts
John Adam Street, Adelphi, London WC2N 6EZ
(01-839 2366)

Royal Society of Ulster Architects
2 Mount Charles, Belfast BT7 1N
(0232) 223760

Royal Town Planning Institute
26 Portland Place, London W1N 4BE
(01-636 9107)

Society of Architectural and Associated Technicians
397 City Road, London EC1V 1NE
(01-278 2206)

Society of Chief Architects of Local Authorities
County Architect, Durham County Council, Architects Department, County Hall, Durham
(0385 64411)

Society of Chief Quantity Surveyors in Local Government
Secretary, Architects Department, Nottinghamshire County Council, County Hall, West Bridgford, Nottingham NG2 7QP
(0602 823823)

Society of Civil Engineering Technicians
c/o Institution of Civil Engineers, 1–7 Great George Street, London SW1P 3AA
(01-222-7722)

Society of Engineers (Inc)
21–23 Mossop Street, London SW3 2LW
(01-589 6651)

Society of Industrial Artists and Designers
12 Carlton House Terrace, London SW1Y 5AH
(01-930 1911)

Society of Surveying Technicians
Aldwych House, Aldwych, London WC2B 4EL
(01-242 4833)

TRADE AND INDUSTRY ORGANISATIONS

Aggregate Concrete Block Association
60 Charles Street, Leicester LE1 1FB
(0533 536161)

Aluminium Federation
Broadway House, Calthorpe Road, Birmingham B15 1TN
(021-455 0311)

Aluminium Windows Association
26 Store Street, London WC1E 7EL
(01-637 3578)

The Architectural Aluminium Association
193 Forest Road, Tunbridge Wells, Kent TN2 5JA
(0892 30630)

Asbestos Cement Manufacturers' Association
16 Woodland Avenue, Lymm, Cheshire
(092575 569 2833)

Association of British Manufacturers of Mineral Insulating Fibres
St Paul's House, Edison Road, Bromley, Kent
(01-466 6719)

Association of British Plywood and Veneer Manufacturers
c/o British Plywood Manufacturers Ltd, Wharf Road, Ponders End, Enfield, Middlesex EN3 4TS
(01-804 2424)

Association of British Roofing Felt Manufacturers
69 Cannon Street, London EC4N 5AB
(01-248 4444)

Association of British Solid Fuel Appliance Manufacturers
Fleming House, Renfrew Street, Glasgow G3 6TG
(041-332 0826)

Association of Builders' Hardware Manufacturers
5 Greenfield Crescent, Birmingham B15 3BE
(021-454 21778/8)

Association of Building Component Manufacturers Ltd
26 Store Street, London WC1E 7BT
(01-580 9083)

Association of Gunite Contractors
45 Sheen Lane, London SW14 8AB
(01-876 4415)

Association of Lightweight Aggregate Manufacturers
c/o Butterley Aglite Ltd, Wellington Street, Ripley, Derbyshire DE5 3DZ
(0773 43661)

Association of Manufacturers of Domestic Unvented Supply Systems Equipment
c/o Maurice Thomson, Sawsley, Snailwell Road, Newmarket, Suffolk CB8 9AJ
(0638 662279)

Association of Structural Fire Protection Contractors and Manufacturers Ltd
45 Sheen Lane, London SW14 8AB
(01-876 4415)

Autocleaved Aerated Concrete Products Association
c/o Celcon House, 289–293 High Holborn, London WC1V 7HV
(01-242 4803)

Bituminous Roofing Council
PO Box 125, Haywards Heath, West Sussex RH16 3TJ
(0444 416681)

Boiler and Radiator Manufacturers Association Ltd
Fleming House, Renfrew Street, Glasgow G3 6TG
(041-332 0826)

Box Culvert Association
60 Charles Street, Leicester LE1 1FB
(0533 536161)

The Brick Development Association
Woodside House, Winkfield, Windsor SL4 2DX
(03447 5651)

British Adhesive Manufacturers Association
2A High Street, Hythe, Southampton SO4 6YW
(0703 842765)

British Aggregate Construction Material Industries
25 Lower Belgrave Street, London SW1W 0LS
(01-730 8194)

British Bath Manufacturers Association
Fleming House, Renfrew Street, Glasgow G3 6TG
(041-333 0826)

British Blind and Shutter Association
5 Greenfield Crescent, Edgbaston, Birmingham B15 3BE
(021-454 2177)

British Ceramic Tile Council Ltd
Federation House, Station Road, Stoke-on-Trent ST4 2RU
(0782 45147)

British Combustion Equipment Manufacturers' Association
The Fernery, Market Place, Midhurst, Sussex GU29 9DP
(073081 2782)

British Constructional Steelwork Association Ltd
92–96 Vauxhall Bridge Road, London SW1V 2RL
(01-834 1713)

British Electrical and Allied Manufacturers' Association Ltd
Leicester House, 8 Leicester Street, London WC2H 7BN
(01-437 0678)

British Electrical Systems Association
Granville Chambers, 2 Radford Street, Stone, Staffordshire
(0785 814023)

British Flue and Chimney Manufacturers' Association
Nicholson House, High Street, Maidenhead, Berkshire SL6 1LF
(0628 34667/8)

British Furniture Manufacturers Federated Associations
30 Harcourt Street, London W1H 2AA
(01-724 0854)

British Hardware Federation
20 Harborne Road, Edgbaston, Birmingham B15 3AB
(021-454 4385)

British Hardware and Housewares Manufacturers Association (BHHMA)
35 Billing Road, Northampton NN1 5DD
(0604 22023)

British Independent Steel Producers' Association
5 Cromwell Road, London SW7 2HX
(01-581 0231)

British Foundry Association
14 Pall Mall, London SW1
(01-930 7171)

British Laminated Plastic Fabricators' Association
5 Belgrave Square, London SW1X 8PH
(01-235 9483)

British Lead Manufacturers' Association
68 High Street, Weybridge, Surrey KT13 8BL
(97 56621)

British Lock Manufacturers' Association
91 Tettenhall Road, Wolverhampton WV3 9PE
(0902 26726)

British Non-Ferrous Metals Federation
Crest House, 7 Highfield Road, Birmingham B15 3ED
(021-454 7766)

British Plastics Federation
5 Belgrave Square, London SW1X 8PH
(01-235 9483)

British Precast Concrete Federation Ltd
60 Charles Street, Leicester LE1 1FB
(0533 28627/9)

British Pump Manufacturers' Association
3 Pannels Court, Chertsey Street, Guildford, Surrey GU1 4EU
(0483 37997)

British Ready Mixed Concrete Association
Shepperton House, Green Lane, Shepperton, Middlesex TW17 8DN
(Walton-on-Thames 43232)

British Refrigeration and Air Conditioning Association
Nicholson House, High Street, Maidenhead, Berkshire SL6 1LF
(0628 34667/8)

British Reinforcement Manufacturers' Association
15 Tooks Court, London EC4A 1CA
(01-831 7581)

British Rubber Manufacturers' Association
90–91 Tottenham Court Road, London W1P 0BR
(01-580 2794)

British Timber Merchants' Association
Blackburn House, 1 Warwick Street, Leamington Spa, Warwickshire CV32 5LW
(0926 29905)

British Wood Preserving Association
Premier House, 150 Southampton Row, London WC1B 5AL
(01-837 8217)

British Woodworking Federation
82 New Cavendish Street, London W1M 8AD
(01-580 5588)

Building Materials Export Group
33 Alfred Place, London WC1E 7EN
(01-636 5345)

Building Societies Association
34 Park Street, Mayfair, London W1Y 3PF
(01-629 7233)

Calcium Silicate Brick Association
11 White Lion House, Town Centre, Hatfield, Hertfordshire AL10 0JL
(30 71580)

Cement Admixtures Association
2A High Street, Hythe, Southampton SO4 6YW
(0703 842765)

Cement and Concrete Association
Terminal House, 52 Grosvenor Gardens, London SW1W 1AQ
(01-235 6661)

Cement Makers' Federation
Terminal House, 52 Grosvenor Gardens, London SW1W 0AH
(01-730 2148)

Chipboard Promotion Association
Stocking Lane, Hughenden Valley, High Wycombe, Buckinghamshire HP14 4NU
(0240 24 5265)

Clay Pipe Development Association
Drayton House, 30 Gordon Street, London WC1E 0HN
(01-388 0025)

Clay Roofing Tile Council
Federation House, Station Road, Stoke-on-Trent
(0782 416256)

Concrete Brick Manufacturers' Association
60 Charles Street, Leicester LE1 1FB
(0533 536161)

Concrete Block Association
60 Charles Street, Leicester LE1 1FB
(0533 536161)

Concrete Lintel Association
60 Charles Street, Leicester LE1 1FB
(0533 536161)

Concrete Pipe Association of Great Britain
60 Charles Street, Leicester LE1 1FB
(0533 536161)

Contract Flooring Association
23 Chippenham Mews, London W9 2AN
(01-286 4499)
Contractors' Plant Association
28 Eccleston Street, London SW1W 9PY
(01-730 7117)
Copper Cylinder and Boiler Manufacturers
St James's Buildings, 89 Oxford Street, Manchester M1 4LT
(061-236 3668/9)
Copper Tube Fittings Manufacturers' Association
Crest House, 7 Highfield Road, Birmingham B15 3ED
(021-454 7766)
Cork Industry Federation
56 Lagham Park, South Godstone, Surrey
(034 285 3363)
Council of British Ceramic Sanitaryware Manufacturers
Federation House, Station Road, Stoke-on-Trent ST4 2RT
(0782 48675/57074)
Decorative Lighting Association
Bishop's Castle, Shropshire SY9 5LE
(058 84 658)
Decorative Paving and Walling Association
60 Charles Street, Leicester LE1 1FB
(0533 536161)
District Heating Association
Bedford House, Stafford Road, Caterham, Surrey CR3 6JA
(0883 42323/4)
The Door and Shutter Association
5 Greenfield Crescent, Edgbaston, Birmingham B15 3BE
(021-454 2177)
Dry Lining and Partition Association
82 New Cavendish Street, London W1M 8AD
(01-580 5588)
Dry Material Cavity Insulation Council
Interpress PR, The Meusings, Church End, Walkern, Stevenage, Hertfordshire
(0438 86347)
Ductile Iron Pipe Association
14 Pall Mall, London SW1Y 5LZ
(01-930 7171)
Environmental Protection Equipment Manufacturers' Association
136 North Street, Brighton BN1 1RG
(0273 26313)

Export Group for the Construction Industries
Kingsbury House, 15–17 Kings Street, St James, London SW1Y 7QU
(01-930 5377)

Federation of British Hand Tool Manufacturers
Light Trades House, Melbourne Avenue, Sheffield S10 2QJ
(0742 663084)

Federation of Concrete Specialists
60 Charles Street, Leicester LE1 1FB
(0533 536161)

Federation of Resin Formulators and Applicators Ltd
2A High Street, Hythe, Southampton SO4 6YW
(0703 842765)

Federation of Manufacturers of Construction Equipment and Cranes
7–15 Lansdowne Road, Croydon, Surrey CR9 2PL
(01-688 4422)

Federation of Piling Specialists
15 Tooks Court, London EC4A 1LA
(01-831 7581)

Fibre Building Board Development Organisation Ltd
1 Hanworth Road, Feltham, Middlesex TW13 5AF
(01-751 6107)

Flat Glass Manufacturers' Association
Prescot Road, St Helens, Merseyside WA10 3TT
(0744 28882 Ext 2585)

GAMBICA (SIMA) Scientific Instruments Manufacturers' Association
Leicester House, 8 Leicester Street, London WC2H 7BN
(01-437 0678)

Glass and Glazing Federation
6 Mount Row, London W1Y 6DY
(01-409 0545)

Glass Manufacturers' Federation
19 Portland Place, London W1N 4BH
(01-580 6952)

Glassfibre Reinforced Cement Association
Farthings End, Dukes Ride, Gerrards Cross, Buckinghamshire SL9 7LD
(0753 882606)

Glazed and Floor Tile Home Trade Association
Federation House, Station Road, Stoke-on-Trent ST4 2RU
(0782 45147)

Heat Pump Manufacturers Association
Nicholson House, High Street, Maidenhead, Berks SL6 1LF
(0628 34667/8)

Heating, Ventilating and Air Conditioning Manufacturers' Association
Nicholson House, High Street, Maidenhead, Berkshire SL6 1LF
(0628 34667/8)

Industrial Warm Air Heater Manufacturers Association Ltd
Business Services Group, Chesterfield House, Bloomsbury Way, London WC1 2TP
(01-242 3366)

Interlock Paving Association
60 Charles Street, Leicester LE1 1FB
(0533 536161)

Kitchen Furniture Manufacturers Section of the British Woodworking Federation
82 New Cavendish Street, London W1M 8AD
(01-580 5588)

Kitchen Specialist Association
31 Bois Lane, Chesham Bois, Amersham, Buckinghamshire
(02403 22287)

Lattice Girder Floor Association
60 Charles Street, Leicester LE1 1FB
(0533 536161)

Lighting Industry Federation
Swan House, 207 Balham High Road, London SW17 7BQ
(01-675 5432)

Machine Tools Trades Association
62 Baywater Road, London W2 3PH
(01-402 6671)

Manufacturers' Association of Radiators and Convectors
45 Sheen Lane, London SW14 8AB
(01-876 4415)

Mastic Asphalt Council and Employers' Federation Ltd
Construction House, Paddock Hall Road, Haywards Heath, West Sussex RH16 1HE
(0444 457786)

Metal Roof Deck Association
Maxwelton House, Boltro Road, Haywards Heath, West Sussex RH16 1BJ
(0444 451835)

Metal Sink Manufacturer's Association
Fleming House, Renfrew Street, Glasgow G3 6TG
(041-332 0826)

METCON
Fleming House, Renfrew Street, Glasgow G3 6TG
(041-332 0826)

Midland Federation of Brick and Tile Manufacturers
Acorn House, 196 High Street, Erdington, Birmingham B23 6QY
(021-373 7445)
Mortar Producers' Association Ltd
74 Holly Walk, Leamington Spa, Warwickshire CV32 4JL
(0926 38611)
National Association of Lift Makers
8 Leicester Street, London WC2N 7BN
(01-437 0678)
National Association of Shopfitters
NAS House, 411 Limpsfield Road, The Green, Warlingham, Surrey CR3 9HA
(08832 4961)
National Brassfoundry Association
5 Greenfield Crescent, Birmingham B15 3BE
(021-454 2177)
National Building and Allied Hardware Manufacturers' Federation
5 Greenfield Crescent, Edgbaston, Birmingham B15 3BE
(021-454 2177)
National Cavity Insulation Association
PO Box 12, Haslemere, Surrey GU27 3AN
(0428 54011)
National Clayware Federation
7 Castle Street, Bridgwater, Somerset TA6 3DT
(0278 58251)
National Council of Building Material Producers
33 Alfred Place, London WC1E 7EN
(01-580 3344)
National Federation of Clay Industries
Federation House, Station Road, Stoke-on-Trent ST4 2TJ
(0782 416256)
National Federation of Housing Associations
30–32 Southampton Street, Strand, London WC2E 7HE
(01-240 2771)
National Federation of Terrazzo-Mosaic Specialists
5th Floor, 3 Berners Street, London W1P 4JP
(01-580 2903)
National Fireplace Council
PO Box 35, Stoke-on-Trent ST4 7NU
(0782 44311)
National Home Improvement Council
26 Store Street, London WC1E 7BT
(01-636 2562)

National Master Tile Fixers' Association
c/o Joliffe Cork & Co., Elvian House, 18–20 St Andrews Street, London EC4 3AE
(01-353 3055)

National Paving and Kerb Association
60 Charles Street, Leicester LE1 1FB
(0533 536161)

Northern Ireland Builders' Merchants' Association
2 Greenwood Avenue, Belfast BT4 6JJ
(0232 650321)

Paintmakers Association of Great Britain Ltd
Alembic House, 93 Albert Embankment, London SE1 7TY
(01-582 1185)

Partitioning Industry Association
1 Lansdale Avenue, Solihull, West Midlands B92 0PP
(021-705 9270)

Patent Glazing Conference
13 Upper High Street, Epsom, Surrey KT17 4QY
(03727 29191)

Phenolic Foam Manufacturers Association
45 Sheen Lane, London SW14 8AB
(01-876 4415)

Pitch Fibre Pipe Association of Great Britain
35 New Bridge Street, London EC4V 6BH
(01-248 5271)

Plastic Bath Manufacturers' Association
Fleming House, Renfrew Street, Glasgow G3 6TG
(041-332 0826/8)

Plastic Pipe Manufacturers' Society
89 Cornwall Street, Birmingham B3 3BY
(021-236 1866)

Plastic Tanks and Cisterns Manufacturers' Association
8 Balmain Close, Grange Road, Ealing, London W5 5BY
(01-579 6081)

Precast Concrete Frame Association
60 Charles Street, Leicester LE1 1FB
(0533 536161)

Prefabricated Building Manufacturers' Association of Great Britain Ltd
c/o S. Wernick & Sons Ltd., Russell Gardens, Wickford, Essex S11 8BL
(03744 5544)

Prestressed Concrete Association
60 Charles Street, Leicester LE1 1FB
(0533 536161)

Refined Bitumen Association
c/o Construction House, Paddockhall Road, Haywards Heath, Sussex RH16 1HE

Sand and Gravel Association Ltd
48 Park Street, London W1Y 4HE
(01-499 8967/9)

Scottish Plant Owners' Association
12 St. Vincent Place, Glasgow G1 2EQ
(041-332 0021)

Scottish Pre-Cast Concrete Manufacturers' Association
c/o 9 Princes Street, Falkirk FK1 1LS
(0324 22088)

Sealant Manufacturers' Conference
2A High Street, Hythe, Southampton SO4 6YW
(0703 842765)

Sectional Chamber Association
60 Charles Street, Leicester LE1 1FB
(0533 536161)

Society of British Gas Industries
36 Holly Walk, Leamington Spa, Warwickshire CV32 4LY
(0926 34357/9)

Solid Smokeless Fuels Federation
York House, Empire Way, Wembley, Middlesex HA9 0PA
(01-902 5405)

Stainless Steel Fabricators' Association of Great Britain
14 Knoll Road, Dorking, Surrey RH4 3EW
(0306 884079)

Steel Lintel Manufacturers' Association
PO Box 10, Newport, Gwent
(0633 272281 Ext 4198)

Steel Window Association
26 Store Street, London WC1E 7JR
(01-637 3571/2)

Stone Federation
82 New Cavendish Street, London W1M 8AD
(01-580 5588)

Structural Fire Protection Contractors and Manufacturers Ltd
45 Sheen Lane, London SW14 8AB
(01-876 4415)

Structural Insulation Association
45 Sheen Lane, London SW14 8AB
(01-876 4415)

The Surveying Equipment Manufacturers' and Dealers' Group of DOMMDA
6th Floor, 25/27 Oxford Street, London W1R 1RF
(01-734 2971)

Suspended Access Equipment Manufacturers Association
82 New Cavendish Street, London W1M 8AD
(01-580 5588)

Suspended Ceilings Associations
29 High Street, Hemel Hempstead, Hertfordshire HP1 3AA
(0442 40313)

Swimming Pool and Allied Trades Association
Faraday House, 17 Essendene Road, Caterham, Surrey CR3 5PB
(22 40110)

Thermal Insulation Manufacturers' and Suppliers' Association
45 Sheen Lane, London SW14 8AB
(01-876 4415)

Timber Research and Development Association
Stocking Lane, Hughenden Valley, High Wycombe, Buckinghamshire HP14 4ND
(0240 24 3091)

Timber Trade Federation
Clareville House, Whitcomb Street, London WC2H 7DL
(01-839 1891)

UK Particle Board Association
Stocking Lane, Hughenden Valley, High Wycombe, Buckinghamshire HP14 4NT
(0240 24 2381)

Vitreous Enamel Development Council
New House, High Street, Ticehurst, Wadhurst, Sussex TN5 7AL
(0580 200152)

Wallcovering Manufacturers' Association of Great Britain Ltd
Alembic House, 93 Albert Embankment, London SE1 7TY
(01-582 1185)

Wood and Solid Fuel Association of Retailers and Manufacturers
PO Box 35, Stoke-on-Trent ST4 7NU
(0782 44311)

Wood Wool Slab Manufacturers' Association
33 Alfred Place, London WC1E 7EN
(01-580 3344)

Woodworking Machinery Importers' Association
Cobbins House, 7 Broadstrood, Goldings Manor, Loughton, Essex IG10 2SE
(01-508 9095)

Published by Building Trades Journal

Titles available:

Arbitration for Contractors
Builders' Reference Book
Building Regulations 1976 in Detail with amendments 1, 2, and 3
Buyers' Guide 1986
Construction Case Law in the Office
Construction Technology Guide (volume 2)
Contract Joinery
Estimating for Alterations & Repairs
Guide to Estimating Building Work 1985/86
Site Carpentry
Techniques of Routing (Completely revised edition)
The Small Contractors' Guide to the Computer
Builders' Detail Sheets (Consolidated)
Drainage Details
Hot Water Details
Sanitation Details (Consolidated)

Practical Guide Series

Alterations & Improvements
Basic Bookkeeping
Builders' Questions and Answers
Estimating Day Work Rates 1985/86
Setting Out on Site
Subcontracting

For further details on the above titles or for a booklist on prices etc. please ring the BTJ Books Department, telephone number 01-404-5531.

BUILDING TRADES JOURNAL

Britain's largest selling building weekly

This is what you have been missing. The multi-useful **Building Trades Journal** every week. There's no other magazine like it in the United Kingdom.

Packed with current news, trade news, product news, features and jobs.

Building Trades Journal caters especially for the needs of Britain's building industry, going behind the scenes with pages of insight into the people and events that make the news.

Current Prices—every week **Building Trades Journal** offers its readers the most up-to-date prices available for small quantities of materials collected from a builders' merchant. The Prices quoted are average and are compiled by computer from information supplied by merchants throughout the country immediately prior to publication. Readers should make adjustments to suit their particular conditions.

Estimating Guide—every month comprehensive up-dated estimating service, covering material, labour and plant prices (fourth week of month).

Monthly features—reviewing a wide range of subjects such as bricks, drainage, timber frame, housing and many more to provide the reader with factual details on new innovations and building systems as soon as they come onto the market.

These frequent collect-and-keep supplements on specialist topics are widely sought after.

Undoubtedly you can benefit professionally from **Building Trades Journal** by taking out a subscription.

Further details and a subscription form can be obtained by either ringing 01-404-5531—Circulation Dept.—or writing to the Circulation Manager, BTJ, 23/29 Emerald Street, London WC1N 3QJ.

NOTES

NOTES